FUNGAL GENETICS

Botanical Monographs

Edited by

W.O.JAMES F.R.S.

Professor of Botany
Imperial College, London, England

VOLUME FOUR

FUNGAL GENETICS

FUNGAL GENETICS

J. R. S. FINCHAM
Ph.D.
John Innes Institute, Bayfordbury, Herts

AND

P. R. DAY
Ph.D.
John Innes Institute, Bayfordbury, Herts

SECOND EDITION

F. A. DAVIS COMPANY
PHILADELPHIA · PA

Printed in Great Britain by
ADLARD & SON LTD, THE BARTHOLOMEW PRESS, DORKING
and bound by
THE KEMP HALL BINDERY, OXFORD

CONTENTS

v

PREFACE TO SECOND EDITION

The growth of the fungal genetical literature is such that the time is approaching when it will be unreasonable even to attempt to survey all the more important work in the field in a single book. However, in this second edition we have tried to take account of the most significant of the advances which have been made during the last three years. Chapters 3, 7 and 8 have been rewritten to a large extent, and less extensive alterations have been made in Chapters 2, 4, 5, 9 and 10. Over a hundred new references have been included, mostly to new publications but some to older work which had been overlooked in the first edition.

J.R.S.F.

John Innes Institute

P.R.D.

Connecticut Agricultural Experiment Station
September 1964

PREFACE TO FIRST EDITION

Although the great development of work on fungal genetics, which has occurred during the last few decades, has revealed a great deal which should be of interest to mycologists, most of this work, and this certainly applies to our own, has not been motivated primarily by an interest in fungi for their own sake. Fungi happen to be uniquely suitable and convenient organisms for experimental attacks on a number of important genetic problems. With their rapid life cycles and suitability for fine-structure genetic analysis, they share many of the advantages which have permitted such sensational progress in the fields of bacterial and viral genetics. At the same time, they have chromosomes which, though small, are easily visible and apparently orthodox in behaviour, and their system of sexual reproduction, with its alternation of regular haploid and diploid phases, is similar to that prevailing in higher plants and animals. Thus the fungi offer a good vantage point from which to

ix

survey a number of key problems of genetics, such as the mechanisms of genetic recombination and the nature of gene action. It is with these broader implications of fungal genetics that we are principally concerned in this book, although in Chapters 9 and 11 we have attempted to cover some of the more specifically fungal aspects also.

A frequent source of difficulty for beginners in genetics, and even sometimes for experienced geneticists, is the tendency for genetic terms like *gene, exchange,* and so on to shift their meanings with the rapid developments of the subject. With the hope of minimizing confusion we have appended a Glossary, which, while it may not command universal agreement, will, at any rate, explain our own usage.

We have been greatly helped by the co-operation of many friends and colleagues. In particular, Drs Raymond Barratt, L. S. Olive, David Perkins, J. R. Raper, Georges Rizet and David Wilkie have been generous in supplying data and manuscripts in advance of publication, and many others, too numerous to mention individually, have helped with advice on points of detail. We are particularly indebted to Dr Edward Barry, of Yale University, who generously loaned us the negatives from which Fig. 20 was made. We owe a special acknowledgment to our colleague Robin Holliday who has read and given us detailed and constructive criticism of many of the chapters. He has done his best to improve the book, but neither he, nor any of the other friends whose help we have acknowledged, can, of course, bear any responsibility for such faults as still remain.

We are also indebted to Miss Maria Shipton for the care which she has given to typing the manuscript, and to Mr L. S. Clarke who undertook all the photographic work in expert fashion. Permission to reproduce Fig. 14 from *Genetical Research*, Fig. 44 from *Zeitschrift für Vererbungslehre*, Fig. 45 from *Science*, Fig. 46 from the *Proceedings of the Royal Society* and Fig. 47 from *Heredity* is gratefully acknowledged. Finally, we should like to thank Dr K. S. Dodds, Director of this Institute, for his support of our project.

J.R.S.F. P.R.D.

John Innes Institute
July 1962

CHAPTER 1

THE CHROMOSOME THEORY AS ILLUSTRATED BY THE GENETICS OF *NEUROSPORA*

The chromosome theory forms the main theoretical basis of fungal genetics, as of the genetics of all other groups of organisms. Different groups of fungi have many different kinds of life history, and some of the more important of these will be surveyed in the next Chapter. The fundamentals of the chromosome theory are, however, best illustrated by reference to a single type of organism, and *Neurospora crassa* is certainly the fungal species which has been the most thoroughly studied, both from the genetical and the cytological points of view.

VEGETATIVE STRUCTURE AND REPRODUCTION

Neurospora is a genus belonging to the Ascomycetes, sub-class Pyrenomycetes. Its nutrition is relatively simple, and it grows well on a fully defined medium containing only a simple carbon source (sucrose or glycerol), the vitamin biotin, and inorganic salts. Among its other outstanding advantages are a rapid rate of vegetative growth, a short generation time (the sexual cycle occupies only 10 days under optimal conditions), and self-sterility which permits the making of controlled crosses.

A growing culture of Neurospora consists of branched filaments (hyphae) some 5µ or less in diameter. The growth of a hypha occurs at the tip, although as Zalokar [605] has shown, the bulk of the protoplasm is synthesized some distance behind the tip and transported to it by a vigorous protoplasmic streaming. At 30°C the rate of extension of a hyphal tip may exceed 5 mm an hour. As a hypha grows side branches are formed, each one having the same potentialities for growth as the primary hypha. The whole hyphal system, or mycelium, is subdivided by crosswalls, or septa. The resulting compartments may be called cells, although they are not closely comparable with the cells of higher organisms since each contains numerous nuclei, of the order of a hundred, rather than a single one. Furthermore, the septa do not form a continuous barrier, but are each pierced by a central pore which allows the passage of streaming protoplasm, including nuclei [487].

As in all other organisms, the nuclei are self-propagating; that is to say, each nucleus can divide to give two daughter nuclei and this is the only way in which new nuclei can arise. The details of nuclear division (mitosis) in vegetative hyphae are rather hard to make out [15, 503] but in essentials the process seems to resemble the more easily studied mitotic divisions in the ascus, to be described below. Briefly, what is seen at mitosis in the ascus [495] is that each nucleus consists of seven-rod-shaped stainable bodies, called chromosomes, and that each chromosome divides longitudinally, one of each pair of daughter chromosomes passing into each daughter nucleus. Thus the self-propagation of the nucleus is due to the individual capacity for self-propagation of its seven constituent chromosomes.

Mycelium growing at the surface of the nutrient medium will send up aerial branches which develop into conidiophores bearing the asexual spores, or conidia, which are the principal means by which the fungus is propagated in nature. The conidiophores are richly branched, and the tips of the branches become subdivided into short segments which round off and are easily detachable as air-borne spores. The conidia, which are bright orange in colour when seen in mass, are quite variable in size and may contain anything from one up to ten or more nuclei [247]. In addition to the more obvious macroconidia, uninucleate microconidia may be formed. Microconidia are only about 1µ in diameter, and are usually far outnumbered by the macroconidia; however special genetic varieties of Neurospora will produce microconidia only, and in great abundance [18].

When placed on the surface of an adequate nutrient medium, both kinds of conidia germinate readily to form a hyphal system. Except in the special case of heterocaryosis, where two or more genetically distinct kinds of nuclei are present in the same mycelium, conidia can only propagate the kind of mycelium from which they came, and represent a vegetative, or clonal, form of reproduction.

The sexual cycle and meiosis

The sexual life cycle of Neurospora was first described by Shear and Dodge [489], and its potentialities as an experimental genetic organism were made clear in subsequent papers by Dodge [e.g. 122, 123].

A pure strain of Neurospora is unable to undergo sexual reproduction;

the sexual fruiting bodies (*perithecia*) are only formed when two mycelia, of different mating type, are brought together. Neurospora like other Ascomycetes, has only two mating types which are usually symbolized by *A* and *a*. There is no morphological difference between *A* and *a* strains, and both can form abundant female reproductive structures, the protoperithecia, when grown on solid (agar) medium of suitable composition [570]. A protoperithecium consists of the ascogonium, which is a coiled multicellular hypha, enclosed in a knot-like aggregation of hyphae. The tip of the ascogonium is extended as a branching system of very slender hyphae, called the trichogyne, which projects beyond the sheathing hyphae into the air. Fertilization occurs when a cell of opposite mating type, which may be a macroconidium, a microconidium, or even a piece of ordinary mycelium, comes into contact with a part of the trichogyne. When such a contact is made, fusion may occur, and one or more nuclei from the fertilizing cell will then migrate down the trichogyne and into the ascogonium [13] (cf Fig. 8).

The details of the nuclear events within the ascogonium after fertilization are not fully understood, but from the work of Colson [86] on the closely related *Neurospora tetrasperma*, as well as from what is known about Ascomycetes generally we can assume that what happens is as follows. No fusion of nuclei of different mating types occurs at this stage. Instead, a pair of nuclei, one from the ascogonium and one from the fertilizing cell, become associated and begin to divide synchronously. The products of these divisions pass, still in pairs of unlike mating type, into numerous *ascogenous hyphae* which now begin to grow out of the ascogonium. While this is happening, the mycelial sheath which enveloped the ascogonium begins to develop as the wall of the perithecium, and becomes heavily impregnated with melanin. The mature perithecium is a flask-shaped structure with a narrow beak-like neck. Occasional perithecia include ascogenous hyphae which have evidently arisen from more than one pair of parental nuclei; such perithecia may well be due to the inclusion of two adjacent ascogonia in a single perithecial wall [372].

The later stages of development of the ascogenous hyphae, and the whole course of development of the asci which arise from them, are well known, thanks to the work of McClintock [333] and Singleton [495] who studied squash preparations of perithecial contents stained with aceticorcein. Each ascogenous hypha bends to form a hook (or *crozier*) at its tip and the two nuclei of opposite mating type within the crozier divide synchronously (Fig. 1 *a*, *b*). Septa now form to divide the crozier into three cells, the central one, in the curve of the hook, containing two nuclei of unlike mating type. This binucleate cell is the ascus initial, and the two uninucleate cells on either side of it commonly fuse to reconstitute a binucleate cell which can grow on to form a further crozier.

Almost immediately after the formation of an ascus initial the two nuclei within it fuse together (Fig. 1c). The chromosomes at this time are in a relatively contracted state, and it can be clearly seen that a nucleus with 14 chromosomes is formed from two nuclei with seven each. The

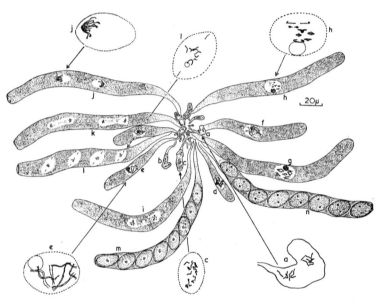

FIG. 1. Stages of ascus development in *Neurospora crassa*. The various asci shown here in the same cluster are based on photographs of several different squash preparations by McClintock & Singleton, using either aceto-carmine or acetic-orcein to stain the chromosomes. The drawing is slightly idealized in the interest of clarity. (a) Pro-metaphase of conjugate mitotic divisions in the crozier; (b) late mitotic anaphase in the crozier; (c) early post-fusion nucleus; (d) zygotene to early pachytene; (e) mid-pachytene; (f) late pachytene; (g) diplotene; (h) metaphase I; (i) anaphase I; (j) late telophase I; (k) anaphase II; (l) metaphase III (mitotic); (m) metaphase of mitotic division in the ascospore; (n) telophase of ascospore division. Note nucleoli in (e)–(h) and (l) and centrioles at the spindle poles in (l).

fusion nucleus is described as *diploid*, since it contains a double set of chromosomes. It is, in fact, the only diploid nucleus in the entire life history of the fungus, and its formation is immediately followed by two nuclear divisions of a special kind which between them constitute the process of *meiosis*, and the effect of which is the formation of four nuclei which once again have the *haploid* (i.e. single) chromosome complement. Meiosis occurs as an essential part of the life cycle of all sexually reproducing organisms, and, in its main features, it seems to be the same process wherever it occurs. The events seen in the *Neurospora* ascus are quite typical of meiosis generally, though not all the stages can be seen in as much detail as is visible in organisms with larger chromosomes. The account which follows is based on the excellent study of Singleton [495].

During meiosis, and especially during the first division, the ascus initial grows rapidly to form an elongated sac, which remains attached to the ascogenous hypha at the base. In the early stages of ascus growth, the chromosomes are seen to undergo lengthwise pairing and subsequently they become progressively more elongated. As a result of this pairing, which occurs during the *zygotene* stages, and which is complete at *pachytene*, it is possible to see that there are seven kinds of chromosome; while different chromosome pairs differ from each other in length, and in the pattern of short deeply staining segments or *chromomeres*, members of a pair are to all appearances identical, with a close point-for-point association all along their lengths (Fig. 1e) (Fig. 20).

At the end of pachytene the chromosomes begin to contract again, and there follows a rather ill-defined stage which appears to correspond to the *diplotene* which is clearly visible in some other organisms (Fig. 1g). So far as they can be seen, diplotene chromosome pairs in the Neurospora ascus appear to have fallen apart along most of their lengths, remaining joined only at a few points, called *chiasmata*. There is also an indication that the individual chromosomes have divided longitudinally at this stage, but with the two daughter strands still closely associated. Pairs of chromosomes joined by chiasmata are referred to as *bivalents*. Further contraction leads to the *diakinesis* stage, during which the bivalents and their chiasmata are comparatively easy to make out.

Diakinesis is followed by the *metaphase* of the first meiotic division (metaphase I, Fig. 1h). The feature of metaphase which distinguishes it sharply from the preceding stages is the appearance in the nucleus of a structure called the *spindle*, consisting of a system of more or less parallel fibres orientated along the long axis of the ascus. The bivalents come to lie in a plane, the spindle equator, which is at right angles to the spindle fibres. Each bivalent lies symmetrically about the equator with one chromosome appearing as if pulled towards each spindle pole. The pull seems to be transmitted through a single point on each chromosome which is called the *centromere* (or sometimes the *spindle fibre attachment*), and which is flanked by two apparently kinetically inert arms.

At *anaphase I* the two chromosomes of each bivalent pull apart towards the two poles of the spindle (Fig. 1*i*), until at *telophase I* the two resulting groups of seven chromosomes form two daughter nuclei, one in each half of the ascus. The separation of the two nuclei is due principally to the elongation of the spindle, which appears to push the two groups of chromosomes apart. During telophase the chromosomes lose their contraction and during the succeeding *interphase* appear as slender threads which are sometimes visibly double (Fig. 1*j*).

The second division of meiosis is initiated by a *prophase*, during which the chromosomes become contracted once again. This is followed by metaphase II, with the formation of a longitudinally orientated spindle within each nucleus and with the chromosomes coming to lie on the spindle equators. At metaphase II, and also at anaphase I, the chromosomes can sometimes be seen to be divided but with the two halves still held together at least at one point. It is generally accepted, on the basis of analogy with meiosis as seen in higher organisms, that the undivided point on each chromosome is the centromere, though this conclusion could hardly have been reached from a study of the Neurospora ascus alone. Genetic evidence, to be considered below, confirms that the chromosomes must already be divided along most of their length before anaphase I.

At anaphase II each chromosome completes its longitudinal division, and a movement of chromosomes essentially similar to that of anaphase I (with sister centromeres passing to opposite poles) results in the formation of four nuclei arranged in a row (Fig. 1*k*). Since the second division spindles seldom or never overlap, the two nuclei in one half of the ascus derive from the same interphase nucleus. Each nucleus now contains seven undivided chromosomes, one of each kind.

At this point the process of meiosis has been completed. It is most simply regarded as two divisions of the nucleus accompanied by only one division of the chromosomes, the chromosomal division occurring partly during the first nuclear division (division of the chromosome arms) and partly during the second (division of the centromere).

During telophase II the chromosomes become elongated once more, but soon contract again for a third nuclear division. This is a mitotic division, essentially similar in effect to the nuclear divisions which occur in vegetative hyphae. By analogy with higher organisms, although this detail can hardly be seen in Neurospora, the chromosomes are believed already to be divided, except at the centromere, before the onset of prophase, and they certainly separate completely into identical daughter chromosomes at the end of metaphase. Anaphase and telophase follow in the usual way. The third division spindles are orientated somewhat obliquely, but do not overlap (Fig. 1*l*) and so the resulting eight nuclei remain in four adjacent pairs, each pair the product of one mitotic division. Around each nucleus the contents of the ascus become organized to form an ellipsoidal ascospore, which, as it matures, develops a thick ribbed wall, impregnated with melanin. During ascospore development yet another mitotic division occurs (Fig. 1*m, n*) so that each ascospore contains two nuclei at maturity.

Although, as we shall see, the chromosomes are of primary importance in genetic theory, they are not the only structures to be seen in preparations of dividing nuclei in the ascus. Each nucleus contains a more or less spherical body, the nucleolus, which hardly stains at all with orcein but is very prominent in preparations stained with carmine. The nucleolus is not self-propagating during meiosis; it tends, in fact, to dwindle in size and vanish as division proceeds, and, if it survives to anaphase, is normally left behind by the chromosome movement. A new nucleolus is formed in each telophase nucleus, apparently always at the tip of the short arm of the second largest chromosome. The nucleolus also differs from the chromosomes in varying greatly in bulk between one stage of ascus development and another; it is particularly well developed during the first division of meiosis, and at pachytene to diakinesis it is far larger and more prominent (in carmine-stained preparations) than the chromosomes themselves (Fig. 1, *e.g.*) The size of the nucleolus is probably connected with the amount of protein synthesis within its sphere of influence; during the first division of meiosis there is, of course, only one nucleolus in the ascus, which is an exceptionally large and rapidly growing cell.

Another interesting feature of ascus development is the presence, during the first post-meiotic division, of prominent centrioles at the poles of the spindles (Fig. 1*l*). The centrioles are triangular plates, commonly seen edge-on, of the same order of size as a chromosome. It is probable that they are present at the spindle poles at all nuclear divisions in Neurospora, and other fungi, but they seem to be seldom observed except at mitosis in the ascus; Singleton has suggested that their special prominence at this time indicates that they have a role in ascospore delimitation. Centrioles in animals are known to be self-propagating, and the same appears to be true in Neurospora. The two centrioles present at anaphase are derived from a single centriole observable at the preceding prophase. This capacity for self-propagation might appear to qualify the centriole as a

2

carrier of genetic factors, but no evidence for such a function seems to exist. Like the nucleolus, but unlike the chromosomes, the centrioles stain poorly with orcein, but are readily seen in carmine-stained preparations.

GENETIC ANALYSIS BY ISOLATION OF ASCOSPORES

When ripe, each ascus in turn elongates until its tip reaches the neck of the perithecium, the tip ruptures, and the eight ascospores are discharged violently. Ascospores can be recovered from the side of the culture tube opposite the perithecia, spread on the surface of medium solidified with agar, and induced to germinate, either by heat shock (30 min at 60° is customary) or by treatment with furfural [138]. Heat shock is the most generally useful method of breaking the dormancy, which otherwise persists for long periods. Single spores, either before or after germination, can easily be isolated into individual culture tubes, and the characteristics of the cultures to which they give rise determined. It is, however, more informative in some respects to express the cluster of asci from a perithecium on to an agar surface before the ascospores have been discharged, and to isolate sets of eight spores in order from individual asci. Methods for the dissection and isolation of ascospores from asci have been described by Beadle [23] and by Emerson [144]. If one is content to isolate the eight spores from an ascus without regard to order, one can dispense with ascus dissection. Strickland [522] showed that if the ascospores discharged from the ripe perithecia are collected on an agar surface, a high proportion of them are in groups of eight which seem nearly always to have come from single asci. Separating the members of such groups is very much easier than ascus dissection, and since the ascospores will be riper than undischarged ones they tend to germinate better.

Genetic segregation in the ascus

The parent strains of any Neurospora cross necessarily differ in mating type; of each pair of nuclei fusing at ascus initiation, one is derived from the A and one from the a strain. When the eight spores from any ascus are germinated individually, and the resulting cultures tested for mating type, four are found to be A and four a. A second important regularity is that members of a spore pair 1 and 2, 3 and 4, 5 and 6 or 7 and 8 (numbering from the top of the ascus to the base) always give cultures of the same mating type. With these restrictions all possible patterns of mating type distribution within asci occur, though not all with equal frequency. The numbers of the six possible arrangements in 274 asci analysed by Lindegren [314] are shown in Table 1.

In the light of our knowledge of the processes of nuclear fusion and meiosis, this kind of inheritance of mating type finds a fairly obvious explanation, namely that the two mating types are determined by a pair of mutually exclusive genetic factors, or *alleles*, as they may conveniently be termed, associated with a pair of homologous chromosomes. Then, provided we assume that each chromosome propagates its own type when it divides, the various ascospore arrangements are exactly what we would expect. During the first division of meiosis each member of a chromosome pair divides (except at the centromere), so that at anaphase I there will be present in the ascus two *A* and two *a* half-chromosomes (chromatids). At the end of the second division each of these four chromatids has been separated into a different nucleus. The four identical pairs of ascospores which are ultimately formed are derived from the four nuclei, which are the immediate products of meiosis, by mitotic divisions in which each nucleus simply reproduces its own kind.

TABLE 1. Arrangements of mating type alleles in *Neurospora crassa* asci

Spore pairs	Ascus types						
	1	2	3	4	5	6	
1 and 2	*A*	*a*	*A*	*a*	*A*	*a*	
3 „ 4	*A*	*a*	*a*	*A*	*a*	*A*	
5 „ 6	*a*	*A*	*A*	*a*	*a*	*A*	
7 „ 8	*a*	*A*	*a*	*A*	*A*	*a*	
Number of asci	105	129	9	5	10	16	Total: 274

Data from Lindegren [314].

The preceding argument shows how the occurrence of two *A* and two *a* spore pairs in every ascus may be explained. It does not, however, explain the differing frequencies of the six possible spore arrangements shown in Table 1. If the four strands of a metaphase I bivalent were distributed at random to the four products of meiosis, one would expect the six arrangements to occur with equal frequencies. In actual fact, types 1 and 2, in which the four spores of each type are together in one half of the ascus, that is, in which *A* factors were segregated from *a* factors at the first meiotic division, together constitute 84 per cent of the total. This means, accepting the hypothesis of a chromosomal determination of mating type, that the two chromatids of each chromosome of the pair determining mating type, or at least the relevant parts of them, tend to be held together at anaphase I. This conclusion can be related to the cytological observations which indicate that there is a point on each chromosome (the centromere) at which the chromatids are held together at anaphase I.

The problem is, in fact, not so much to explain the predominance of asci in which *A* and *a* chromosomes are segregated as wholes at the first division, as to show how the 16 per cent of second division segregation asci can occur. Given that the second division spindles never overlap (which seems to be very nearly true for *Neurospora crassa*), the only way out is to postulate that, in about 16 per cent of asci, an exchange of homologous segments can occur between chromatids in the region between the centromere and a locus determining mating type. The effect of such an exchange is shown in Figure 2.

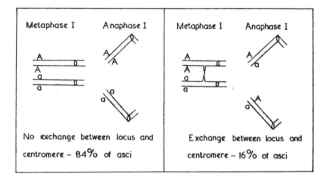

Fig. 2. Explanation of segregation of the mating type alleles *A* and *a* at the first or at the second division of meiosis, on the basis of a certain probability of exchange between homologous non-sister chromatids in the region between the mating type locus and the centromere.

Another aspect of the behaviour of the chromosome pair determining mating type, which is shown by Table 1, is that the orientation of the centromeres at metaphase I is a matter of chance (classes 1 and 2 not significantly different in frequency), and that the same applies to the orientation of chromatids at anaphase II (approximate equality of classes 3, 4, 5 and 6). To put this in another way, there is no tendency for one allele to pass towards one end of the ascus rather than towards the other. This absence of polarization is, in fact, found for all those pairs of alleles which have been studied in Neurospora; as we shall see, this implies non-polarized segregation for all seven chromosome pairs.

Nearly all inherited differences in Neurospora (the exceptions fall under the heading of extra-nuclear inheritance, which is discussed in Chapter 10) can be analysed in terms of differences between pairs of

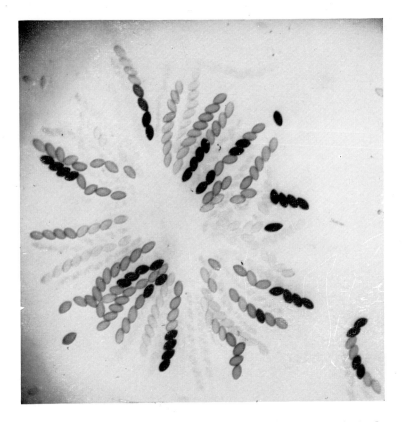

FIG. 3. Asci from the cross *lys-5* × wild-type in *Neurospora crassa*. The *lys-5* mutation prevents normal ripening of ascospores carrying it. The most mature asci in the photograph show four dark wild-type and four pale *lys-5* ascospores, in various arrangements indicative of first or second segregation. Many immature asci have all eight spores still pale; a few asci appear to have lost one or two spores. See Stadler[510].

chromosomal factors, or alleles, each of which segregates 4:4 in asci just as do the *A–a* pair determining mating type. The only difference between the behaviour of different pairs of alleles is in the characteristic frequency of second division segregation which each shows; this frequency can be anything between zero and about 67 per cent and, as will be explained below, can be used as a measure of the distance of the locus of the allelic difference from the centromere of the chromosome. The characteristic behaviour of segregating alleles can be shown most graphically where the genetic difference affects the appearance of the ascospore itself. Several such cases are known in different species of Ascomycetes; an example from Neurospora is illustrated in Fig. 3.

Independent assortment and linkage

When the diploid nucleus of the ascus initial contains a pair of unlike alleles, such as the *A–a* pair determining mating type, it is said to be *heterozygous* with respect to the chromosomal locus in question. When two identical alleles are present the ascus is said to be *homozygous*. The mating type locus is the only one for which Neurospora asci are invariably heterozygous, but a great number of allelic differences at other loci are available for experiment, mainly as a result of artificially induced mutation (see Chapter 3). It is thus possible to study the types and arrangements of ascospores obtained from asci heterozygous at two or more loci simultaneously. As an example we will take a sample of 1161 asci dissected from a single cross by Howe [244]. The parents of this cross actually carried different alleles at four different loci, but for the moment we will restrict our attention to two of them. One parent was of mating type *A* and required riboflavin for growth because of the presence of the allele *r*, while the *a* parent had only the wild-type nutritional requirements and carried the corresponding wild-type allele *r⁺*. It may be noted that the + superscript is generally used to denote alleles normally present in the wild-type. The data for the mating type and *r* loci are shown in Table 2.

The main point which emerges from Table 2 is that the segregation of *A–a* is independent of that of *r–r⁺*. It will be seen that the frequency of second division segregation of *r–r⁺* is even lower than that shown by mating type (2 per cent as compared with 12·5 per cent). Consequently the most frequent classes of asci are those in which both allelic pairs segregated at the first division. Among such asci it seems to be a matter of chance whether *r* is segregated with *A* or with *a* at the first division. The difference between the frequencies of the two cases (473 as against

521) is hardly significant statistically (the appropriate statistical test for significance, the χ^2 test, is outlined in Chapter 4). The frequencies of the other classes of asci are also reasonably close to what would be expected, given the frequencies of second division segregation of the two loci, on the assumption that the two allelic pairs were behaving quite independently.

It is clear that this kind of result is easily interpreted by supposing that A-a are associated with one homologous pair of chromosomes, and r-r^+ with a different pair, since visual observation of meiosis shows no evidence of any interaction between different bivalents.

TABLE 2. Independent segregation of mating type difference (A vs. a) and riboflavin requirement versus independence (r vs. r^+) in asci from the cross $ar^+ \times Ar$. (Data from Howe [244].)

Score pairs		Ascus type						
		1	2	3	4	5	6	7
1,2 and 3,4 *or*		Ar	Ar^+	Ar	Ar^+	Ar	Ar^+	Ar
5,6 „ 7,8		Ar	Ar^+	Ar^+	ar^+	ar^+	ar	ar^+
5,6 and 7,8 *or*		ar^+	ar	ar	Ar	Ar	Ar^+	Ar^+
1,2 „ 3,4		ar^+	ar	ar^+	ar	ar^+	ar	ar
Number of asci		473	521	21	143	2	0	1
Total: 1161								

If this sort of independent segregation depends on the loci concerned being in non-homologous chromosomes, and since the haploid chromosome number is seven, there can obviously be only seven genetic loci, or groups of loci, which can be inherited independently. Some hundreds of genetic differences have been investigated in Neurospora and it is now clear that they can indeed be ascribed to loci in seven groups. Members of different groups show independent assortment, while members of the same group show linked inheritance.

The principles of linked inheritance can best be illustrated by the analysis of asci heterozygous for at least three linked loci. Howe's data can again be used as an example here. His parent strain of mating type A carried not only r but two other mutant alleles at different loci: ad, causing a requirement for adenine in the culture medium, and v, causing a visible change in morphology associated with slow growth. The parent of mating type a carried the corresponding wild-type alleles ad^+ and v^+. Thus, neglecting r, which, as we have already seen, is inherited independently of mating type, we can write the cross as $A\ ad\ v \times a\ ad^+\ v^+$, or, for simplicity, $A\ ad\ v \times a++$. The classification of Howe's 1161 asci with respect to these three allelic differences is shown in Table 3.

The first point to notice is that all three loci are clearly linked, in the sense that the parental combinations of alleles occur more frequently among the progeny ascospores than recombinant types; indeed, in this particular example, ascospores of parental type outnumber all other kinds put together. The second important feature is that, in the more common classes of asci which do contain recombinant spores (classes 2, 3 and 4 of Table 3), only two of the four spore pairs are recombinant, the other two having parental combinations of alleles. Thirdly, in these same common classes of asci, recombination between A and ad is always accompanied by recombination between A and v, but not between ad and v (class 2); and recombination between ad and v is always accompanied

TABLE 3. Linked segregation of three allelic differences in asci from the cross A ad $v \times a$ ad^+ v^+*

Ascus type

	1	2	3	4	5
⎰	A ad v	A ad v	A ad v	A ad v	A ad v
⎱	A ad v	a ad v	a+v	A ad+	A+v
⎰	a++	A++	A ad+	a+v	a ad+
⎱	a++	a++	a++	a++	a++
Number of asci	888	85	43	126	0
	6	7	8	9	10
⎰	A ad v	A+v.	A ad+	A ad v	A ad v
⎱	a+v	a ad v	A++	a++	a++
⎰	A++	A ad+	a ad v	A ad v	A ad+
⎱	a ad+	a++	a+v	a++	a+v
Number of asci	0	0	1	2	2
	11	12	13	14	15
⎰	A ad+	A ad v	A ad v	A ad+	A ad+
⎱	a+v	a ad+	a ad+	a ad v	a ad v
⎰	A ad+	A+v	A++	A+v	A++
⎱	a+v	a++	a+v	a++	a+v
Number of asci	2	3	5	3	1

* ad = adenine requirement, v = 'visible' slow growth; wild-type alleles of these mutants represented by + in the body of the Table.

N.B. Spore pair types occurring in the same half ascus are bracketed. Asci which differ only in the reversal, in one relative to the other, of the two halves of the ascus, or of the two spore pairs in one-half of the ascus, are lumped together. (Data from Howe [244].)

by recombination between A and v, but not between A and ad (classes 3 and 4). Postponing for the moment consideration of the much less frequent classes 5–15, we can account for classes 2, 3 and 4 by postulating that the three loci are linearly arranged along a chromosome in the order A–ad–v, and that reciprocal exchanges can occur between chromatids of homologous chromosomes during the first division of meiosis. On this interpretation, asci of class 1 result when no exchange occurs in the A–v

region, asci of class 2 result from single exchanges between A and ad, and those of classes 3 and 4 from single exchanges between ad and v. We have already seen that exchanges of the kind just postulated will account for segregation of alleles at the second, rather than the first division of meiosis; second division segregation will occur as a result of a single exchange between the locus in question and the centromere. In classes 2 and 3, mating type segregates at the second division and v at the first. In class 4, on the other hand, v segregates at the second division and mating type at the first. Since these are by far the most common classes in which exchanges occur at all, it is apparent that mating type and v must be on opposite sides of the centromere.

Asci of class 3 result from exchanges which fall between ad and the centromere, and those of class 4 are due to exchanges which fall between the centromere and v. The completed map must be A–ad–centromere–v.

The interpretation of asci of classes 1 to 4 in terms of single exchanges is shown in Table 4. Classes 5–15 have to be explained as due to various kinds of double exchange, which are specified in Table 4. It will be noted that it is necessary to suppose that all four chromatids can participate in exchanges. There are, in fact, four kinds of double exchange: 'two-strand' doubles, in which the same two strands are involved in both exchanges, 'four-strand' doubles in which two chromatids participate in the first exchange and the other two in the second, and two kinds of 'three-strand' doubles in which one chromatid is involved twice and two others once each. The data are not sufficiently extensive to show whether the four types occur with nearly equal frequency, as would be expected if the strand composition of one exchange is without effect on that of the other, but they are consistent with this being the case.

It is pertinent to inquire whether the frequencies of these postulated double exchanges are about what would be expected on the simple hypothesis that the probability of a double is the product of the individual probabilities of the two component single exchanges. Taking the doubles into account, we can calculate the probability of an exchange occurring in a given region. Thus, from the data of Table 4 there are 98 exchanges between A and ad (region I) in 1161 asci, a frequency of 8·4 per cent. The corresponding frequencies for the intervals ad–centromere (region II) and centromere–v (region III) are 4·3 per cent and 12·4 per cent respectively. The products of these frequencies give, on the basis of our simple hypothesis, expected frequencies of 0·36, 0·53 and 1·04 per cent for doubles in regions I and II, II and III, and I and III respectively. The

TABLE 4. Explanation of the ascus types listed in Table 3 in terms of exchanges between chromatids

(chromosome regions indicated by Roman numerals)

Chromatids

```
 1    A            ad           v
 2    A    I       ad    II  III   v
 3    a    +       +
 4    a    +       +
                   centromere
```

Ascus type	1	2	3	4	5	6	7	8
Simplest explanation-type of exchange	No exchanges	Single in I	Single in II	Single in III	Double in I and II (2-strand)	Double in I and II (3-strand-2,3-2,4)*	Double in I and II (3-strand-2,3-1,3)*	Double in I and II (4-strand)
Number of asci	888	85	43	126	2	0	0	1

Ascus type	9	10	11	12	13	14	15	
Simplest explanation-type of exchange	Double in II and III (2-strand)	Double in II and III (3-strand-2,3-2,4 or 2,3-1,3)*	Double in II and III (4-strand)	Double in I and III (2-strand)	Double in I and III (3-strand-2,3-2,4)*	Double in I and III (3-strand-2,3-1,3)*	Double in I and III (4-strand)	All others. Triples of various kinds
Number of asci	0	2	2	3	5	3	1	0

Total exchanges

I	98
II	50
III	144

Total asci 1161

Double exchanges

	Observed	Expected
I–II	3	4·2
II–III	4	6·2
I–III	12	14·4

* In distinguishing between the two kinds of three strand double exchange, the exchange to the left is arbitrarily designated as involving chromatids 2 and 3. The two possible kinds of three strand double under Class 10 are indistinguishable.

corresponding observed frequencies are approximately 0·3, 0·4 and 1·2 per cent, a very close fit to expectation considering the relatively small numbers of doubles in the data. On the same basis, the expected frequency of asci due to triple exchanges in regions I, II and III is only 0·04 per cent, so it is not surprising that no asci occur which have to be explained in this way. It will be seen that the whole of the data can be satisfactorily accounted for on the basis of definite probabilities of exchange between linked loci without the necessity of assuming any kind of interference between different exchanges. We shall return to consider this question of interference in more detail in Chapter 4.

The explanation of the ascus data which has just been outlined would clearly be on a firmer basis if one could actually see exchanges between chromatids. In Neurospora the relevant stages of meiosis cannot be seen in sufficient detail for this to be possible, but in some higher plants and animals the appearance of chiasmata at diplotene is exactly what would be expected if chiasmata represented chromatid exchanges. Whether it is safe to conclude that visible chiasmata at meiosis *all* represent genetic exchanges is still a matter of controversy, but there seems to be little doubt that such an interpretation of chiasmata is usually correct (see, for example, the study of chiasmata in the lily made by Brown and Zohary [60]; the whole question is fully discussed by Swanson [534]).

In the diagrammatic representation of linkage groups, it is customary to space the loci according to the amount of recombination which occurs between them. The unit in which linkage map distances are expressed is 1 per cent of recombination among the meiotic products. The map based on the data of Table 3 is thus as follows:

$$\longleftarrow 4\cdot3 \longrightarrow \quad \longleftarrow 2\cdot15 \rightarrow \quad \longleftarrow 6\cdot1 \rightarrow$$

$$A \qquad\qquad ad \qquad\qquad\qquad v$$

It will readily be seen that a map distance of 50 units corresponds to an average incidence of one exchange per ascus in the chromosome region in question, since one exchange involves two out of the four chromatids and thus leads to 50 per cent recombination. Hence the length of a linkage group in map units should be equal to 50 times the average number of chiasmata in the corresponding bivalent at meiosis, if chiasmata represent exchanges. In maize, where both accurate chiasma counts and rather complete linkage maps are available, the correspondence between the two is in quite good agreement with this prediction [534]. In Neurospora chiasma counts of comparable accuracy are not available, but the indications are [495] that bivalents usually have two or three chiasmata, which is in good accord with the present linkage maps (see Fig. 21) which indicate a probable length of 100 to 150 units for several of the chromosomes.

THE CHROMOSOME THEORY

We will now summarize the main points of the chromosome theory at which we have arrived. Most inherited differences can be analysed in terms of differences between homologous chromosomes, which thus seem to be largely responsible for the transmission and determination of hereditary characteristics. Pairs of character differences determined by different (non-homologous) chromosomes are inherited independently of each other; those determined by the same chromosome show linked inheritance. Limited recombination between linked factors is possible as a result of exchanges between homologous chromosomes. These exchanges which lead to genetic *cross-overs*, vary in their frequency and position from one cell to another; they occur after the paired chromosomes at the first division of meiosis have divided into chromatids, and they each involve only one chromatid from each chromosome, though different chromatids may participate in successive exchanges. Through the analysis of patterns of crossing-over, it is possible to construct linkage maps in which the loci of genetic differences are arranged in linear order, each corresponding to a particular site on a chromosome.

If a change at a particular chromosomal locus can determine an inherited difference in the organism, there must clearly be physiologically important structures in the chromosome which can undergo change (mutation). Such structures, which are really only recognizable through their capacity for mutation, may conveniently be called *genes*. The more precise definition of the gene is quite a complicated matter, and will be discussed at much greater length in later chapters. For the present it will suffice to point out that alleles (mutually exclusive genetic factors located in corresponding segments of homologous chromosomes) may conveniently be defined as different forms of the same gene, and that many more than two alleles of one gene can exist, although, of course, no more than two of them can be present in any one diploid cell.

CHAPTER 2

OUTLINE OF THE BIOLOGY OF FUNGI
OF GENETIC INTEREST

The purpose of this chapter is to outline the more important features of some of the fungi either in current use in experimental genetics, or which seem to demand genetical investigation. It is intended primarily as an introduction for the reader who is not a specialist in mycology. Genetical studies with fungi are of necessity almost entirely conducted in the laboratory, and so the practical experimental techniques which have been successfully employed with the various species are emphasised here, and the natural or ecological aspects have to a large extent been neglected.

NUTRITION

All fungi have in common a basic inability to use carbon dioxide as a major source of carbon, and are therefore saprophytic or parasitic. For reasons of experimental convenience geneticists have, on the whole, chosen to study fungi which are able to complete their life cycles on non-living media, that is to say, saprophytically. A variety of culture media may be employed, all of which provide mineral salts, a source of organic carbon, and possibly organic nitrogen and vitamins. The solution containing these components is frequently made into a gel by the addition of agar. For large scale cultures liquid medium in aerated vessels of up to 15,000 gallons capacity may be employed [168], as in the production of penicillin from species of Penicillium. The nutritional requirements of the majority of fungi with which we shall be concerned are known with precision, so they may be cultured on completely defined nutrient media. The simplest chemically defined medium on which the wild type of a species will grow well is often called *minimal medium* for that species. Sometimes a fungus is grown on several different media to enable it to complete its life cycle satisfactorily in culture. This is true of *Neurospora crassa* and *Coprinus lagopus*, for example.

Table 5 lists culture requirements of some ten different fungi of importance in experimental genetics.

TABLE 5. Cultural requirements of fungi which have been used in genetics

	Opt. temperature for vegetative growth	Special nutrient requirements	Opt. pH	Special fruiting requirements	Max. growth rate mm/24 hrs	References
Allomyces arbuscula	25°	thiamin, methionine	7	media poor in carbohydrate	ca. 12 hrs*	72, 140, 331
Saccharomyces cerevisiae	30°	none	5–6		2·5 hrs* at 25°	315, 582
Ophiostoma multiannulatum	25°	thiamin, pyridoxin	?	malt extract	?	177
Venturia inaequalis	25°	thiamin	5–6	malt extract and apple leaf decoction at 8°	?	282
Neurospora crassa	35°	biotin	5–6	pH 6·5 25° limiting nitrogen	120·0–144·0	470, 542, 570
Ascobolus immersus	? (25° used)	thiamin, biotin	? (6·5 used)	urea replacing asparagine as N-source; high C/N ratio	?	604
Aspergillus nidulans	37°	none	6·5	none	6·5	424
Ustilago maydis	30°	none	5·8	host plant	2·5–3·0 hrs*	232
Coprinus lagopus	37°–40°	thiamin	6·8	light, dung at 28°	6·0	176, 335, 536
Schizophyllum commune	32°	thiamin	6·8	light	5·3	442

* Approximate doubling time in liquid culture measured as mass for *Allomyces* and cell number for *Saccharomyces* and *Ustilago*.

LIFE CYCLES AND BREEDING SYSTEMS

The fungi have a variety of life cycles which can be classified roughly according to the relative importance of the haploid and diploid phases. In this respect they are like other organisms which show an alternation between haploid and diploid phases but in the majority of the fungi the diploid phase is very short lived and is not even reproduced mitotically. Many fungi can form *heterocaryons* in which genetically different haploid nuclei may co-exist and multiply in the same cell or cytoplasmic system. Heterocaryons can originate by mutation in a homocaryon, in which all the nuclei are identical, or by the migration of nuclei through points of fusion between different hyphae. They are of several different kinds, ranging from the transient condition prior to fusion of haploid nuclei after sexual fusion of gametes, through forms which arise casually and are readily resolved into their components by the production of uninucleate spores, to the regularly binucleate cells of the ascogenous hyphae of Ascomycetes or the stable dicaryon of the higher Basidiomycetes.

Although systematic mycologists rarely agree on any one detailed scheme of classification of the fungi most accept a division into four main classes: Phycomycetes, Ascomycetes, Basidiomycetes and Fungi Imperfecti. The examples we shall examine are drawn from all four classes. Although the true Fungi Imperfecti have no sexual or so-called perfect stage some of them are amenable to parasexual genetical analysis (Chapter 5).

Phycomycetes

This is a large and complex group of fungi of very varied forms. In some the organism is reduced to a single cell. The mycelial species are *coenocytic*, that is they form hyphae without cross walls which contain many nuclei in a common cytoplasm. Genetical analysis has been almost entirely limited to two orders, the Blastocladiales and the Mucorales.

Blastocladiales

The Blastocladiales are a small order of water moulds and soil fungi. The smaller forms (Blastocladiella) consist at first of a single cell which cuts off a terminal cell forming a sporangium. The larger mycelial forms (Allomyces) have a tree-like organization with branches on which the reproductive structures are borne. The members of this order in general appear to have an extended diploid phase, and at least two genera can be subdivided on the basis of the relative importance of a haploid gamete-forming phase in relation to the diploid. *Allomyces arbuscula* has equally important haploid and diploid phases and is representative of the sub-genus Euallomyces; its life cycle is shown in Fig. 4.

The diploid phase of *A. arbuscula* reproduces by motile uniflagellate spores produced in a colourless spore case, or *sporangium*. Since the nuclear divisions which give rise to these spores are mitotic, Emerson [140]

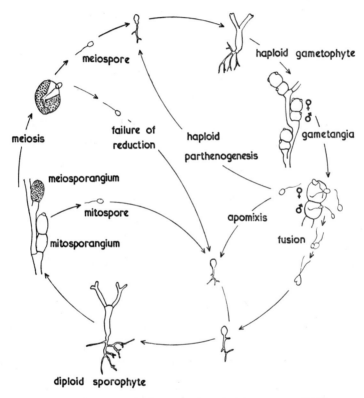

FIG. 4. Life cycle of *Allomyces arbuscula* (after Emerson[139]).

has called the sporangia mitosporangia and the spores mitospores. The sporophyte, or spore forming diploid phase, also produces dark brown meiosporangia. As their name implies, the nuclei of a meiosporangium undergo meiosis, but this is deferred until the meiosporangium germinates. A mature meiosporangium contains from eight to twenty diploid nuclei. When it is placed in water, meiosis and the formation and release of from twenty to ninety-six uniflagellate, motile meiospores occur within 100 to 130 minutes [141, 583]. The meiospores germinate to form haploid hermaphrodite mycelia bearing male and female gametes differentiated by size and produced in separate *gametangia*.

In *A. cystogena* (sub-genus Cystogenes) the haploid phase is reduced to a single gametangium in a cyst formed directly by the meiospore. Each cyst gives rise to four undifferentiated gametes, and these fuse in pairs to reconstitute the sporophyte. In *A. anomala* (sub-genus Brachy-allomyces) there is no cyst formation and no sexuality. The uniflagellate spores produced by the dark brown sporangia germinate directly to form mycelia bearing dark brown sporangia. By analogy with the other forms only the diploid phase is represented in Brachyallomyces. Species of the genus Blastocladiella also show three types of life cycle similar to those in Allomyces.

Crosses between *A. arbuscula* and *A. macrogynus* were made by Emerson and Wilson [142]. The paired male and female gametangia in each species were separated. The male gametangia in both species are orange, sub-terminal in *A. arbuscula* and terminal in *A. macrogynus*. When the uniflagellate gametes have emerged from the isolated gametangia they are mixed and transferred to hanging drops with a fine pipette. After three hours the biflagellate zygotes which are present may be transferred to nutrient agar where they grow into sporophytes. In some strains the female gametes may develop without fertilization (i.e. apomictically) giving rise to nonhybrid sporophytes. Studies of chromosome numbers in strains of the two parents isolated from the wild revealed a polyploid series in each of them. Haploid chromosome numbers of 8, 16, 24, and 32 occurred in *A. arbuscula* and 14, 28 and 50+ in *A. macrogynus*. The hybrids studied by Emerson and Wilson were from parental strains with 16 and 28 chromosomes respectively in their gamete nuclei. The segregations of parental characters were complicated by aneuploidy (see Chapter 5). Some of the hybrids were practically unisexual, with 1 per cent or fewer gametangia of the minority sex, and were later used to demonstrate that a sex hormone is produced by the colourless female gametangia [329, 330].

Undoubtedly the occurrence of polyploidy in Allomyces accounts for some of the difficulties encountered in genetical analysis and the isolation of mutants. When each type of chromosome is represented several times, a mutation in one chromosome will often have no obvious effect. The use of strains with the basic haploid chromosome number should facilitate the investigation of the genetic control of sexual and morphological differentiation for which the group seems so well suited. The physiological and morphogenetic studies of Cantino and his associates [70, 72] will form a valuable background to an approach of this kind.

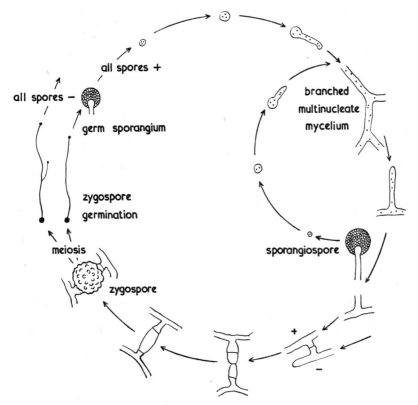

Fig. 5. Life cycle of *Mucor hiemalis*. (Partly after Sjowall [497]).

Mucorales

This is a large order mostly made up of soil or dung inhabiting saprophytes which form extensive coenocytic mycelia with sporangia containing non-motile, usually multinucleate spores. The life cycle of *Mucor hiemalis* is shown in Fig. 5.

Sexual reproduction is by fusion of two multinucleate hyphal cells or gametangia, with the formation of a thick-walled, usually dormant *zygospore*. At germination the zygospore forms a germ sporangium which is similar in appearance to the vegetative sporangium. Many of the Mucorales are heterothallic with two mating types called + and —. The greatest obstacle to genetical work is the difficulty of germinating zygospores. Another difficulty appeared to be the apparent lack of hyphal anastomosis between strains of the same mating type preventing the formation of heterocaryons. However, recently the synthesis of heterocaryons from morphological or nutritional mutants has been claimed in *Mucor rammanianus* (E. H. Evans, personal communication) and *Rhizopus javanicus* [355]. Among the species which have been investigated there are variations in the events following the fusion of gametangia and leading to the release of spores from the germ sporangium. The inheritance of mating type is a good example of this variation. Blakeslee [40], who discovered heterothallism, defined four types of zygospore germination. These were: (1) pure germinations in which all germ sporangiospores were of the same mating type either + or —; (2) mixed germinations in which both + and — spores were present; (3) mixed germinations in which +, — and homothallic (i.e. self-fertile) spores were present; (4) pure germinations in which all the spores were homothallic, this last type being found in the homothallic species. It is generally agreed that during its initial development the zygospore contains many nuclei from the gametangia and that these appear to fuse in pairs. What is not clear, however, is how many of the diploid nuclei thus formed undergo meiosis and at what stage. In those species which give mixed germinations it is not known whether all the products of each meiotic division survive and can be recovered.

Since Blakeslee defined the four types of germination many workers have studied the cytology of zygospore development. In the heterothallic species Cutter [95] defined three basic types of nuclear behaviour according to whether nuclear fusion and meiosis occurred before the resting period, at germination or in the germ sporangium. Sjowall [497] however recognized only two types of behaviour. In the first group, which included *Mucor hiemalis*, *M. mucedo* and *Absidia glauca*, nuclear fusion and meiosis occurred before the resting period. In the second group nuclear fusion was delayed until zygospore germination. Meiosis took place in the germ sporangium and involved many diploid nuclei in *Phycomyces blakesleeanus* but only one nucleus in *Rhizopus nigricans*.

The cytological findings do not completely account for the genetic evidence. Thus Kohler [289] found that in *M. mucedo* not only are all the spores from a germ sporangium of the same sex but in crosses involving two pairs of alleles only one of the four possible genotypes was found in any one germ sporangium. However Sjowall claimed to have observed degeneration of nuclei in zygospores during the resting stage which would account for their each having only one or a few nuclei of the same genotype at the time of germination. In *P. blakesleeanus* Burgeff [62] found that in crosses heterozygous at three loci no more than four spore genotypes were produced in each germ sporangium. In each case the four types can be interpreted as the complementary meiotic products of a single diploid nucleus. The zygospores can thus be scored as ditype or tetratype for each pair of loci (cf. Table 10). There is no indication from cytological studies in this species how all except one of the diploid nuclei are eliminated. Recently Gauger [192] germinated zygospores of *Rhizopus stolonifer* (= *R. nigricans*) and found that they gave both pure and mixed germ sporangia with respect to the mating type of the spores. No spores were found which were homothallic. For the present it would seem that the cytological evidence must await detailed genetical studies before it can be evaluated.

The Mucorales would seem to offer excellent opportunities for the transfer of nuclei from one strain to another by Wilson's [584] micro-injection technique (see p. 233). The hyphae are wide and, judging by their capacity to survive mechanical injury in culture transfer, should be able to heal after puncture with a needle. If artificial nuclear transfers are possible, the general failure of members of the order to form hetero-caryons naturally need be no great hindrance to experiment.

Peronosporales

This order includes some important plant pathogens among which is *Phytophthora infestans*, the cause of late blight in potatoes. While the my-celium is similar to that of the Mucorales, the thin walled sporangia are borne on richly branched structures, the *sporangiophores*. The tips of the sporangiophore branches can continue growth for an indefinite period and produce a succession of sporangia. Germination in water generally gives rise to biflagellate zoospores which swim away and, after coming to rest, produce an infection hypha which penetrates the host tissues through a stoma. Many members of the order are heterothallic. Heterothallism was only recently discovered in *Phytophthora infestans;* the populations occurring in Europe and North America consist of only one mating type. Both mating types occur in central Mexico, where the fungus grows on a wide range of primitive cultivated potatoes and wild Solanum species [186]. Sexual reproduction involves the fertilization of an *oogonium* containing a single egg by a nucleus from an adjacent male hypha or *antheridium*. Cytological evidence has been put forward suggesting that meiosis occurs in the oogonia and antheridia of *Pythium debaryanum* and *Phytophthora cactorum* [474, 475, 477]. If correct, this would mean that the greater part of the life cycle of these two species, and possibly other Peronosporales, is diploid. The only clear demonstration of meiosis in the Phycomycetes is in *Allomyces arbuscula* and *A. macrogynus*, where it conveniently occurs when meiososporangia are placed in water. In view of the general difficulty of interpreting division figures in other Phyco- mycetes the most crucial test for meiosis remains a genetic one, the characteristic segregation and recombination of markers which it brings about.

Germination of the oospores of the Peronosporales is difficult. Some success has been achieved in *Phytophthora erythroseptica* and *P. cactorum* by passing the oospores through the digestive tract of the snail *Helix aspersa* [202]. Oospores of *P. infestans* have been germinated on dung decoction media [502].

Another order of water moulds, the Saprolegniales, includes the genera Dictyuchus and Achlya. There is cytological evidence that meiosis takes place in the antheridia and oogonia of Achlya [477]. This genus has been the object of intensive study of the hormonal control of sex organ development (see p. 228), but genetical investigation has been entirely thwarted by failure to germinate oospores. Oospore germination was readily obtained by Couch [88] in Dictyuchus, and many problems concerning the genetics of sex organ determination may well be solved by future work with this genus.

Ascomycetes

The Ascomycetes form the largest class of the fungi and provides most of the species which have been widely used in genetics. The principal feature of this group is the ascus, which is a special type of sporangium enclosing the products of meiosis, the ascospores; these are either four or, following an extra mitotic division, eight in number. We adopt here the widely accepted system of classification which recognizes two sub- classes: the Hemiascomycetes in which the asci are formed singly and not in a fruit body, and the Euascomycetes in which the asci are borne on ascogenous hyphae usually enclosed within a fruit body.

Endomycetales

This is the most important order of Hemiascomycetes, and includes the yeasts (Saccharomycetaceae) which are unicellular fungi with no true mycelium. The yeasts are of great economic importance in baking and brewing, and are a valuable source of vitamins for the food industry. A great many genetic studies have been made with members of the genus Saccharomyces. The life cycle of *S. cerevisiae* is shown in Fig. 6.

Haploid cells of opposite mating type fuse readily to form the diploid phase. Pure haploid lines are generally established from individual ascospores dissected from intact mature asci with a micromanipulator. A sporulating colony may be freed of vegetative cells by suspending the

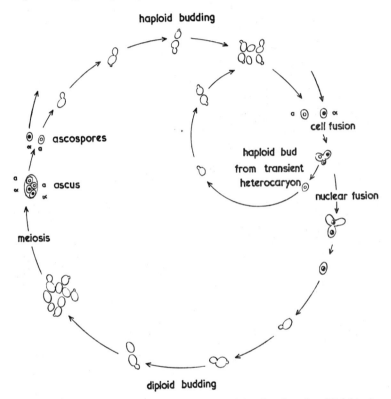

FIG. 6. Life cycle of *Saccharomyces cerevisiae* (partly after Wright & Lederberg[595]).

mixture of ascospores and vegetative cells in water, and holding at 58°C for 2 to 4 minutes [315]. The vegetative cells are killed but the ascospores, which are more heat resistant, survive and will grow when plated on solid media. Some of the resulting colonies are likely to be diploid because of copulation between adjacent spores, of opposite mating type, which are still held together in intact asci. Both haploid and diploid cells increase in number by budding. Each dividing cell produces a small outgrowth which increases in size until it has reached about half the size of the parent cell, when it is cut off as a daughter cell. This growth as separate cells is experimentally convenient in several ways. Cell numbers can be estimated by measurements of the optical densities of suspensions, and replica plating (see p. 51) of colonies is easily possible. Furthermore, the free cell growth means that samples of uninucleate cells can be obtained from growing cultures at any time. None of these technical advantages is readily available in fungi with a mycelial organization.

Matings in Saccharomyces are carried out by pairing single cells [84], by isolating mating pairs from mass matings, or by selecting prototrophic (i.e. nutritionally wild type) diploids from mass matings of complementary auxotrophic parental strains spread on minimal medium [417]. Fused haploids of *S. cerevisiae* usually bud off diploid cells only, but occasionally form haploid buds which are not recombinant for parental markers. Thus the product of fusion of haploid cells appears to be a transient heterocaryon [169, 595]. The diploid cells are larger than haploids, but are otherwise very similar to them. Meiosis and ascus formation take place when diploid cells are placed on a starvation medium which is usually a gypsum block [320] or sodium acetate agar [332].

The cytology of the yeast cell has long been a subject of dispute. Indeed, the evidence for haploid and diploid states in *S. cerevisiae* is largely genetical except for the demonstration by Ogur et al [385] that diploid cells possess twice as much deoxyribonucleic acid as haploids (cf. p.171). Lindegren [315] at one time maintained that the central vacuole commonly seen in stained preparations is the nucleus, containing six or seven chromosomes in the haploid phase. A darkly staining body to one side of the vacuole he believed to be the centrosome (a structure containing the centriole). Other investigators, notably DeLamater [114], claim that this body is the nucleus, and that it is not attached to the vacuole. More recent accounts by Lindegren et al. [323, 602] suggest some measure of agreement with DeLamater's interpretation. No reliable chromosome counts are yet available. The most recent linkage map of Hawthorne and Mortimer [220] has ten groups of markers linked to different centromeres, four of which have markers in both chromosome arms.

The fission yeast *Schizosaccharomyces pombe* multiplies by a process of equal cell division rather than by budding. The diploid cell formed by the fusion of two haploids usually undergoes meiosis without any intervening growth or cell division. The ascus is dumb-bell shaped and has four haploid spores arranged in a row. The experimental selection of diploids may be carried out by a method devised by Leupold [304]. Cells from a mating mixture are spread on nutrient agar. Diploid colonies, identified by their production of ascospores which stain brown when briefly exposed to iodine vapour, are then selected. Non-sporulating cells are isolated from the diploid colonies and a proportion of these cells are stable diploids. The diploids can be distinguished from haploids by their larger cells.

Some other members of the Endomycetales have a mycelial organization, and a few have a well-defined diploid phase like that of the yeasts. In *Ascocybe grovesii* the haploid mycelium produces an erect diploid branch, called an *ascophore*, with clusters of four-spored asci at its tip. When dissected off the parental haploid mycelium the diploid ascophore will grow indefinitely in liquid culture, retaining its unbranched morphology [121]. Mutants with specific growth requirements have been isolated by Dixon (unpublished), but attempts to produce ascophores of hybrid origin have been unsuccessful so far.

The Euascomycetes may be divided into three sub-groups: Plectomycetes, Pyrenomycetes and Discomycetes, which are separated by differences in the structure of their fruit bodies.

Plectomycetes

The fruit body of the Plectomycetes is called a *cleistothecium*, and the enclosed ascospores are released by rupture of the cleistothecium wall. The most important genera are Aspergillus and Penicillium. Some of the species usually included in these genera, on the basis of vegetative morphology, have no known sexual stage. The species which have been studied genetically include *Aspergillus nidulans* [424], *A. niger* [423], *A. oryzae* and *A. sojae* [251], and *Penicillium chrysogenum* [425]. All of these have been studied by exploiting parasexual recombination (see Chapter 5), but only *A. nidulans* has also been subjected to more orthodox genetic analysis utilizing sexual reproduction. The life cycle of *A. nidulans* is shown in Fig. 7.

The following brief account of *A. nidulans* is largely based on the authoritative paper of Pontecorvo et al [424]. The colourless mycelium is made up of multinucleate cells and bears conidiophores, each of which arises from a foot cell and ends in a globose multinucleate vesicle (cf Fig. 7). From the surface of the vesicle uninucleate primary finger-like *sterigmata* protrude, and each of these gives rise to two or more secondary sterigmata. Each secondary sterigma remains uninucleate, but buds off a chain of uninucleate conidia by repeated division of its nucleus; the youngest conidia are at the base of the chain. The conidia are dark green to dull-grey -green in colour in wild strains.

Hyphal fusions within the same mycelium or between different mycelia occur frequently. Heterocaryons are fairly easily produced by either of two methods. A balanced heterocaryon may be obtained by inoculating a supplemented liquid medium with large numbers of conidia of two mutants requiring different nutritional supplements, both present in the

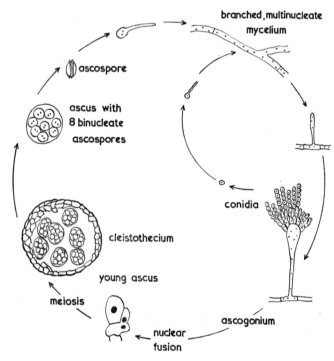

FIG. 7. Life cycle of *Aspergillus nidulans* (partly after Elliott[137]).

medium. After incubation for 24 hours, or long enough to produce a thin mesh of mycelium, the latter is removed, washed in liquid minimal medium, and teased out over the surface of solid minimal medium. After approximately four days, growth starts from a few points, and single hyphal tips may be isolated. The majority of these are balanced heterocaryons capable of vigorous growth on minimal medium. Alternatively, the mixed conidial suspensions may be streaked or stabbed onto the surface of solid medium. If the heterocaryon is more vigorous than its parent monocaryons it will grow out from some of the inocula and can be isolated.

The separate components of a heterocaryon may be recovered by isolating single conidia. Where the vesicle of a conidiophore is heterocaryotic, all the conidia in one chain are of the same type, although adjacent chains may differ. This is most strikingly shown when the different nuclei present in the heterocaryon determine different kinds of pigmentation in the conidia carrying them.

Although *A. nidulans* is homothallic, crosses can readily be made with the aid of suitable genetic markers. Two methods may be used. In the first, strains with complementary nutritional requirements are paired on minimal medium to form a balanced heterocaryon in which each type of nucleus supplies the function lacking in the other. The second method is similar except that it does not require auxotrophic mutants. Mixtures of equal proportions of conidia of the two parental strains are plated at high densities, not less than 5×10^6 per plate, so that most of the limited growth which can occur before the development of cleistothecia, is heterocaryotic. Both methods commonly yield more than 50 per cent of asci of hybrid origin [224]. In certain crosses the proportion of hybrid asci approaches 100 per cent. It seems clear that, in such crosses, the development of asci which are heterozygous rather than homozygous for the mutant markers contributed by the parents must be strongly favoured.

Even under the most favourable circumstances, however, the proportion of asci resulting from self-fertilization will be appreciable, and some method must be used which permits the exclusion of spores from such asci from the analysis of a cross. Several methods are available. In one, known as recombinant selection, only spores which are recombinant for two genetic markers, usually unlinked, are selected for further analysis. A second technique, called cleistothecium (or 'perithecium') analysis, followed from the finding that individual cleistothecia practically always produce asci which are either all selfed or all crossed. Thus, if a sample of intact asci from a single cleistothecium are found to be of crossed origin, the remaining ascospores from the same fruit body can, with a high degree of confidence, be regarded as genuine progeny of the cross. Analysis can then proceed using these random ascospores. It seems likely that, as in Neurospora, only two nuclei, which multiply by conjugate division, usually give rise to all the nuclei of the ascus initials of one fruit body. In a third technique, intact asci are dissected, and the eight spores cultured individually. Asci of selfed origin, recognized by their failure to show segregation of the markers distinguishing the parents, are rejected. In a study of this kind Strickland [520] examined 1,842 asci from nine different crosses; of these 1,773, or 96.2 per cent, were hybrid. Mature cleistothecia contain up to 100,000 asci each. The ascospores are hat-shaped and red-brown in colour. The cleistothecia show no interrupted stage of development comparable to the protoperithecium of Neurospora. The proper title of the Aspergillus fruit body is a matter of some disagreement. Although the term cleistothecium is mycologically correct, geneticists working on the fungus almost always use perithecium. Elsewhere in this book we shall fall in with the customary genetical usage.

The chromosomes of *A. nidulans* are small and not very easy to study. However, the haploid number of eight is very clearly established [137], and agrees with the number of known linkage groups [272].

The synthesis of diploid lines of Aspergillus, and the methods used in parasexual analysis, are described in Chapter 5.

Other species of Aspergillus, and *Penicillium chrysogenum*, are, except for the absence of a sexual stage, similar in general to *A. nidulans*. The conidiophore of Penicillium differs from that of Aspergillus in having no foot cell and, instead of a terminal vesicle, a brush-like structure which consists of whorls of branches bearing sterigmata. Each sterigma forms a chain of uninucleate conidia. In *P. chrysogenum*, which has a slower growth rate than *A. nidulans*, the growth on minimal medium given by pairs of complementary auxotrophic mutants is normally due to mutual stimulation by diffusion of metabolites (syntrophism), rather than to heterocaryosis. In order to overcome this difficulty, double mutants, which are much less readily stimulated syntrophically, are used in the synthesis of balanced heterocaryons [425].

Pyrenomycetes

In the Pyrenomycetes the asci are borne in a flask shaped fruit body called a perithecium. This has a small opening, or *ostiole*, through which the ascospores are discharged when they are mature. The most intensively studied species in this group, or indeed among all fungi, is *Neurospora crassa*.

A detailed account of development in Neurospora has already been given in Chapter 1. The life cycles of the other Pyrenomycetes are very similar to that of Neurospora which is shown in the diagram in Fig. 8.

Most Pyrenomycetes produce eight haploid ascospores in each ascus, and a few (e.g. *Chromocrea spinulosa*) even have sixteen. Some, however, produce four-spored asci (e.g. *Neurospora tetrasperma, Podospora anser-*

ina and *Gelasinospora tetrasperma*) each spore containing, at the time of its formation, two nuclei of opposite mating type (see Chapter 9). The form of the ascospores varies from a single uninucleate cell (*Glomerella cingulata*), a binucleate cell (*Neurospora crassa*), or two uninucleate cells (Venturia), to long filiform multiseptate threads coiled round each other

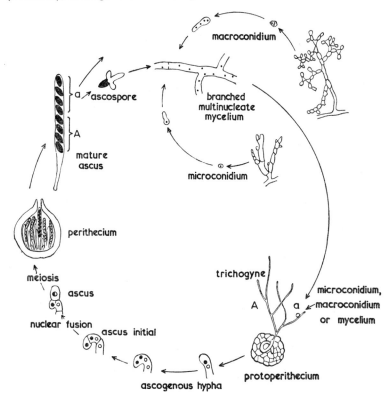

FIG. 8. Life cycle of *Neurospora crassa*.

in the ascus (Cochliobolus). The vegetative mycelium is septate with from one to two nuclei per cell, as in Venturia, to many per cell as in Neurospora. Pores in the cross walls between adjacent cells which allow the passage of nuclei and cytoplasm have been observed in *N. crassa* [487] and Gelasinospora [127] and other fungi [365], and are probably of widespread occurrence. The mycelia generally form vegetative spores (conidia), and some species (e.g. *Podospora anserina*) form microconidia which function as male fertilizing elements, fusing with the trichogyne at fertilization. Sometimes, as in Neurospora, both micro- and macro-conidia can function in either capacity. Hyphal fusion by anastomosis has been recorded in several species and is probably common. In Neurospora, provided the anastomosing strains are heterocaryon compatible (see p. 233), heterocaryon formation is easily achieved in the laboratory by superimposing inocula on agar. In *Venturia inaequalis* heterocaryons could not be synthesized by the techniques employed with *Aspergillus nidulans* and *Penicillium chrysogenum*.

The mycelia, conidia and ascospores of many species of Pyrenomycetes are pigmented which makes possible isolation of mutants with altered pigmentation. The pigments of a number of species of Helminthosporium have been identified by Raistrick [434]. Some of these species have known perfect stages in the genus Cochliobolus and should be well suited to mutational studies of pigment biosynthesis with a concurrent genetical analysis.

Methods for making crosses vary within the group. In Neurospora conidia of one mating type are dusted or poured in suspension over a culture of the other mating type which has already formed protoperithecia after six days' growth on a medium favouring sexual reproduction [570]. Crosses are made at 25°C and ascospores are ripe in about a week after fertilization. In *Venturia inaequalis* conidia and mycelia of the two mating types are mixed and added to cooled molten nutrient agar containing a decoction of apple leaves. After a short period of incubation the cultures are held at 8°C for two to three months during which time perithecia form [282]. The laboratory method simulates the period of overwintering which Venturia normally undergoes in nature.

Many Pyrenomycetes are homothallic, and methods for selecting perithecia of crossed origin have to be employed, as in *Aspergillus nidulans*. In *Sordaria fimicola* cultures are placed some distance apart on the surface of nutrient agar and form a line of perithecia where they grow into contact. If this method fails, co-inoculation of the two stocks at the same point may be successful [387]. Ascospore colour mutations are used to recognize asci of crossed origin. Another method, which is discussed in Chapter 9, is to use mutants in which the development of normal perithecia is blocked at different points. Such mutants are normally sterile, but when paired form perithecia, some of which may, however, be of selfed origin.

Discomycetes

In this class the asci and interspersed sterile cells form a *hymenium*, or fertile layer. The fruit body, called an *apothecium*, may be cup-shaped, with the hymenium lining its inner surface, or flat or club-shaped with the hymenium covering its exposed surface. Very few species have been investigated genetically except to determine whether they are homothallic or heterothallic. The best known species include *Ascobolus stercorarius* and *A. immersus*, both of which grow on dung (i.e. they are *coprophilous*), and *Sclerotinia trifoliorum*, which is a parasite of clover. The essential details of the life cycles of these species are similar to those of the other Euascomycetes.

In Ascobolus crosses are made by inoculating mycelia of the two mating types to nutrient agar, when ripe apothecia are produced after seven to twelve days. Alternatively apothecia all at the same stage of development may be obtained by spreading conidia of one mating type over a three-day old culture of the other. Mature asci are present after six days at 24–25°C. Most studies have been concerned with mutant ascospore characters. Bistis [37] followed segregation in the linear asci of *A. stercorarius* using intact and dissected asci. Working with *A. immersus* Lissouba and Rizet [326] studied segregation patterns for light and dark spores in discharged sets of eight spores collected on agar surfaces placed over the apothecia and periodically renewed (see pp. 150–154).

Sclerotinia trifoliorum produces stalked apothecia from mycelial aggregates called *sclerotia* which normally over-winter in the soil. The ascospores infect clover plants causing a disease known as clover rot. The fungus is homothallic but may produce heterocaryotic sclerotia. Using a combination of a small ascospore strain and a large ascospore strain Carr [73] has shown that crossed asci may be found in apothecia developing from heterocaryotic sclerotia. Crossed asci were also recovered from apothecia formed by a heterocaryon between a non-sclerotial line and wild type.

Basidiomycetes

The Basidiomycetes form a clearly defined group in which the products of meiosis are borne on special cells called basidia. Meiosis takes place in each basidium and is followed by the production of four basidiospores externally. Many Basidiomycetes have two phases in their life cycles, a haploid homocaryotic phase, which is frequently monocaryotic, with a branched septate mycelium propagated by asexual spores, and a heterocaryotic phase in which the two nuclear components occur in a one-to-one ratio, each cell containing two nuclei which are of different mating types. The second phase is called a dicaryon and in many species is characterized by the possession of clamp connexions, which are formed during cell division at each new septum of the mycclium. The two nuclei of a terminal cell of the mycelium divide synchronously and at the same time the clamp connexion is formed as a small outgrowth in the mid-region of the cell into which one of the four daughter nuclei passes. Its sister nucleus remains in the terminal cell with one of the products of the other divided nucleus. The other product, the fourth daughter nucleus, becomes separated from these two nuclei by a cross wall which thus cuts off a uninucleate subterminal cell. The side branch is also cut off from the new binucleate terminal cell and curves to fuse

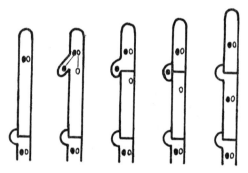

FIG. 9. The stages in the formation of a clamp connexion in a dicaryotic mycelium.

with the subterminal cell close to the septum. Its nucleus then passes into the latter cell which thus becomes binucleate. The steps in this process are illustrated in Fig. 9. This mode of division is an effective, though apparently cumbersome, way of ensuring that each cell of the mycelium contains two nuclei of opposite mating type. The dicaryotic phase is terminated by the fusion of the two nuclei in the basidium to give a diploid nucleus which undergoes meiosis immediately.

Classification is based primarily on the form of the basidium which in the Heterobasidiomycetes is either transversely or longitudinally septate and in the Homobasidiomycetes is one celled. Members of two orders of the Heterobasidiomycetes the Uredinales (rusts) and the Ustilaginales (smuts) have been investigated genetically. All members of both orders are parasitic on vascular plants.

Recent electron microscopic studies [365, 50, 193, 199] have shown that the septal pores of several Basidiomycetes are remarkable for a complexity which has so far been encountered in no other group of fungi. The edge of the pore in the centre of the septum is greatly thickened, forming a tube open at either end. On either side of the septal pore is a cup-shaped discontinuous membrane, the pore cap, which is continuous with the cell endoplasmic reticulum. In *Rhizoctonia solani* the passage of mitochondria, and objects as large as nuclei, during cytoplasmic streaming appears to be accomplished by stretching of the pores [51] (see also p.214).

Uredinales

All rusts are obligate parasites; attempts to culture them on artificial media have, so far, been unsuccessful. While a number of different rusts have been investigated genetically two species, *Puccinia graminis tritici*, black stem rust of wheat, and *Melampsora lini*, flax rust, merit particular attention. Both species have two phases in their life cycles. In stem rust

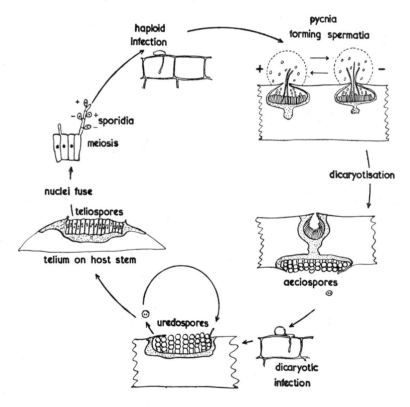

FIG. 10. Life cycle of *Melampsora lini* (flax rust) (partly after Allen[2]).

the monocaryon occurs on a so called alternate host *Berberis vulgaris* and only the dicaryon is able to infect wheat. In flax rust both the monocaryon and dicaryon grow on *Linum usitatissimum*. The life cycle of flax rust is shown in Fig. 10. It is very similar to that of stem rust.

In the spring a teliospore present on a dead stem germinates to give a short promycelium in which meiosis occurs. The promycelium then becomes septate and each cell produces a sterigma on which a basidiospore or sporidium develops. The sporidia are discharged and, if they land on a host, can initiate monocaryotic infections. These appear as pycnia within approximately eight days. Each pycnium is a hollow flask-shaped structure submerged in the host tissue and opening by a stoma. Inside it uninucleate pycniospores, sometimes called spermatia, are produced. These pass out of the pycnium into a drop of fluid on the leaf surface. *Melampsora lini* is heterothallic and each pycnium is of + or − mating type. The spermatia are transferred from one pycnium to another, probably by small insects. In experimental studies this transfer is made with the aid of a wire loop [160]. According to Allen [2] the spermatia invade the host tissue either directly by puncturing the epidermis or through a pycnium, when they may come in contact and anastomose with a compatible mycelium. At the point of anastomosis the spermatial nucleus migrates into the established mycelium and by division and further migration converts it into a dicaryon. Aecial pustules appear, nearby on the same or the other surface of the leaf some three or four days after transfer of spermatia. In the pustule chains of binucleate aeciospores are formed. In damp conditions they imbibe water and round off suddenly, becoming detached from the chain. The aeciospores give rise to dicaryotic infections which are repeated by the orange pigmented dicaryotic uredospores which arise in patches under the host epidermis. Towards the end of the season teliospores are produced on the stems of the host, These are sessile and are not liberated. They germinate *in situ* in the following spring.

For stem rust Craigie and Green [91] recently gave an account of nuclear migration following the fusion between spermatia and the hyphae protruding from the ostioles or openings of the pycnia (flexuose hypae). This migration led to the dicaryotization of protoaecia, or aecial rudiments, near to the pycnia which then developed into aecia. In flax and stem rust dicaryotization may be brought about by the fusion of adjacent, compatible pycnia. In stem rust aeciospores and uredospores, both of which are dicaryotic, may be used to fertilize pycnia [87].

The aeciospores of stem rust are produced on Berberis but are unable to reinfect this host. Instead they infect wheat plants these infections giving rise to uredospores which as in flax rust bring about unlimited clonal reproduction and dissemination of the dicaryon.

As far as the authors are aware rust monocaryons cannot be propagated clonally. The spermatia are unable to initiate new infections. This fact severely limits the number of tests which can be made with any single monocaryon.

The cereal rusts have always constituted a threat to food crops which has partly been met by the development through plant breeding of resistant host varieties. This has resulted in the selection of races of many rusts which can no longer be controlled by host varieties which were formerly resistant. Most of the markers used in genetic studies with the rusts have been genes controlling the capacity to infect and produce rust symptoms on certain so-called differential host varieties possessing various genes for rust resistance. Pigment mutants which affect the colour of the spores have also been used.

Ustilaginales

In contrast to the rusts many of the smuts can be cultured in the haploid monocaryotic condition. Most smuts are obligate parasites in the dicaryotic phase, but they have no provision for the clonal reproduction of the dicaryon comparable to the uredospore stage of a rust. *Ustilago maydis* is probably the most extensively investigated species and its life cycle is shown in Fig. 11.

Germination of the diploid brandspore commences with a meiotic division in the germ tube to form a short septate promycelium from which haploid basidiospores are budded off. Each basidiospore germinates immediately by budding off a sporidium. The sporidia are uninucleate and multiply by yeast-like budding, forming compact colonies on solid media. *Ustilago maydis* is heterothallic, and mating type is controlled by two genes a and b. Only sporidia which have different alleles at both loci can fuse and form a pathogenic dicaryon. Mating is carried out by mixing suspensions of two compatible monosporidial lines and inoculating the mixture to young seedlings of *Zea mays* with a hypodermic syringe or by merely stabbing the cut end of the coleoptile successively into sporidial colonies of the two desired types [233]. In the host plant a dicaryotic mycelium develops which causes hypertrophied growth of the maize tissue. Galls containing mature brandspores appear near the point of inoculation in eight to ten days in 70–80 per cent of the inoculated seedlings. Sporidia are also present in the gall tissue.

4

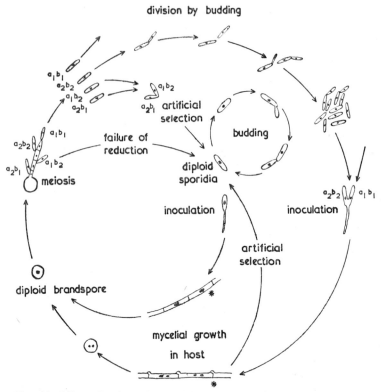

FIG. 11. Life cycle of *Ustilago maydis* (partly after Holliday[233]). * The presence or absence of clamp connexions on the mycelium formed in the host tissues has not been established conclusively.

Diploid sporidial lines sometimes arise by failure of reduction at meiosis when the brandspore germinates [233, 466] or they may be synthesized from pairings of compatible and complementary auxotrophic mutants by the method described on p. 106.

Among the Homobasidiomycetes there are two groups: the Hymenomycetes in which the hymenium is exposed before maturity and the Gasteromycetes in which it is enclosed in a fruit body until mature.

Hymenomycetes

The most extensively investigated group among the Hymenomycetes is the family Agaricaceae. The members of this family possess macroscopic fruit bodies, in the shape of a toadstool or bracket, the under surfaces of which bear gills covered by the hymenial surface. Some species are homo-thallic the remainder having either bipolar or tetrapolar systems of heterothallism (see Chapter 9). In the bipolar forms mating type is determined by multiple alleles at one locus, and in the tetrapolar forms there are two mating type loci also with multiple alleles. In heterothallic forms

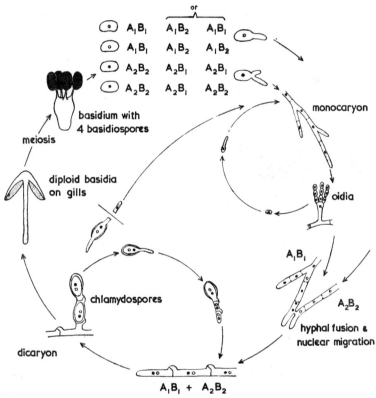

FIG. 12. Life cycle of *Coprinus lagopus*.

the fruit bodies are composed of dicaryotic mycelium. These mating systems are discussed in detail in Chapter 9. The life cycle of a representative tetrapolar form, *Coprinus lagopus* is shown in Fig. 12.

The black basidiospores of Coprinus are borne in tetrads which can be picked off intact with a micromanipulator. A germinated basidiospore forms a branched septate mycelium composed mainly of cells with one, or more rarely two, nuclei. Short branches called oidiophores bud off

uninucleate oidia which accumulate in a drop of fluid. The oidia germinate on nutrient media to repeat the monocaryotic phase.

Matings are made by placing two compatible inocula together on a nutrient agar surface. When the two mycelia touch anastomosis occurs, and
is followed by rapid migration of nuclei through both mycelia, resulting
in the formation of dicaryotic hyphae, easily distinguished by their clamp
connexions, which grow out from both homocaryons (Fig. 44). The dicaryon will produce fruit bodies on a defined medium [335], but fruits
more readily when inoculated into sterile dung or farmyard manure.
Light is necessary for fruit body development. The first stage in the development of a fruit body is an accumulation of hyphae which forms a
small spherical structure. The cap first appears as a dome-shaped mass
of dense tissue in the upper part of the rudiment. The whole is by this
time covered by an outer layer of tissue called a veil, which is ruptured as
expansion commences. On the under side of the cap radiating plate-like
gills develop, orientated vertically. Meiosis takes place when the rudiment is between 10 and 15 mm long. The closely packed gills are held
apart by long bridging cells, the *cystidia*, which grow out from the hymenial surfaces. Apart from these cells, the hymenium is made up of other
sterile cells and basidia of two lengths, so that the tetrads occur at two
levels. The total number of spores produced by an average fruit body of
Coprinus lagopus is as high as 10^9. Rapid expansion of the stalk or stipe
ruptures the veil and the cap expands and commences to shed basidiospores. All the basidia of one fruit body stem from a single pair of nuclei.
At maturity the fruit body undergoes autolysis forming an inky fluid
coloured by the basidiospores. The haploid components of the dicaryon
may be recovered by dissecting off the occasional homocaryotic germ
tubes with no clamp connexions which are produced by germinating
dicaryotic *chlamydospores* [310]. These chlamydospores are formed on
the older parts of dicaryotic mycelia. Heterocaryosis cannot easily be
considered apart from the function of the mating type genes and is
discussed in Chapter 9.

The life cycle of *Schizophyllum commune* is broadly similar to that of
Coprinus. The fruit body consists of a small fan or kidney-shaped cap
attached to the substrate at one point. The gills on the under side of the
cap radiate from this point and when dry are seen to be split longitudinally; hence the generic name. Dried fruit bodies when moistened
discharge viable basidiospores even after storage for several years. The
homocaryotic mycelium does not produce asexual reproductive spores
but this disadvantage may be overcome by using macerated mycelium

instead of a spore suspension in, for example, experiments to induce mutations [442].

The resolution of the Hymenomycete dicaryon into its component haploid types is a valuable technique in studies of cytoplasmic inheritance (see Chapter 10). It can be achieved in most clamp-forming species by using a surgical method developed by Harder [214]. The dicaryotic hypha chosen for the operation must be one in which the terminal cell has only just been cut off by its septum. The method is to kill by puncturing with a fine needle both the terminal cell and its clamp before it has fused with the subterminal cell, and to recover the hyphae which later grow out of the subterminal cell.

In *Collybia velutipes* recovery of the component nuclei of a dicaryon is very simple since uninucleate oidia are formed by both the monocaryon and the dicaryon. In both the oidia are formed endogenously in chains by the fragmentation of aerial hyphae [57]. The oidia in each chain in a dicaryon are uniform in mating type. Different chains may be of different mating types.

The complex fruit bodies of the Hymenomycetes present many problems in morphogenesis which parallel those in higher plants but which in some respects are more amenable to analysis. Plunkett [415, 416] has analysed the role of light, humidity and carbon dioxide concentration in the formation of fruit bodies of Collybia and *Polyporus brumalis*. The use of mutants blocked at different steps of development may become a promising addition to this type of study.

Gasteromycetes

No Gasteromycetes have been investigated genetically beyond studies in heterothallism. *Cyathus stercoreus* is important in this respect since it was in this fungus that the different roles of the two genes controlling tetrapolar heterothallism were first recognized [183].

Fungi Imperfecti

The members of this class are divided up on the basis of the arrangement and form of their asexual spores into so called *form genera*. Since they normally lack sexual methods of reproduction their genetical interest is determined largely by the possibilities of mutation studies, heterocaryosis, artificial synthesis of diploids and parasexual recombination. The extent of these possibilities have hardly begun to be investigated in this group except for work on Penicillium and Aspergillus, genera already discussed, and the plant pathogens: *Fusarium oxysporum* f. sp. *pisi*,

Cladosporium fulvum, Verticillium albo-atrum and *Colletotrichum lagen-arium. F. oxysporum pisi* and *V. albo-atrum* both attack their host plants, peas and hops (Humulus) respectively, through the roots from inoculum present in the soil and infect the vascular tissues producing wilting symptoms. Both organisms have septate hyphae composed of multinucleate cells, and readily form heterocaryons. *Cladosporium fulvum* infects the leaves of tomato plants. Germ tubes formed by the conidia penetrate through stomata producing lesions which give rise to many conidiophores and conidia. The hyphal cells are uninucleate except for the tip cells which contain several nuclei. Attempts to produce balanced hetero-caryons of *C. fulvum* were unsuccessful (Day, unpublished). *Colleto-trichum lagenarium* infects members of the Cucurbitaceae from spores carried on the seed coats or present on plant debris. It produces lesions on leaves and fruits which develop black spore masses.

CHAPTER 3

THE INDUCTION, ISOLATION AND CHARACTERIZATION OF MUTANTS

A mutation is a more or less abrupt and stable change in a cell which can be transmitted in a hereditary fashion to daughter cells. In this chapter we will deal only with mutations in the chromosomal genetic material, and will postpone consideration of extra-chromosomal genetic change until Chapter 10.

Almost the whole modern development of fungal genetics is based on the study of mutations at specific chromosomal loci; without mutations the loci could never be identified nor their functions studied. Although wild populations of fungi probably carry a variety of different alleles at most loci, most naturally occurring allelic differences are not sufficiently sharp in their effects on the phenotype to furnish good material for experiment. Most kinds of genetic investigation are made possible by the occurrence and preservation in the laboratory of mutants which could hardly survive in nature. The development of efficient methods of isolation and characterization of mutants is thus of great importance. Some of the available techniques will be considered below.

Mutations occur spontaneously and apparently unavoidably in all cell populations, but usually with very low frequency. Most programmes for isolating mutants include procedures which are known to increase the mutation rate. The most generally used mutagenic treatments are irradiation with ultraviolet light or with X-rays, although various kinds of chemical treatment are also effective. The process of the induction of mutation will be considered in more detail later in this chapter. We will note here merely that most treatments which induce a useful increase in mutation rate will also result in the death of many of the cells treated. A balance must be struck between having too few surviving cells, at a high dose of the mutagen, and having many survivors with rather few mutants among them, at a low dose.

A convenient distinction can be made between 'forward' and 'reverse' mutations. Forward mutations are those *from* the wild type to a non-wild, or 'mutant' condition, while reverse mutations are those from a mutant condition back to wild, or something close to it.

Leaving aside the class of lethal mutations, which are referred to later in this chapter, forward mutations may be broadly classified into *visibles*, that is those with visible effects on morphology or colour, and *biochemicals*, or those with effects on the capacity of the organism to carry out definable metabolic reactions. Two of the more striking visible mutants which have been isolated following mutagenic treatment of *Coprinus lagopus* are shown in Fig. 13.

In Neurospora studies much use has been made of visible mutants, particularly those which have vigorous but abnormally compact growth, such as the various *colonial* mutants. The usefulness of morphological variants is, however, limited by the difficulty of scoring two or more morphological effects in combination, and by the loss of viability usually encountered when several such mutants are introduced into the same strain. Generally only one morphological character can conveniently be used in any one cross. In most fungi only a few types of clear-cut difference in pigmentation are likely to be found. Those which do occur are generally extremely valuable because of their normal vigour and ease of scoring in practically any gene combination. Good examples are provided by the *albino* mutants of Neurospora, with white conidia and mycelium in contrast to the normal orange, and the *white* and *yellow* mutants of *Aspergillus nidulans* which contrast strikingly with the dark green of the wild type. However, the greater part of the development of fungal genetics since 1945 has been based on the use of biochemical mutants, first studied in Neurospora by Beadle and Tatum [25].

The type of biochemical mutant which it is most convenient to study is the *auxotroph*. Auxotrophs fail to grow on a medium (minimal medium) containing the minimum nutrients essential for the growth of the wild type, but will grow if one or more specific substances are added to the medium. When supplied with their growth requirements, auxotrophs commonly resemble wild type in appearance, although this is not invariably the case. A strain which can grow on minimal medium may be referred to as a *prototroph*.

METHODS FOR ISOLATING AND CHARACTERIZING AUXOTROPHIC MUTANTS

Total isolation procedures

The most straightforward, though the least efficient, procedure for isolating mutants is to take a large random sample of nuclei from a population treated with a mutagen and to test them all individually for ability to support growth on minimal medium.

FIG. 13. Wild type (top) and two morphological mutants of *Coprinus lagopus*; left *lacy* (2203) and right *poodle* (2205).

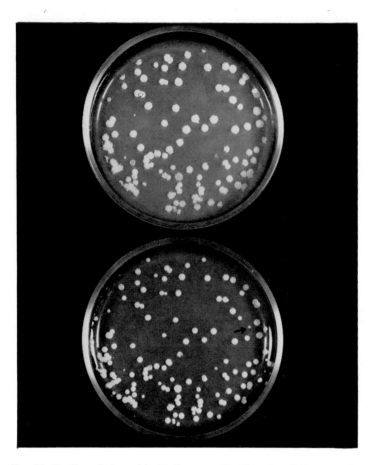

FIG. 14. Replica plating with *Ustilago maydis*. The plate below contains complete medium; the replica plate above contains minimal medium. The colony indicated by the arrow is auxotrophic (photograph by courtesy of R. Holliday).

In fungi with uninucleate conidia which germinate to form compact colonies, such as *Aspergillus nidulans,* the method consists simply in spreading treated conidia, at a suitable dilution, on plates of agar medium, and isolating the resulting colonies into individual tubes. The visible mutations, which may have been distinguishable already on the original plate, can then be identified by eye, and the auxotrophs by testing for growth on minimal medium. The medium used in the plates and the derived tube cultures is generally a 'complete' one, containing every kind of supplement which auxotrophs might reasonably be expected to require. Recipes for complete media vary, but most contain casein hydrolysate (to supply amino acids), hydrolyzed yeast nucleic acid (to supply purine and pyrimidine bases and/or nucleosides), a synthetic mixture of water-soluble vitamins, and perhaps yeast extract or malt extract or both.

In Neurospora, a complication is introduced by the fact that the ordinary conidia are predominantly multinucleate. This means that in a simple plating method like the one outlined above many mutant nuclei will not be detected because the presence of one or more viable non-mutant nuclei in the same conidium will permit normal growth. If one is not interested in quantitative relationships between dose of mutagen and yield of mutants this may not matter since, if more than 95 per cent of the conidia are killed by the mutagenic treatment, as is commonly the case, the chance of more than one nucleus surviving in any one conidium is probably small. However, many workers in the field have adopted some method which enables them to be sure that they are sampling single nuclei. Two such methods have been used with Neurospora. The first depends on the use of strains of Neurospora which produce only the uninucleate microconidia. The Lindegrens [318] used a mutant *fluffy* which is unable to form macroconidia, but which will generally produce microconidia in fair quantity after the surface of the culture has been wetted with sterile water. Tatum and his colleagues at Stanford University [543] preferred to use another mutant *peach-microconidial* (*pe^m*) which has the virtue of forming abundant microconidia under ordinary cultural conditions.

Most Neurospora strains have a very rapidly spreading growth and do not form the sort of compact colonies following spore germination which are necessary if the mycelia from individual spores are to be isolated after plating. One can overcome this difficulty either by using *colonial* mutant strains which will form compact colonies on any medium, or by using special media which will induce a colonial habit in morphologically normal strains. Tatum *et al.* adopted the first solution; in some experiments they used the double mutant *pe^m col-1* which is a microconidial strain which grows colonially on any medium, and in other experiments the double mutant *pe^m inos*, which has the property of colonial growth when the medium contains limiting concentrations of inositol. Strains carrying *inos* have the special advantage that they can be made to grow either colonially or in the normal spreading fashion merely by adjusting the level of inositol in the medium. A disadvantage in the use of either of these double mutant strains for the induction of further mutations is that after a new mutant has been obtained one has the additional task of ridding it of the genes for microconidiation and colonial growth habit which are undesirable because of the reduction in vigour which they bring about. In current work colonial growth is most usually achieved by the use of a medium in which sucrose is partly replaced by the non-utilisable sugar sorbose [542]. Most strains of *Neurospora crassa* show a very restricted growth rate on such a medium.

In the original work of Beadle and Tatum [25], the problem of the isolation of single nuclei following mutagenic treatment was solved in a different way. Ordinary macroconidia were irradiated and used to fertilize protoperithecia of the opposite mating-type, and a sample of the ascospores formed were isolated into tubes of complete medium and germinated by heat shock. Assuming that each perithecium results from fertilization by a single nucleus, mutant ascospores, where they occur, will generally constitute one-half of all the ascospores in the perithecium. To avoid isolating the same mutant more than once, Beadle and Tatum took only one ascospore from each perithecium. This procedure is certainly more laborious than the plating methods that were subsequently developed.

Characterization of auxotrophic mutants obtained by total isolation

In *Neurospora crassa*, one particular kind of auxotroph can be identified by eye among the colonies obtained by plating irradiated conidia on sorbose medium. This is the *adenine-purple* type of mutant, which is due to mutation in the *ad*-3 region, and which has a defect in a particular step in adenine synthesis which results both in a growth requirement for adenine and in the accumulation in the medium of a purple pigment. The purple pigment is evidently a derivative of an accumulated adenine precursor. De Serres and Kolmark [119] found that it was possible to pick out *ad*-3 mutants by eye when irradiated conidia were allowed to grow into compact colonies in liquid medium containing sorbose and adenine and then poured into large white porcelain dishes, even though there were only a few of them among a million non-purple colonies. This is perhaps the only kind of auxotrophic mutation in Neurospora which can be identified by eye at the colony stage. A similar situation exists in yeasts, where certain classes of adenine-requiring auxotrophs, for example *ad*-6 and *ad*-7 mutants of *Schizosaccharomyces pombe*, develop a striking red colour. This pigmentation has permitted the identification and isolation of a great many such mutants; the collection of *ad*-7 mutants in *S. pombe* made by Leupold [307] and Gutz [212] is one of the most extensive series of independently occurring allelic mutations yet obtained in the fungi. Furthermore, starting with *ad*-6 or *ad*-7 one can easily detect further mutations blocking the adenine pathway *before* the intermediate which is converted to the red pigment, since these give white colonies (Heslot [227]). These secondarily derived mutants fall into four genetic and biochemical classes, *ad*-1, *ad*-3, *ad*-4 and *ad*-5.

In general, the identification of auxotrophs requires the separate culture of each mutant type and tests for ability to grow on various differential media. It is customary to use a series of tests, each one narrowing the range of possibilities. Beadle and Tatum [25] isolated their mutants on a complete medium, which, in addition to the constituents of minimal medium, contained yeast extract, malt syrup, glucose, hydrolyzed casein, hydrolyzed nucleic acid, and a mixture of vitamins. Mutants which grew on this medium, but which failed to grow on minimal medium containing sucrose as carbon source, were transferred to (*a*) minimal plus vitamins plus hydrolyzed nucleic acid, (*b*) minimal plus hydrolyzed casein, (*c*) minimal with glucose as carbon source, and (*d*) complete medium. All but a few mutants grew either on (*a*) or on (*b*), as well as on complete medium; none were found which required glucose rather than sucrose as carbon source. Those growing on (*a*) were then tested on all the vitamins and the purine and pyrimidine bases or nucleosides individually, while those growing on (*b*) were tested on all the more likely individual amino acids. The great majority of auxotrophs responded to a single supplement, although a few with multiple requirements were found (see Chapter 8). One category of mutants found by Beadle and Tatum, those growing on ammonium salt as nitrogen source but failing to utilize nitrate, could not be distinguished when standard minimal medium (containing ammonium tartrate) was used, but revealed themselves when additional tests were made on a medium containing nitrate as sole nitrogen source.

When one knows what kinds and relative numbers of mutants to expect, one can often rationalize the screening procedure so as to save labour. For example Pontecorvo [424] found that the auxotrophs obtained by the 'starvation' technique, outlined below, included up to 60 per cent non-utilizers of sulphate which responded to methionine, cystine or inorganic thiosulphate. It thus became worthwhile to separate this numerous class at the outset by an initial test on minimal plus thiosulphate. The general principle to be followed in devising a series of diagnostic tests is that each test should sub-divide the unidentified mutants into two more or less equal groups.

A considerable saving of labour may be achieved by means of the method suggested and used by Holliday [231], and subsequently by Takahashi [538]. This method utilizes a series of media each containing a number

TABLE 6. Design of media for distinguishing fifteen different kinds
of growth requirement

Medium no.	Supplements added					Auxotroph grows on		Substance required
1	A	B	C	D	E	1 only		A
2	F	B	G	H	I	2 „		F
3	J	C	G	K	L	3 „		J
						4 „		M
4	M	D	H	K	N	5 „		O
5	O	E	I	L	N	1 and 2		B
						1 „	3	C
						1 „	4	D
						1 „	5	E
						2 „	3	G
						2 „	4	H
						2 „	5	I
						3 „	4	K
						3 „	5	L
						4 „	5	N

N.B. The analogous design with six media each with six supplements will cater for 21 different substances, the 7×7 design for 28, and the 8×8 for 36.

of substances according to a pattern which provides for each substance occurring either in one medium only, or in two media, the two being different for each substance. The most efficient procedure is to have n different media with n different substances in each; in this case requirements for $n(n+1)/2$ different substances can be distinguished. In Table 6 a design differentiating between fifteen kinds of growth requirement through the use of five different media is illustrated. With a larger number of substances to be differentiated, it might be an advantage to adopt a more complicated pattern in which each substance could occur either in one, two or three different media, but for the 20 to 30 different requirements one is likely to find among auxotrophs the simpler design seems better. The labour involved in testing many strains on a number of different media can be greatly reduced if it is possible to use the replica plating technique of Lederberg [299], which was originally designed for experiments with bacteria. A sterile piece of velvet, supported by a circular wooden block of a size to fit the surface of a plate, is used to transfer cells from the colonies on one plate to one or more other kinds of medium. If all the cells are able to grow on all kinds of medium, the pattern of colonies after incubation will be identical on all plates. If, however, a certain strain growing on the master plate is unable to grow on one of the test media, then the corresponding colony will be absent from that medium. In its application to the identification of auxotrophs, the method involves growing colonies from mutagenically treated cells on a plate of complete medium, and the replication of these colonies on to an appropriate series of differential media. A hundred or more colonies can thus be tested within a few minutes. This technique is well adapted for use with yeasts [538], or with any fungus which can form moist yeast-like colonies of single cells which will adhere readily to the pad. Its application to the screening of auxotrophs in *Ustilago maydis* is shown in Fig. 14. Unfortunately, one cannot replicate colonies of a filamentous fungus in the same way unless they produce abundant conidia while still relatively small.

In *Aspergillus nidulans*, which fulfils this requirement, Roberts [455] has used a replicator consisting of an array of closely-spaced steel pins mounted points outward in a sheet of 'Perspex'. If the points of the pins are moistened by pushing them lightly into the surface of the agar plate, conidia of any Aspergillus colonies will adhere to them. Mackintosh and Pritchard [334], however, obtained very good results with the same species using a velvet replicator, provided the velveteen surface was damp and the pile was short; these conditions minimized scattering of the conidia. These authors were able to replicate up to 600 colonies at a time by restricting the colony size through incorporating 0·08 per cent sodium desoxycholate into the plating medium. Maling [339] has successfully used either a velvet or a filter paper surface for replicating Neurospora colonies carrying the two morphological mutants *crisp* and *ragged*. Such double mutant strains produce compact and abundantly conidiating colonies very suitable for replication, but their usefulness is limited in other respects by their slow growth, and their sexual infertility when crossed among themselves. Neurospora colonies with the morphology of wild type can be replicated on sorbose medium by the method devised by Reissig [445] which consists of spreading spores on top of thin paper ('pipe paper') resting on the surface of the master plate. After the spores have grown sufficiently to have penetrated through the paper into the medium below, the paper is peeled off and transferred to the test plate. The strands of mycelium left in the original plate will continue growth to form colonies, and that remaining in the paper is capable of initiating a corresponding pattern of colonies in the test plate.

When isolates have been identified as auxotrophs by testing on minimal medium their growth requirements can often be identified very easily by the 'auxanographic' technique of Pontecorvo [418]. The method consists in pouring a layer of molten agar minimal medium containing a suspension of cells of the auxotroph all over the surface of a minimal plate, and then introducing small crystals of various substances at well separated marked points on the surface of the agar. After incubation, growth-promoting compounds are indicated by zones of growth. Inhibitory substances, and double requirements, can also be graphically revealed. This method is applicable to any organism which is capable of showing a compact and limited zone of growth; in Neurospora the growth rate has to be restricted by the use of sorbose.

An indication of the kinds and relative frequencies of auxotrophs obtainable in different fungi by total isolation is given in Table 7.

Visual selection of auxotrophs—'rescue' methods

The principle of 'rescue' may be illustrated by the layer plating technique used on yeast by Reaume and Tatum [444]. Irradiated haploid cells were plated at a suitable dilution on minimal medium and incubated for several days so as to allow the prototrophic survivors to grow and form colonies the positions of which were marked. Then a layer of supplemented agar medium, cooled to just above the setting temperature, was poured on to the plate, and allowed to solidify. Colonies which grew up only after this layering were isolated, and, on testing often proved to be auxotrophic. The layer plating idea has found more recent application in the 'starvation' methods for selection of auxotrophs, outlined below.

An alternative to layering is to identify non-growing or retarded spores by microscopic examination and to transfer them to enriched medium in the hope that a proportion of them will now grow normally. Such a method has been used with success by Lein and the Mitchells on Neurospora [301], and by Boone et al. on Venturia [48].

For the isolation of specific mutants it may be an advantage to add only one or a few substances to the 'rescue' medium, since some auxo-

TABLE 7. Auxotrophic mutants obtained in different species

Growth requirement	Penicillium chrysogenum[42]	Ophiostoma multiannulatum[178]		Aspergillus nidulans[424]		Neurospora crassa[17]	Ustilago maydis[232]	Venturia inaequalis[48]
	T	T	F	T	S	T	T	T, R
Carboxylic acids:								
Acetic acid						1		
Krebs cycle acid						2		
Vitamins:								
Biotin	5	2	3	4		*		1
Thiamin	18			3	2	1		
Nicotinamide	5	6	2	7	2	5	7	1
Pantothenic acid		6		1		1	15	3
p-Aminobenzoic acid	6		4	2	3	4	2	
Riboflavin				1	1	1	2	6
Pyridoxin	6			1	7	3	2	
Inositol	6	18					2	3
Choline	9		2	1		1	5	5
Nucleic acid bases:								
Adenine	} 19	5	50	10	134	21	17	} 17
Guanine			4	1			6	
Uridine or cytidine		26	15			4		19

Growth requirement	Penicillium chrysogenum[42]	Ophiostoma multiannulatum[178]		Aspergillus nidulans[424]		Neurospora crassa[17]	Ustilago maydis[232]	Venturia inaequalis[48]
	T	T	F	T	S	T	T	T, R
Amino acids:								
Methionine	32	9	8	1		24	7	4
Methionine or cysteine	50†			1	2	19	7	
Arginine	52	42	56	6		2	12	18
Arginine or proline	13			1	14			
Proline	14				2			2
Lysine	54	32	22	3	7	19		1
Histidine	16		6					3
Serine or glycine						1	1	
Isoleucine plus valine	1					2		
Isoleucine					1			
Leucine	12					6	4	
Phenylalanine	4					1		
Tryptophan	2	6			1	4		
Complete mixture of aromatic metabolites						1		
Tryptophan plus nicotinamide				3	1	3		
Anthranilic acid plus nicotinamide				4	1			
Thiosulphate or other reduced S		34	40	7	298		4	16
Ammonia or other reduced N	31			17	3			
Unknowns	43	14	12	24	7	18	6	1

Notes: * Wild type requires biotin. † Including 'thiosulphate' mutants. T=Total isolation; F=Filtration enrichment; S=Starvation method; R=Rescue method.

trophs, while responding well to their specific requirement, grow poorly or not at all on a complete medium. For example, Neurospora and Venturia mutants requiring histidine fail to grow on the standard complete medium because the uptake of histidine is antagonized by the presence of some other amino acids [350].

Filtration enrichment

In recent years total isolation procedures, and methods in which mutants are picked out by eye, have tended to be replaced by methods which result in the selective elimination of the prototrophic survivors of the mutagenic treatment but in the retention of many or most of the desired types of auxotrophs.

The principle of the *filtration enrichment method* is to suspend treated conidia from a wild type strain in liquid minimal medium, and to remove prototrophic growth as it appears by repeated filtration through some kind of sterile filter in which the hyphae of germinated spores are held. Since Fries' original trial of the method in Ophiostoma [178], Woodward and co-workers [593] and Catcheside [79] have independently applied it to Neurospora. Woodward used filters of muslin, while Catcheside used cotton, but the essential features of the method were the same in both cases. When it appeared that nearly all conidia capable of growth on minimal medium had been removed, the remainder were spread on agar plates of medium supplemented with one or a few substances for which nutritionally deficient mutants were being sought. The plating medium also contained sorbose to induce colonial growth. In practice, many of the colonies appearing on the supplemented plates were prototrophic, either because different auxotrophs had anastomosed to give prototrophic heterocaryons, or because the conidia giving rise to them had been delayed in germination as a result of the mutagen treatment. The method in general yields many mutants which grow slowly on any medium. Under proper conditions, however, a fair proportion (10 per cent or more) of the colonies will be of the desired auxotrophic types.

The success of the filtration enrichment method depends on a number of factors, the most important of which is the time of incubation in the liquid minimal medium before the final filtration. Too short a period will lead to too large a number of delayed prototrophs being recovered, while too long a delay will result in the death of most of the auxotrophic mutants. As a rule, filtrations are continued for so long as prototrophic growth visible to the eye appears, but no longer. The period is usually two to four days depending to some extent on the mutagenic agent; ultraviolet light causes a considerable delay in growth and thus necessitates a rather longer course of filtrations. It is important that prototrophic growth should not be allowed to build up in the liquid medium, since this will tend to lead to loss of auxotrophic conidia by anastomosis, or by causing them to be trapped in the felt of mycelium held by the filter. Woodward recommended a first filtration after 18 hours at 25°C followed by further filtrations at 3 to 6 hourly intervals during the second 18 hour period, and at 6 to 12 hourly intervals thereafter. It is also essential to keep the conidia in suspension, since if they are allowed to settle they will readily fuse together to form prototrophic heterocaryons. It is usual either to shake or to aerate the conidial suspension continuously during the incubation period. One might expect the efficiency of the method to be increased by the use of uninucleate, rather than multinucleate conidia. Catcheside, in his original method, used microconidial strains, while Woodward used a macroconidial strain with a high proportion of uninucleate conidia. More recently, however, Catcheside [81] has achieved very successful results with an ordinary macroconidial strain. It seems probable that, when a high proportion of the conidia are killed by the mutagenic treatment, most of the survivors will, in any case, have only one viable nucleus.

The filtration enrichment method is extremely efficient in the isolation of certain types of auxotrophs, but is not so good for others. In general, mutants which are enabled to grow by the presence of minute traces of vitamins, such as may well be present in the suspension medium as a result of breakdown of dead spores, will not be recovered. Neither will any type of auxotroph, such as that requiring inositol, which tends to die rapidly in minimal medium (see below under 'starvation methods'). On the other hand, auxotrophs requiring a relatively high concentration of, for example, an amino acid, will be readily obtained. Woodward was able to select large numbers of arginine requiring mutants, but was unsuccessful in attempts to obtain nicotinamide requiring ones.

Although the success of filtration in removing prototrophs normally depends on the formation of mycelium, the method can, in special cases, be applied to organisms with yeast-like growth. Takahashi [538] has applied the method a 'flocculent' strain of *Saccharomyces cerevisiae* in which budding cells tend to remain together in clumps. He removed multicellular clumps from his initial suspension by filtration through sintered glass, irradiated the resulting single cells, suspended them in minimal medium, and after incubation removed most of the prototrophic survivors by a series of further filtrations. The final suspension of single

cells, consisting partly of the occasional fully separated buds from prototrophs, but containing a fair proportion of auxotrophic mutants, was
plated on complete medium.

There seems to be no reason why filtration enrichment should not be
applied far more widely in the isolation of auxotrophs in filamentous
fungi. Apart from its use on Neurospora and Ophiostoma, Day and
Anderson [111] have obtained good results with the method in work on
Coprinus lagopus.

Selective elimination of prototrophs by antibiotics

The standard method for the isolation of auxotrophic mutants in the
bacterium *Escheria coli* was devised by Davis [97], who made use of the
fact that penicillin only kills growing bacteria. Fungi are insensitive to
penicillin, but Moat *et al.* [364] showed that it was possible to apply the
same principle to Saccharomyces using certain other antibiotics, of
which amphotericin B, endomycin and nystatin were the most effective.
Populations of yeast cells containing a small proportion of auxotrophs
were suspended in minimal medium plus antibiotic and, after a period
of incubation, the antibiotic was removed by centrifuging and the cells
were spread on antibiotic-free supplemented medium. The proportion
of auxotrophs among the surviving cells growing to give colonies was
over 50-fold greater than in the original population, owing to the selective killing of the prototrophs. The efficiency of the method was limited
by the fact that a high proportion of the survivors of the antibiotic treatment tended to be cytoplasmically determined respiration-deficient cells
(cf. Chapter 10) rather than auxotrophs. However, the antibiotic actidione has been used with success for the isolation of auxotrophs by
Pittman *et al.* [414].

Selective advantage of double mutants over single mutants

The *starvation method* for the selection of auxotrophs was also suggested
by work of Fries on Ophiostoma [179]. Fries showed that double mutants
with two nutritional deficiencies sometimes survived longer in minimal
medium than the single mutants from which they had been derived.
This observation was used by McDonald and Pontecorvo [424] as the
basis for an effective method for obtaining auxotrophs in *Aspergillus
nidulans.* Conidia of a biotin-requiring auxotrophic mutant of Aspergillus
were irradiated with ultraviolet light to about 5 per cent survival, spread
on minimal agar and covered with a further layer of the same medium.

After incubation of the plates for at least 96 hours at 37°C, during which some 99 per cent of biotin requiring single mutant conidia were shown to die, a further layer, this time of complete medium, was poured on to the plate. Colonies which grew only after this supplementation were picked off and tested for growth requirements for substances in addition to biotin. Up to 60 per cent of the isolates turned out to be double auxotrophs. The second mutations obtained are listed in Table 7; it will be seen that they were of many different kinds with a notably high proportion showing loss of ability to reduce sulphate. It is clear that different second requirements had widely differing effects on the survival of biotin requireing conidia, a requirement for reduced sulphur being among the most effective in improving viability. The choice of the biotin requiring mutant as the starting strain was a fortunate one, since it turned out to have an exceptionally high death rate in minimal medium.

Essentially the same principle has been applied by Lester and Gross [302] to *Neurospora crassa*. In this case the single mutant in which further mutations were selected was one requiring inositol. The procedure was as follows. A washed conidial suspension was irradiated with ultraviolet light to about 10 per cent survival, incubated briefly in medium supplemented with inositol (this with a view to keeping the double auxotrophs alive through any lag in the phenotypic expression of their second requirement) washed by repeated centrifugation and resuspension in water, and plated on minimal sorbose agar. After two to four days at 34°C, during which almost all the inositol requiring single mutant spores died, the plates were supplemented with a layer of complete medium, and the colonies which grew up were isolated and tested. Up to 80 per cent of double auxotrophs were recovered in some experiments. As in the Aspergillus experiments, the method tended to be selective for certain kinds of second mutation; many leucine requiring mutants were always recovered, though many other types of auxotroph occurred with lesser frequencies. One can, of course, make this method selective for certain specific types of mutant by supplementing the second layer of agar with only one or a few substances.

In the two cases just considered the physiological basis for the improved viability of double auxotrophs as compared with the original single auxotroph is not understood. It seems likely, however, that they are examples of the phenomenon of unbalanced growth, and its prevention by a second nutritional requirement. Some cases of this type are well understood in bacteria. For example, Bauman and Davis [21] have made use of the fact that *Escherichia coli* mutants requiring diaminopimelic acid die rapidly in minimal medium supplemented with lysine, but that viability can be greatly improved by the induction of a further growth requirement. Diaminopimelic acid is required as a precursor of lysine, and is also used directly for the specific purpose of cell wall synthesis. If lysine is supplied to this mutant, protein can be synthesized, and all parts of the cell can grow except the cell wall material. Under these conditions the cells grow and lyse for lack of cell walls. The superimposition of a requirement for a substance with a less specific function can stop virtually all growth and the cell can be prevented from committing suicide. It seems likely that a similar type of explanation may be applicable to the fungal examples just considered; there is evidence that inositol starvation, for example, leads to a specific failure to form cell membranes [488].

In special cases, the viability of an auxotroph, even when it is supplied with its growth requirement, may be improved by the addition of a second metabolic block in the same biosynthetic pathway. In Neurospora, Mitchell and Mitchell [359] reported an instance where a culture of an *adenine-purple* (*ad-3*) mutant was found to have acquired a second mutation blocking adenine synthesis at a stage earlier than the formation of the precursor of the purple pigment. This had evidently occurred as a result of spontaneous mutation followed by a growth of the double mutant more vigorous than that of the parental single mutant. In an analogous case in Ophiostoma, Fries [181] noted a tendency for old cultures on complete medium of a guanine auxotroph to be overgrown by a double mutant of spontaneous origin which required hypoxanthine or adenine in addition to guanine.

A situation of this kind has been utilized by Roman [457] for the isolation of adenine auxotrophs in *Saccharomyces cerevisiae*. As in Neurospora, some adenine requiring mutants in yeast accumulate a reddish pigment. Roman showed that cultures of such mutants tend to be overgrown by cells which, when plated on agar medium, form white or pink, rather than red colonies. This overgrowth occurs most readily in relatively anaerobic cultures in liquid yeast extract-peptone medium. The white or pink variants were shown to consist predominantly of double mutants carrying a second genetic block in adenine synthesis. The second block was at some stage in adenine synthesis before the formation of the intermediate which, when accumulated as a result of the first block alone, led to the production of the pigment. It appears that, in yeast strains which are, in any case, unable to synthesize adenine, there is a selective advantage in the biosynthetic pathway being blocked earlier rather than late.

The position of the second block was not always the same; the second mutation could, in fact, be in any one of five physiologically distinct genes. Single mutants for each of these five genes were easily recovered by outcrossing the double mutants to the wild type. The pink colonies were shown to be due to mutation to 'intermediate' alleles at one or other of these same five loci, causing incomplete blockage of the pigment formation.

MUTATIONS FROM AUXOTROPHY TO PROTOTROPHY

The best kind of situation for quantitative work is one in which the desired type of mutant can grow on a certain type of medium on which the parental strain fails to grow at all. This is the case with reverse mutations from auxotrophy to prototrophy; such mutations, even the very rare spontaneous ones, can generally be detected, and their frequencies measured, by the simple procedure of plating cells from the auxotrophic mutant at high density on minimal medium. Extensive studies of reversion to prototrophy have been made by, among others, Giles [194] working on Neurospora, and Heslot [228] working on Schizosaccharomyces. To confirm that all prototrophic revertants are able to grow and express themselves, it is necessary to show in reconstruction experiments that added prototrophic cells in known numbers do, in fact, appear as colonies against the background of non-growing or feebly growing auxotrophs. That this is not necessarily the case was emphasized by Grigg [205], who demonstrated situations in which prototrophic Neurospora conidia were suppressed by the presence of high concentrations of auxotrophic conidia. In Grigg's experiments the competitive power of the non-growing auxotrophs was probably exercised through their ability to respire and hence to reduce the level of sucrose, which was already limiting in the sorbose plating medium used to induce colonial growth. Where reconstruction experiments indicate that the 'Grigg effect' is operating, it should be possible to overcome the difficulty by suitable adjustment of the conidial density in relation to the nature of the medium.

In many cases in Neurospora [194, 599] reversions to prototrophy have proved to be due, not to a reversal of the original mutation, but to the suppression of its effect by a second mutation, at a different locus. It is occasionally found that a particular kind of auxotrophic mutation has the effect of suppressing the effect of another. Mitchell and Mitchell [360] were the first to show that in Neurospora some *pyr*-3 mutants acted as suppressors of *arg*-2; that is to say, the double mutant *pyr*-3 *arg*-2 required pyrimidine but not arginine. Reissig [446] exploited this situation to isolate *pyr*-3 mutants. Starting with *arg*-2 strains, he plated mutagenically treated conidia on minimal medium supplemented with pyrimidines but lacking arginine. He obtained two kinds of mutant capable of growing on this medium, those due to reversion at the *arg*-2 locus and those due to a 'forward' mutation at *pyr*-3. Some of the *pyr*-3 mutations which suppressed the arginine requirement did not cause an absolute pyrimidine requirement. While the mechanism of the suppression here is not yet completely clear, this system allows large scale and relatively trouble-free isolation of mutations in a specific gene. Reversions of *pyr*-3 to *pyr*-3$^+$ can, of course, be selected on minimal medium so both 'forward' and 'back' mutations can readily be selected at this locus.

Some reversions in auxotrophic strains are due to 'partial reverse mutation' giving new alleles at the original locus with effects intermediate between the presumably negative allele in the auxotroph and a fully wild type allele [196, 394]. Mutants with 'intermediate alleles' permitting slow growth on minimal medium have been termed *bradytrophs*. Both suppressor mutations and mutations to intermediate alleles are of importance for the theory of gene action, and will be discussed more fully in Chapter 8.

MUTATIONS TO DRUG RESISTANCE

Some kinds of 'forward' mutation result in a gain of ability to grow in a given environment. 'Suppressor' mutations, discussed, above, afford one example, and another is provided by mutations resulting in gain of virulence in pathogenic fungi, which are dealt with specifically in Chapter 11. Mutations to drug resistance have been selected in a number of fungi by plating cells at high densities on medium containing a high enough concentration of the drug in question to inhibit wild type growth completely. Mutations have been selected in Aspergillus [462] and in Neurospora [245] to acriflavin resistance, in Neurospora [245, 246] and in yeast [354] to actidione resistance, in Neurospora [245] to sodium azide resistance and in *Venturia inaequalis* [296] to antimycin resistance, to give only a few examples.

Mutants resistant to analogues of normal metabolites are particularly interesting since they may show modifications of normal metabolic regulatory mechanisms. Mutants of Aspergillus have been obtained resistant to 8-azaguanine and to p-fluorophenylalanine [366], while in yeast mutants resistant to canavanine (an analogue of arginine) are easily selected [506]. A case of particular interest, since it is one of the few in which the mechanism of resistance is known, is that of 4-methyl-tryptophan resistant mutants of Neurospora, isolated by Stadler [513]. These mutants were shown to lack an uptake mechanism ('permease') specifically concerned with concentrating aromatic and perhaps some other amino acids from the growth medium, and all were mutant in the same short genetic segment. A double mutant, carrying both the gene for 4-methyltryptophan resistance and a second mutation causing a tryptophan requirement, grows only with difficulty on low concentrations of tryptophan owing to its very slow uptake of the amino acid. From such a strain revertants with normal permease production can easily be selected. This is one of the few systems (the one developed by Reissig, cf. p. 61, is another) in which both forward and reverse mutations at a specific locus can be selected for. Lewis's [311] investigation of ethionine-resistant mutants of *Coprinus lagopus* has drawn attention to yet another possible mechanism of resistance to amino acid analogues. Ethionine is a methionine analogue, and part of its inhibitory effect is probably due to its being incorporated into protein in place of methionine. The evidence suggests that one kind of resistant mutant in Coprinus has an altered methionine-activating enzyme (cf. p. 278, Appendix I) with a more restricted specificity such that it no longer promotes the incorporation of ethionine to a significant extent.

RECESSIVE LETHAL MUTATIONS

In diploid organisms, of which the fly *Drosophila melanogaster* is the best studied, a large proportion of induced mutations are recessive lethals; that is to say they result in death when homozygous but not when accompanied in the same nucleus by a corresponding normal allele. In a haploid fungus a comparable mutation would always be lethal in a cell containing only one nucleus, but it would be able to survive in a heterocaryon, sheltered by nuclei wild type for the locus in question. Atwood and Mukai [11] have made an ingenious attempt to assess the frequency of mutations to recessive lethal conditions, in comparison with that of mutations to auxotrophy, in *Neurospora crassa*. They produced a heterocaryon which carried *arg-6* in one kind of nucleus, and *me-7* in the other. These mutant alleles, in homocaryons, cause growth requirements for arginine and methionine respectively, but the heterocaryons can grow well on minimal medium since each kind of nucleus supplies the process deficient in the other. The *me-7* nuclei also carred a 'visible' mutation called *amycelial*, which, in a homocaryon, causes an exceedingly slow, almost yeast-like growth with little or no recognizable mycelium. Conidia from this hetero-caryon, which was morphologically normal, were plated on minimal sorbose agar. All colonies which grew were necessarily from heterocary-otic conidia, since neither type of homocaryon is able to grow on minimal medium. Such colonies were isolated and grown in tubes of minimal medium, and conidia from them were spread on plates of sorbose-mini-mal supplemented with methionine. Such a medium permits both hetero-caryotic conidia, and those which happen to contain only *me-7* nuclei, to grow and form colonies. In most cases, then, some of the colonies which grew on the methionine-supplemented plates were normal in mor-phology, while others were *amycelial*. Sometimes, however, the expected *amycelial* colonies failed to appear, and it was established in these cases that although the heterocaryon possessed nuclei which were capable of supplying the *arg*-6+ function necessary for growth, these nuclei were incapable of promoting growth on their own, even in the presence of externally supplied methionine. These mutant nuclei could be carrying 'incurable' recessive lethal mutations or mutations giving additional nutritional requirements. Atwood and Mukai found that, out of twenty-six distinguishable spontaneous mutations which were identified in this way, only two could be made to grow in homocaryotic condition on standard complete medium. The rest were either 'incurable' lethals, or failed to find the substances they needed for growth in the medium used. One type of lethal mutation occurred with particularly high frequency, and continued to do so, even when the heterocaryotic stock was repeat-edly reisolated from single conidia [11].

On the basis of their results, Atwood and Mukai argued that auxo-trophic mutants in Neurospora, which can mostly be made to grow norm-ally by the addition of a single substance, and thus presumably represent single metabolic lesions, may be much less frequent than the lethal class, which may represent more complicated derangements of metabolism. They suggested that many genes might have functions more complicated, or less accessible to experiment, than one might suppose from a study of auxotrophs alone. We will return to this point in Chapter 8.

The induction of recessive lethal mutations is a common effect of X-irradiation (cf. p. 77).

SPONTANEOUS MUTATION

The determination of spontaneous mutation rates in a growing mycelium is a matter of some difficulty. Ideally, one would like to measure the

nique (see above) for the detection of recessive lethal mutations in Neurospora, has obtained evidence that mutation occurs in dry macroconidia. She found that as conidia aged at 32°C the proportion carrying recessive lethal mutations detectable by the method used increased at the rate of 0·3 per cent per week. The rate at 4°C was much lower. Assuming that no synthesis or turnover of DNA, the presumed genetic material, occurs in 'dry' conidia, one must conclude that some type of irreducible chemical instability of the gene substance is involved here.

The frequencies measured by Auerbach represent mutation at an unknown, but presumably large, number of loci. Only very scanty information is available on the mutation rates of individual genes. Tatum and his colleagues [543] found only a single spontaneously auxotrophic N. crassa microconidium among 3,000 tested, and, since auxotrophy can result from mutation in any one of at least some hundreds of genes, this suggests that the forward mutation rate can hardly be more than of the order of 10^{-6} per nucleus per gene. De Serres and Kølmark [119] found no spontaneous ad-3 (adenine-purple) mutants of N. crassa among $7·6 \times 10^6$ viable conidia, though as many as 180 per 10^6 surviving conidia were obtained following X-irradiation. Woodward [592] attempted to measure the frequency of spontaneous auxotrophic mutants of Neurospora by the filtration enrichment method, making corrections for loss of mutants during the incubation in minimal medium by adding known numbers of auxotrophic conidia in control experiments. He estimated that there were 5 mutants responding to citrulline per 10^6 conidia, and 80 per 10^6 responding to glutamic acid. Since macroconidial strains were used these figures were presumably low estimates of the proportions of mutant nuclei, since an auxotrophic nucleus in a conidium also containing wild type nuclei would go

TABLE 8. Frequencies of spontaneous reverse mutants in *Neurospora crassa*

Gene	Mutant No.	Number of revertants per million macroconidia	Reference
ad–3	38701	0·008–0·29	Jensen, Kirk, Kolmark & Westergaard[257]; Kolmark & Westergaard[292]; Kolmark[291]
inos	JH5202	15·0	
	37401	0·02	
	37102	0·01	
	64001	0·10	Giles[153]
	46316	0·02	
	JH2626	0·01	
	89601	0·02	
	46802	None	

undetected. On the other hand, the two classes of auxotrophs looked for may well each have represented several genes; the figures obtained were certainly surprisingly high for mutation rates of single genes.

Spontaneous reverse mutation in auxotrophic mutants is very much easier to measure, since mutants of this kind can be efficiently selected on minimal plates, and counted as colonies against the background of non-mutant ungerminated or barely germinated conidia. Some representative values for the fraction of reverse-mutant conidia in cultures of various Neurospora auxotrophs are listed in Table 8. Here the rates per conidium are presumably overestimates of rates per nucleus, since one proto-trophic nucleus in a conidium will probably be enough to promote growth on minimal medium. It will be seen that although many of the reversion frequencies are much the same, one *inositol* allele reverted some thousand times more frequently than most of the others. There is no doubt that especially highly revertible mutations do occur.

Some very remarkable observations by Barnett and De Serres [16] indicate that where a mutant is particularly highly revertible the in-stability may be a function of the genetic site rather than of the particular genetic element (base pair?) occupying it. They found a Neurospora mutant, number 137 of the *ad-3B* series (see p. 138), which reverted to prototrophy with a frequency of 3 to 4 per 10^6 conidia, which is 10^2 to 10^4 times more frequent than the reversions at other known *ad-3* mutant sites. Remarkably enough, the apparent wild types obtained by reversion of 137 all showed an extremely high spontaneous frequency of adenine-purple mutants—about 15 per 10^6 conidia, which is some 50 times the normal rate. Almost all the secondary *ad-3* mutants proved to be members of the *ad-3B* rather than the *ad-3A* class, and all of these showed the same high reversion rate as the original mutant 137 to which they seemed, in fact, to be identical. Thus there seemed to be a condition of persistent instability at a particular genetic site, causing it to mutate back and forth between a mutant and an apparently wild type condition. No explanation can at present be given for this very curious result.

In general, the frequency of reversion that one is likely to find in an auxotrophic mutant depends on the genetic nature of the revertants. In some cases reversion to the prototrophic condition can result from 'suppressor' mutation at any one of a large number of genetic sites within one or more genes other than the one which originally mutated. In such cases, of which Heslot's [228] study of an *arg* mutant in *S. pombe* provides an example, the 'reversion' frequency can be higher than the frequency of forward mutation within most individual genes. In many other cases only further mutation within the originally mutant gene will restore prototrophy, but even so it is not necessarily exactly the same *site* which is involved in the reversion as was altered in the primary mutant. Examples of 'second site reversion' are known in bacteria [598] and are strongly suspected in Neurospora (cf. p. 188). Nevertheless it must often be true that only a true reverse mutation, resulting in a restoration of the original wild genotype, will restore the ability to grow on minimal medium. Thus, on the whole, reversions to prototrophy will represent mutation at only a few sites, often only a single site, whereas forward mutation to auxotrophy can occur at any one of a large number of sites within the wild type gene; many different changes will stop the normal gene from functioning but it requires a rather specific change to repair a damaged gene. It is, in fact, the general experience where both forward and reverse mutation rates for one gene can be measured [16, 56] that the former are some orders of magnitude higher than the latter.

MUTAGENS

Mutation rates can be very greatly increased by certain kinds of radiations and chemical treatments. Among radiations, ultraviolet light is effective, and so is any kind of ionizing radiation such as X-rays, gamma rays, α-particles, or fast neutrons. Many kinds of chemicals are effective as mutagens.

Radiations

In the case of X-rays, it is commonly found that the fraction of mutants among surviving nuclei is nearly proportional to the dose of radiation. Such a result has been obtained with *Neurospora crassa*, both for forward [476] and reverse mutations [194], and approximately for visible forward mutations in *Aspergillus terreus* [516] (Fig. 15) and *Glomerella cingulata* [343]. The tendency for mutation to be proportional to X-ray dose, implying as it does that mutations are each induced by 'one hit', *i.e.* by a single ionization track, may be taken as an indication that the effect of X-rays may be a rather direct one. In many cases, though not in that of *Aspergillus terreus* illustrated in Fig. 15, the killing of uninucleate conidia by X-rays proceeds as if the chance of a spore being killed, like the chance of its undergoing a viable mutation, is proportional to radiation dose. This encourages the belief that much of the lethal effect of radiations is due to the induction of lethal mutations.

Klein and Klein [286] found with Neurospora that both the killing and the mutagenic (*ad-3* to *ad-3+*) effect of X-rays could be somewhat enhanced by far-red light, and that this enhancement could be negated by subsequent exposure to red light; this suggests that X-irradiation can have indirect as well as relatively direct effects.

Ultraviolet light may give a more complex dose-mutation relationship; in the low dose range a more than linear increase of mutation with dose has sometimes been found, and some experiments show an optimum and subsequent decline in yield of mutants as the dose is increased in the higher range. The existence of an optimum ultraviolet dose for mutagenesis could

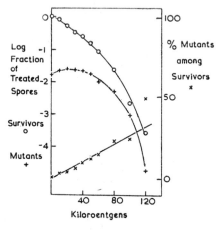

FIG. 15. Lethal and mutagenic effects of X-rays on conidia. Data on *Aspergillus terreus* from Stapleton, Hollaender & Martin[516].

be explained if a fraction of the cell population were resistant both to mutation and to killing, so that the survivors of high doses tended to be non-mutant; Markert has provided evidence in support of this explanation for conidia of *Glomerella cingulata* [342]. The tendency towards an exponential increase in the proportion of mutants with dose in the lower range has been related by Witkin [588], on the basis of her work on the bacterium *Escherichia coli*, to the delay imposed by ultraviolet irradiation on DNA synthesis. Her evidence suggests that the process of mutation depends on the occurrence of protein synthesis in the interval between irradiation and the resumption of DNA synthesis, and that increasing the dose of ultraviolet light both increases the probability of mutation per unit time, and lengthens the time during which it can happen. Evidence of a link between post-ultraviolet mutation fixation and protein synthesis has also been obtained in experiments on yeast [479].

Little can be said in detail about the chemical basis of the mutagenic effect of X-rays, though there is no doubt that they can cause many different chemical changes. There is, however, some rather precise knowledge about at least one major effect of ultraviolet light on cells, namely the linking (dimerization) of pairs of thymine residues in DNA [34, 35]. Microbial cells have various repair mechanisms, some dependent on visible light (photoreactivation mechanisms) [344, 469] and others occurring in the dark [49, 485], which can remove thymine dimers from DNA chains and replace them with normal thymine residues. The effect of visible light in restoring viability to ultra-violet irradiated cells is well established in a number of fungi, and there are some reports that the potential for mutation as well as the lethal effect is subject to photoreversal [59, 200]. It seems very likely that the formation of thymine dimers is responsible for at least a part of the mutagenic effect of ultraviolet light, as well as for much of the killing effect.

Chemical mutagens

Perhaps the most extensive and systematic surveys of the action of chemical mutagens on fungi have been made by Westergaard on *Neurospora crassa* [569] and by Heslot on *Schizosaccharomyces pombe* [228], and much of what follows is based on their work. A very considerable number of substances have been shown to be effective as mutagens, and at first sight they might appear to be a very heterogeneous assemblage. They can, however, be regarded as falling into a few distinct classes on the basis of their presumed or known action on DNA or its replication.

Most chemical mutagens can act on non-growing cells, and probably act by direct chemical modification of the DNA, which is the most general type of genetic material (see Appendix I). Most of these probably act as alkylating agents, reacting with the amino groups of the DNA bases. The most chemically straightforward alkylating agents include such compounds as the ethylating agents diethylsulphate, ethylmethane-sulphonate ($CH_3SO_2OC_2H_5$), and the corresponding methylating agents dimethylsulphate and methylmethanesulphonate. Even more widely used have been various β-chloroethyl amines or sulphides, the so-called mustard compounds, which were the first compounds to be demonstrated as mutagens (in Drosophila) and among the first to be shown to be effective on fungi. They include *bis*(β-chloroethyl)sulphide (mustard gas) [242], *n*-butyl-β-chloroethylsulphide [519], di(β-chloroethyl)methylamine (nitrogen mustard) [569], *tris*(β-chloroethyl)methylamine [444, 543] and di(β-chloroethyl)-L-phenylalanine [569]. Another group of compounds which probably act as alkylating agents are various epoxides, including ethylene oxide ($CH_2{\overset{\displaystyle O}{\diagup\diagdown}}CH_2$), propylene oxide ($CH_3CH{\overset{\displaystyle O}{\diagup\diagdown}}CH_2$), glycidol ($HOCH_2{\overset{\displaystyle O}{\diagup\diagdown}}CH{-}CH_2$), epichlorohydrin ($ClCH_2{\overset{\displaystyle O}{\diagup\diagdown}}CH{-}CH_2$), 1-2,3-4-diepoxybutane ($CH_2{\overset{\displaystyle O}{\diagup\diagdown}}CH{-}CH{\overset{\displaystyle O}{\diagup\diagdown}}CH_2$) and 1-2-monoepoxybutane ($CH_3CH_2CH{\overset{\displaystyle O}{\diagup\diagdown}}CH_2$). Akin to the epoxides are ethyleneimine ($CH_2{\overset{\displaystyle NH}{\diagup\diagdown}}CH_2$) and a number of related compounds including the complex heterocyclic derivative triethylene-melamin. All of these substances have been shown to be mutagenic in Neurospora or Schizosaccharomyces or both. β-Propiolactone [501] and N-nitroso-N-methylurethan [606], both potent mutagens, may also act as alkylating agents.

There are some fairly precise ideas about the mechanism of action of at least the ethylating mutagens, based on studies of their action on deoxyribonucleic acid. It seems that the principal target for ethylation is probably the 7-nitrogen atom of guanine, and that 7-ethylguanine tends to be removed by hydrolysis from the polynucleotide chain. Adenine can also be ethylated, but there is evidence [22] that adenine is not removed from DNA by ethylation to the same extent as guanine; the pyrimidines cytosine and thymine are apparently not readily ethylated. On the basis of the currently accepted view of the mode of replication of DNA (see Appendix I) it is easy to imagine that removal of a base could result in a more or less random incorporation of any of the four bases opposite the gap at the following replication. The new base thus brought into the chain will incorporate its own complementary partner base at the next replication, and a new stably replicating pair will have been established.

Nitrous acid, which is usually generated by dissolving sodium nitrite in a solution buffered at around pH 4·6, is generally effective as a mutagen on non-dividing cells. It acts in quite a different way from the alkylating agents. Through its well known property of reacting with primary amino groups to give hydroxyl groups it will convert adenine to hypoxanthine, guanine to xanthine and cytosine to uracil. Hypoxanthine should have the pairing specificity of guanine (see Appendix I) and should thus bring in cytosine instead of thymine at the next replication, while uracil should, like thymine, pair with adenine. The guanine to xanthine change should cause no change in pairing specificity and is not expected to be mutagenic. Nitrous acid has been used to induce mutations in Neurospora [513], Schizosaccharomyces [212] and *Aspergillus nidulans* [493].

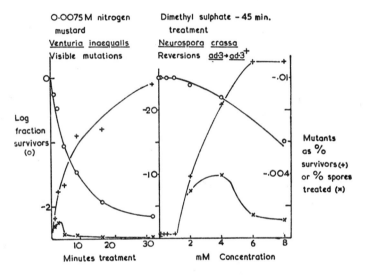

Fig. 16. Lethal and mutagenic effects of chemicals on conidia. Data on *Venturia inaequalis* from Boone *et al*[48]; those for *Neurospora crassa* from Kolmark[291].

Hydroxylamine, though there are no reports of its mutagenicity in fungi, is a potent mutagen for bacteriophage [175]. It probably acts specifically on cytosine, altering its pairing properties.

All the chemical mutagens mentioned above will induce mutations in non-dividing conidia or (in the case of yeasts) in non-dividing suspensions of vegetative cells. The cells to be treated are suspended in a buffered solution containing a suitable concentration of the mutagen and, after a certain period of incubation, are centrifuged down, washed, and plated on a suitable medium for screening for mutants.

Many studies have been made on the quantitative relationships between yield of mutants and dose of mutagen, especially by Heslot (and see Fig. 16). As Heslot points out [228], it is better to keep to a constant time of incubation and to vary the concentration of the mutagen, so as to equalize the effects of instability of the mutagen. Even so one cannot be certain that the effective concentration of mutagen within the cell is proportional to the concentration in the external solution. Thus it is difficult to place much theoretical significance on the shapes of the curves relating mutagen concentration to yield of mutants. In Heslot's experiments on Schizosaccharomyces these curves were sometimes roughly linear, but more often the yield of mutants (expressed as percent of surviving cells) increased in proportion to rather more than

TABLE 9. Yields of mutants obtained with various mutagens.

Mutagenic treatment	Cells treated	Type of mutation	Per cent		Ref.
			Survivors	Mutants among survivors	
Ultraviolet light	*N. crassa* macroconidia	*ad-3→ad-3+*	50	·0012	569
Dimethyl-sulphate	ditto	ditto	44	·015	569
Diethyl-sulphate	ditto	ditto	68	·0026	569
Ethylmethane sulphonate	ditto	ditto	14	·012	569
'Nitrogen mustard'	ditto	ditto	60	·0005	569
Diepoxy-butane	ditto	ditto	56	·015	569
Ethylene oxide	ditto	ditto	63	·0027	569
Ethylmethane sulphonate	*N. crassa* microconidia	*pyr-3+→pyr-3*	50	ca. ·003	447
Nitrous acid	ditto	ditto	50	ca. ·01	447
Diethyl-sulphate	*Schizosaccharomyces pombe*	*pyr→pyr+*	62	·0002	228
'Mustard gas'	*N. crassa* macroconidia*	All identifiable mutants	?	7·6	242
L(+)-diepoxy-butane	*Penicillium multicolor* conidia	All visible mutants	33	20	368

Notes: *Conidia were used to fertilize protoperithecia, and the resulting ascospores were screened.

The *ad-3* mutant (no. 38701) referred to in the upper 7 lines of the Table seems to be especially highly revertible by alkylating agents.

the first power of the mutagen concentration. Moreover, with several mutagens (for example, nitrous acid and diethyl sulphate) he found an optimum mutagen concentration, above which the proportion of mutants among the survivors actually decreased. The reason for this kind of result is unknown, though a formally adequate explanation would be that those cells most prone to mutate have a higher than average chance of being killed at high mutagen concentrations.

It is difficult to give any concise summary of the relative effectiveness of different mutagenic treatments, since much depends on the organism and on the type of mutant being scored, but Table 9 will give some idea of the results which can be obtained with some of the better known mutagens.

The reagents mentioned so far attack non-replicating DNA. A further class of compounds, the base analogues, can become incorporated into DNA at replication 'in mistake' for normal bases, and cause mutations as a result. Two base analogues have been particularly widely used by workers on bacteriophage [174]. 5-Bromouracil (BU), or its deoxy-nucleotide derivative 5-bromodeoxyuridine (BDU), can be incorporated into bacterial or bacteriophage DNA with high efficiency in place of thymine, provided that the endogenous supply of thymine is restricted either by a mutational block in its synthesis or by inhibition of its synthesis by a drug. 2-Aminopurine (AP), which is an adenine analogue, can substitute for adenine to a more limited extent [174]. Both BU and AP are mutagenic, probably by virtue of their somewhat ambivalent pairing properties. While BU is a close analogue of thymine there are chemical grounds for suggesting that it can occasionally, through a tautomeric shift in the position of a double bond, pair as if it were cytosine, and hence either become incorrectly incorporated opposite guanine, or replicate incorrectly, bringing in guanine instead of adenine. In rather a similar way, AP has the possibility of pairing with cytosine as well as with thymine. Since their action depends on DNA replication, base analogues, to be effective as mutagens, have to be supplied to growing cells by incorporation into the growth medium.

The use of base analogues as mutagens for fungi has so far been limited, but Brockman and De Serres [56] have used AP to induce *ad-3* mutants of Neurospora, while Ishikawa [248] reports the use of BDU for the induction of *ad-8* mutants in the same organism. In the latter case 5-fluorodeoxyuridine was incorporated into the medium along with the BDU in order to inhibit the synthesis of normal thymine.

Caffeine, which has been reported to be mutagenic for Ophiostoma

[180], is a methylated purine, and can also be regarded as a base analogue, but there is no evidence that it can be incorporated into DNA.

How many independently mutable strands per chromosome?

Assuming, as we must do (cf. Appendix I), that the genetic material is DNA, we must suppose that the genetic information in each chromosome is carried in *at least* two polydeoxynucleotide strands of one DNA molecule. Mutagens which act by attacking, or substituting for, individual DNA bases would not be expected to cause mutations capable of being transmitted to all the descendants of a treated cell, since only one strand of a DNA molecule should be altered at any given site. The mutation, or potentiality for mutation, should be transmitted to only one of two daughter genes following mutagen treatment. Siddiqi has, indeed, observed [493] that when *Aspergillus nidulans* conidia are treated with nitrous acid to about 30 per cent survival most *white* or *yellow* mutants appear as sectors, often halves or quarters of colonies, rather than as whole colonies; quarter-colony sectors would be expected if the DNA had replicated in preparation for the next nuclear division at the time of treatment. Reissig [447], on the other hand, found no significant frequency of mosaic colonies when selecting for *pyr-3* mutants (cf. p. 61) following treatment of *Neurospora crassa* microconidia with nitrous acid; practically all the mutants tested were pure for *pyr-3*, though *pyr-3+* nuclei should certainly have been able to persist in the selected colonies had they been initially present. The amount of killing in this experiment (36 per cent) was not enough to make it reasonable to argue that each colony represented only one surviving DNA strand out of two originally present. Reissig's conclusion was that nitrous acid could induce mutations affecting both DNA strands simultaneously. Holliday [234] has come to the same conclusion with regard to mutants induced by ultraviolet light. In experiments on *Ustilago maydis* he used the phenomenon of mitotic segregation (see Chapter 5) to identify both daughter strands of ultra-violet-treated chromosomes and has shown that both can transmit the same induced mutation. Holliday suggests that there is some mechanism for correcting mis-matched base pairs in DNA by taking one of the pair out and replacing it with a properly complementary one. Thus a mutational replacement or alteration of a base in one strand might lead to a base replacement in the second strain so as to restore the complementarity of the base pair. This idea is of great interest in connexion with the theory of gene conversion (see Chapter 7) as well as for the understanding of mutation, but it is still quite speculative.

Selective effects of mutagens

There is very little indication that any of the mutagens act preferentially on one gene rather than another. Comparisons between the effects of ultraviolet light and nitrogen mustard on Venturia [48], mustard gas, X-rays and ultraviolet light on Neurospora [25, 242] and ultraviolet light, diethylsulphate, isopropylmethane-sulphonate and ethyleneimine on *Schizosaccharomyces pombe* [227] have revealed no clear differences in the relative frequencies of different kinds of biochemical function affected by mutation. In current theory the gene, in the sense of a short chromosomal segment acting as a unit of function, is pictured as a large array of different kinds of mutable sites, presenting a relatively large and diverse target, and so mutagen specificity as between one gene and another would not be expected.

On the other hand where one is looking specifically for mutation at one or a few sites, as is often the case in studies of induced reversion in auxotrophs, for example, one might well expect some mutagen specificity. Such specificity is in fact often found. Giles [194], for example, showed that of two Neurospora auxotrophs mutant in the *inos* gene, one (98601) can be induced to revert much more frequently with X-rays than with ultraviolet light, comparisons being made at doses giving equal amounts of killing, while for the other (37401) the reverse was true. Kølmark [290] synthesized a double mutant strain carrying both the 37401 *inos* allele and the mutant allele (37801) of *ad-3* which had been used so extensively in other reverse mutation studies by Westergaard's group. It was shown that in the double mutant strain reversions of the mutation within the *ad-3* locus could be induced with a strikingly high frequency with diepoxybutane, but that this mutagen caused almost no reversion of the *inos* mutation. Ultraviolet light induced both kinds of reversion, rather more frequently in *inos* than in *ad-3*.

A similar striking specificity of certain mutagens for certain genetic sites has been shown in studies on forward mutation. Gutz [212], working with *Schizosaccharomyces pombe*, has obtained a very extensive series of *ad-7* (adenine red) mutants. By methods which will be described in Chapter 7 he was able to show that the mutations occurred at many different sites, linearly arranged within the *ad-7* gene. It was evident that, as compared with ultraviolet or X-irradiation, nitrous acid tended to attack only a limited number of sites but caused these to mutate over and over again. Ishikawa [248], in isolating many mutants at the *ad-8* locus of Neurospora, found no very definite site specificity on the part of nitrous acid but a very strikingly specific effect of bromodeoxyuridine. Fifteen independently isolated mutants following bromodeoxyuridine treatment were all at the same site (Fig. 17). The chemical basis of these

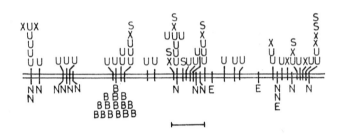

FIG. 17. Map of the *ad-8* locus of *Neurospora crassa* showing the distribution of mutations induced by different mutagenic treatments. Each symbol indicates one independently induced mutation. U=ultraviolet, X=X-ray, S=spontaneous, N=nitrous acid, E=ethylmethanesulphonate, B=5-bromodeoxyuridine. The horizontal scale shows the map distance equivalent to the formation of 0·01 per cent prototrophs from an inter-mutant cross. Adapted from Ishikawa [248].

rather extreme site preferences of mutagens is no clearer in fungi than it is in bacteriophage, where the phenomenon is much better known [29]. Since there are only four kinds of DNA base pair, occurring with very similar frequencies, one must suppose that the intrinsic nature of the individual base pair is not the only factor involved in determining its vulnerability to mutagens, but that the nature of neighbouring bases or base sequences also has a large effect.

Structural effects of radiations

In higher plants and animals X-rays, in common with other ionizing radiations, are known to produce chromosome breaks with consequent rearrangements such as inversions, segmental interchanges and deletions of chromosome segments. Deletions will normally be lethal in a haploid organism, but they can be preserved in one of the two kinds of nuclei in a heterocaryon. De Serres and Osterbind [120] studied adenine-purple mutants induced by X-rays in a Neurospora heterocaryon carrying *ad-3+* in one nucleus and both *ad-3A* and *ad-3B* mutations (cf. p. 63) in the other. Each nucleus carried other (complementary) auxotrophic markers in order to make homocaryotic growth on minimal medium supplemented only with adenine impossible. Two kinds of adenine-purple heterocaryons were obtained. The first had genetic damage confined to either the *ad-3A* or the *ad-3B* gene in the originally *ad-3+* nucleus. In these cases the newly induced adenine-requiring mutants could be recovered as homocaryons on suitably supplemented medium. The second class consisted of cultures from which the newly induced mutant component could not be recovered as a viable homocaryon, and was, in other words, recessive lethal (cf. pp. 62–63). Complementation tests showed that many of these had suffered simultaneous damage to both *ad-3A* and *ad-3B* genes. It was presumed that the non-recoverable *ad-3* mutants carried deletions including parts of the *ad-3A* and/or *ad-3B* genes together with neighbouring genetic material of unknown but essential function. These inter-genic deletions were commoner than the intra-genic changes (some of which may also have been deletions) which were viable on supplemented medium as homocaryons; the former outnumbered the latter by 29 to 16 (see also [118]). Intra-genic deletions can be distinguished from true point mutations by fine structure mapping (cf. Chapter 7) and by their inability to revert to wild type by further mutation. Coincident non-revertibility and abnormal linkage relationships are apparently frequent among X-ray induced mutants (Giles [195]).

Ultraviolet light is generally regarded as being much less likely than X-rays to induce structural chromosome changes. McClintock, however [333], has obtained microscopic evidence for segmental interchanges in two ultraviolet mutants of Neurospora as well as in one X-ray mutant.

CHAPTER 4

CHROMOSOME MAPPING

Linkage maps of chromosomes can be constructed either through the analysis of the products of meiosis, or by the newer techniques of di-ploidization and analysis of mitotic segregation. The second method, which is of comparatively recent introduction, will be dealt with in the next chapter. In this chapter we shall give an account of the classical approach via meiotic analysis.

In all organisms where the process occurs, the primary products of meiosis are tetrads of nuclei, though in egg formation in higher plants and animals, only one nucleus of each tetrad survives. Fungi differ from higher organisms in the greater opportunities which they offer for the re-covery of meiotic products in their original tetrads. In fungal genetics one often has the choice between the analysis of tetrads (of ascospores, ascospore pairs, or basidiospores as the case may be) and the isolation of the same spores after they have been released from their tetrads and mixed in a random fashion. As we shall see, both types of analysis have their special advantages.

TYPES OF TETRAD

The argument in this chapter is based on the assumption that unit differ-ences between homologous chromosomes always show 2:2 segregation during meiosis. While, as is explained in Chapter 7, this is not strictly true, the exceptions are so infrequent as to be negligible in the present context. From this basic assumption it follows that, if the diploid cell undergoing meiosis is heterozygous for two loci, marked by mutations a and b, there are, disregarding the ordering of the spores, only three pos-sible types of tetrad. These are: parental ditype (PD), non-parental di-type (NPD), and tetratype (T). These three types are shown in Table 10.

THE DETECTION OF LINKAGE

In the genetics of higher organisms, where tetrad analysis is practically never possible, the test for linkage has always been the comparison of the frequencies of parental and recombinant types among the products of meiosis. If, from a double heterozygote, one obtains a significant excess of parentals over recombinants (i.e. a *recombination fraction* of less than 0·5), one may conclude tentatively that the two loci concerned are on the same chromosome, although there are other possible explanations which will be mentioned at the end of this chapter. To take as an example

TABLE 10. Possible tetrad types formed by a double heterozygote
from the cross $a+ \times +b$

Parental ditype	Non-parental ditype	Tetratype
$a+$	$a\,b$	$a+$
$a+$	$a\,b$	$a\,b$
$+b$	$++$	$++$
$+b$	$++$	$+b$

The order of the spores within each tetrad is arbitrary.

the cross shown in Table 10, one could test for linkage by comparing the total frequency of the parental types, $a +$ and $+ b$, with that of the recombinant types, $a\,b$ and $+ +$, in a sample of random spores. The appropriate statistical procedure for assessing the significance of any deviation from a 1:1 ratio of parentals and recombinants would be the X^2 test. This test, and other statistical techniques relevant to the detection and assessment of linkage, are well described by Mather [346]. Here it will suffice to point out that, where one is testing for goodness of fit to a 1:1 ratio, the X^2 test is extremely simple. If the numbers observed in the two classes are x and y, $X^2 = (x - y)^2/(x+y)$. From the value of X^2 one can, from tables, determine the probability of getting so large a deviation from a 1:1 ratio merely by sampling error. If X^2 is greater than 3·8, this probability is less than one in twenty, while if it is greater than 6·6, the probability is less than one in a hundred. The test is only applicable if both x and y are greater than 5.

When one is analysing tetrads, however, it is misleading to compare the total frequencies of parental and recombinant types among the component single meiotic products. Such a procedure is wrong because the four components of a tetrad do not constitute independent observations; having, for example, determined the two upper members of any of the three tetrad types shown in Table 10, the other two members are also determined. Thus if one counts all four members of each ascus one will obtain an inflated estimate of the significance of any apparent linkage. The frequencies which can be compared statistically, since they are of independent events, are those of the three tetrad types. It will be seen from Table 10 that all of any excess of parental over recombinant products must be due to an excess of PD over NPD tetrads. Therefore the appropriate test for linkage with tetrad data is the assessment, by the χ^2 test, of the significance of any such excess; in the absence of linkage, of course, one expects a 1:1 PD:NPD ratio.

Although tetratype tetrads, since they contain equal numbers of parental and recombinant spores, can contribute nothing to the evidence *for* linkage, their frequency can, in certain cases, provide critical evidence *against* two loci being on the same chromosome. If one obtains a PD: NPD ratio of approximately one, this may be due either to the loci being on different chromosomes, or to their being far apart on the same chromosome. In the former case, the proportion of tetratypes can take any value depending on the distances of the loci from their respective centromeres (see below). In the extreme case, where both loci are very close to their centromeres and show 100 per cent of segregation at the first meiotic division, the proportion of T tetrads will be zero. If, on the other hand, the loci are far apart on the same chromosome, the fraction of tetratypes should approximate to 67 per cent, and the T:NPD ratio should never fall below 4:1. This will become evident from a study of Table 14 and Fig. 18. Briefly, the argument is that, if there are so many double exchanges as to make PD and NPD tetrads nearly equal in frequency, there will be enough triples to hold the tetratype frequency high (singles as a class give 100 per cent tetratype, doubles 50 per cent and triples 75 per cent, assuming no chromatid interference). To give just one example from *Neurospora crassa* (quoted in Barratt *et al*, [19]), a two point cross involving the loci *iv* and *ad-4* gave 7 PD, 11 NPD and 2 T; here the low frequency of tetratypes furnishes convincing proof of the independence of the two loci. Since it is often as useful to have proof of non-linkage as of linkage, this kind of test finds a good deal of use in mapping work.

In spite of this special use of tetrad data, preliminary tests for linkage are, on the whole, best made with random spores. Random spores contribute more information *per spore* on the recombination fraction than do tetrads. The reason for this is that each random spore, provided it is taken from a large population, can be regarded as an independent observation, while, as already pointed out, the constitutions of the four components of a tetrad are mutually dependent. When linkage is fairly close, so that recombinant spores nearly all come from tetratype tetrads, each tetrad contributes little more than two random spores would do to the estimate of the recombination fraction (Mather and Beale [347]).

THE ESTABLISHMENT OF GENE ORDER

If linkage values could be relied upon to be constant from one cross to another, one could build up a map of a linkage group from a series of crosses each involving only two markers. In Neurospora, at any rate, there is enough genetically determined variability in cross-over frequency to make this an uncertain procedure, and a much sounder method is to build up a picture of the entire gene sequence by a series of 3- or 4-point crosses involving overlapping sets of markers.

Whether one is analysing random spores or tetrads, the basic principle in multipoint crosses is that the most probable gene order is that which requires the postulation of the fewest cross-overs. This principle was implied in the example given in Chapter 1. In Table 11 the various possibilities in a 3-point cross are set out systematically. When the intervals between markers are short, the classes resulting from double exchanges should hardly occur at all and only when the recombination fraction shown by adjacent markers approaches 0·5 (i.e. no detectable linkage) will the frequency of double cross-over products approach that of the least frequent single cross-over products. In the latter situation the establishment of an unambiguous gene order will depend on the use of additional markers to shorten the inter-marker intervals. Among the tetrad types, classes 2, 3 and 4 (Table 11) usually provide most of the evidence for gene order; if the intervals are fairly short one of these classes should be very infrequent relative to the other two. When a higher frequency of multiple crossing-over prevails, classes 9, 10 and 11 may give a good indication of the most probable order, since an order requiring the postulation of a quadruple exchange is much less probable than one requiring only a double.

It was noted above that random spores were generally more efficient than tetrads for the establishment of linkage. The same is generally true for the ordering of genes, though, where linkage is very loose, tetrads may give information not obtainable from random spores because of the greater number of comparisons which may be made between different classes. When one is dealing with auxotrophic markers, and particularly when the recombinant classes are rather infrequent, random spores have a very marked advantage. Thus, in a cross between two auxotrophic mutants, the prototrophic recombinant class can be estimated by spreading spores on minimal medium and counting the number capable of forming colonies. If the two loci concerned are unlinked, 25 per cent of the spores capable of forming colonies on complete medium should do so on minimal medium. Selective media can also be used with great sav-

TABLE 11. Interpretation of a three-point cross

Cross: *a b c* × + + +

Meiotic products		Simplest cross-over explanation with gene order:		
		a–b–c	*b–a–c*	*a–c–b*

A. Random spores

Meiotic products	*a–b–c*	*b–a–c*	*a–c–b*
$\begin{cases} +++ \\ c\,b\,c \end{cases}$	No exchange	No exchange	No exchange
$\begin{cases} +\,b\,c \\ a++ \end{cases}$	Single	Double (2-str. or 3-str.)	Single
$\begin{cases} a\,b+ \\ ++c \end{cases}$	Single	Single	Double (2-str. or 3-str.)
$\begin{cases} a+c \\ -\,b+ \end{cases}$	Double (2-str or 3-str.)	Single	Single

B. Tetrads

Tetrad type with respect to:

	a–b	*a–c*	*b–c*	*a–b–c*	*b–a–c*	*a–c–b*
1.	PD	PD	PD	No exchange	No exchange	No exchange
2.	PD	PD	T	Single	Single	Double (*a–c, c–b*, 2-str.)
3.	T	PD	T	Double (*a–b, b–c*, 2-str.)	Single	Single
4.	T	T	PD	Single	Double (*b–a, a–c*, 2-str.)	Single

Tetrad type with respect to / Simplest cross-over explanation with gene order:

	a–b	a–c	b–c	a–b–c	b–a–c	a–c–b
5.*	T	T	T	Double (a–b, b–c, 3-str.)	Double (b–a, a–c, 3-str.)	Double (a–c, c–b, 3-str.)
6.	NPD	T	T	Triple (b–c, 4-str. double in a–b)	Triple (a–c, 4-str. double in b–a)	Double (a–c, c–b, 4-str.)
7.	T	NPD	T	Double (a–b, b–c, 4-str.)	Triple (b–a, 4-str. double in a–c)	Triple (c–b, 4-str. double in a–c)
8.	T	T	NPD	Triple (a–b, 4-str. double in b–c)	Double (b–a, a–c, 4-str.)	Triple (a–c, 4-str. double in c–b)
9.	PD	NPD	NPD	Double (within b–c, 4-str.)	Double (within a–c, 4-str.)	Quadruple (4-str. doubles within both a–c and c–b)
10.	NPD	PD	NPD	Quadruple (4-str. doubles within both a–b and b–c)	Double (within b–a, 4-str.)	Double (within c–b, 4-str.)
11.	NPD	NPD	PD	Double (within a–b, 4-str.)	Quadruple (4-str. doubles within both b–a and a–c)	Double (within a–c, 4-str.)

* Two kinds of 3-strand doubles distinguishable.

ing of labour for the analysis of 3-point crosses involving auxotrophic mutants. An example from *Neurospora crassa* is shown in Table 12.

TABLE 12. Ordering of loci with the use of selective plating

Neurospora crassa cross *inos*+*sp* × +*am*+. *Inos* grows only in the presence of inositol; *am* grows well with alanine but not on minimal medium*, *sp* is prototrophic but has a characteristically stunted growth habit. A random sample of over a thousand ascospores were classified on each medium; the germination was 98 per cent. Data from Smith [499].

		Supplement to minimal medium		
	1. Alanine+inositol	2. Alanine only	3. Inositol	4. None*
Genotypes expected to grow	*inos*+*sp*		*inos*+*sp*	
	inos+ +		*inos*+ +	
	+*am sp*	+*am sp*		
	+*am*+	+*am*+		
	+ +*sp*	+ +*sp*	+ +*sp*	+ +*sp*
	+ + +	+ + +	+ + +	+ + +
	inos am sp			
	inos am+			
Growing spores as per cent. of those germinated	*sp*+ 51·2	46·1	3·7	0·2
	sp 48·8	6·5	46·1	3·4

Interpretation: The following frequencies can be obtained with fair accuracy: + + + 0·2% (column 4); + + *sp* 3·4% (column 4); *inos*+ + 3·5% (columns 3 and 4, by difference); +*am sp* 3·1% (columns 2 and 4 by difference). Of these the first must be the double cross-over class, and the last three single cross-over classes. Thus the order must be *inos–am–sp*.

* *am* strains tend to grow a little on ordinary minimal medium, but this slow growth is strongly inhibited by glycine. In this experiment glycine was added to plates lacking alanine to inhibit growth of *am* sporelings; this detail is omitted from the Table for simplicity.

MAPPING OF CENTROMERES

In *ordered tetrads*, the arrangement of the spores is such that first and second division segregation can be distinguished. In genera of Ascomycetes with linear asci, such as Neurospora, Sordaria, Venturia, Ascobolus, Bombardia and Chromocrea, the sister-products of each second meiotic division are usually adjacent in the same half of the ascus. Thus if we represent a pair of alleles as + and −, the arrangements (+ + − −) and (− − + +) would indicate first division segregation, and the arrangements (+− +−), (− +− +), (+−− +), (− +− +) second division segregation. The secondarily homothallic species *Neuro-*

chance of mutation occurring per nucleus per unit time or per nuclear generation time. This is possible in the case of an organism growing in the form of free cells, such as a bacterium, a yeast, or the sporidial stage of a smut. Novick and Szilard [382], working with bacteria, devised a 'chemostat' for continuous culture of cells under constant condition in liquid culture. The original inoculum containing no mutant cells, the proportion of cells carrying a mutation conferring resistance to a bacteriophage was determined at intervals by withdrawing samples and plating them in the presence of the virus. Since it was determined that the mutation to virus resistance made no difference to growth rate in the absence of the virus, the mutation rate, in terms either of mutations per cell per hour, or of mutations per cell per generation time, could be calculated from the rate of increase in the proportion of mutant cells in relation to the rate of cell division. Given a suitable screening technique for determining numbers of mutants, the same principle could be applied to fungi with free cell growth.

In filamentous fungi, however, nuclei can be easily sampled only at the stages when spores are produced. Thus one can determine the proportion of mutants among conidia formed late in the development of a culture of Neurospora, but one has no means of knowing whether these mutations occurred in the conidia themselves, or several nuclear divisions before conidiation. Even if one were able to sample nuclei from growing hyphae, it would still be a matter of some complexity to determine the rate of nuclear division in the growing hyphal tips, nor would it be easy to devise a test to determine whether a newly arisen mutant nucleus in a hypha divides more or less rapidly than the original type. All that one can really do is to determine the proportions of mutant conidia in a series of replicate cultures, each started from a small inoculum. A few cultures will contain unusually many mutant conidia; these will be due to the occurrence of mutation early in the growth of the cultures concerned. Many others, however, will contain much the same rather low frequency of mutants, and these will give an approach to a measure of the frequency of mutation per conidium, although the estimate must always be inflated somewhat through the multiplication of mutant nuclei arising a few nuclear generations before conidiation. It must also be borne in mind that the frequency of mutant macroconidia, where the mutation in question is dominant in a heterocaryon, will be an overestimate of the frequency per nucleus.

It has often been supposed that spontaneous mutation only occurs in dividing nuclei. However, Auerbach [12], using the Atwood-Mukai tech-of randomized ascospores from single asci. This method, which makes use of the fact that the spores of one ascus are most commonly discharged as a compact group, was, in fact, first used as early as 1927 by Shear and Dodge [489], and subsequently forgotten.

Second division segregation frequencies

Unordered tetrads, which are all one can get from Aspergillus, many strains of yeast and Basidiomycetes in general, can often be made to yield information about second division segregation frequencies. If, in a particular cross, there is a 'centromere marker' which is known to segregate nearly always at the first division, the segregation of any other markers can be classified also, since a tetratype ascus will indicate second division segregation and a ditype ascus first division segregation of the second marker. This method becomes safer when two or more unlinked centromere markers are present in the cross. Hawthorne and Mortimer [220] have made extensive use of this type of argument in their work on Saccharomyces. They used linear asci to establish which were the centromere markers, and then used the latter to classify segregations of other markers in oval asci.

Even if one has only unordered tetrads and no known and reasonably reliable centromere markers, it still may be possible to calculate second division segregation frequencies even though one cannot decide the type of segregation in each individual ascus. The argument, developed by Whitehouse [576], is as follows.

Suppose we have three loci a, b and c, with frequencies of second division segregation x, y and z. Then the frequencies of first division segregation will be $1-x$, $1-y$ and $1-z$. Provided that the loci are all on different chromosomes, the frequencies of tetratype tetrads in crosses involving any pair of them will depend only on the second division segregation frequencies. Thus in a cross heterozygous at a and b, tetratype tetrads will result in all cases where a segregates at the first divison and b at the second (frequency $(1-x)y$), in all cases where a segregates at the second division and b at the first (frequency $x(1-y)$), and in half the cases where both segregate at the second division (frequency $xy/2$). Thus the total frequency of tetratype tetrads is given by

$$T_{ab} = x(1-y) + y(1-x) + xy/2 = x + y - 3xy/2$$

Similarly, tetratype frequencies from crosses involving heterozygosity at a and c, and at b and c will be

$$Tbc = y + z - 3yz/2 \text{ and}$$

$$Tac = x + z - 3xz/2$$

With three equations and three unknowns it is possible to solve for x, y and z:

$$x = 2/3 \left\{ 1 \pm \sqrt{\left(\frac{4 - 6Tab - 6Tac + 9TabTac}{4 - 6Tbc} \right)} \right\}$$

and corresponding expressions for y and z. In practice there is seldom any doubt as to which of the two solutions to these formulae is the real one, since one solution will be greater than 2/3 and thus improbably high even when it is less than unity.

A good example of the use of the formula is provided by some data of Desborough and G. Lindegren [115] on Saccharomyces. Three loci, α (mating type), *ga* (galactose utilization) and *ur* (uracil requirement) showed no linkage in any combination, and gave the following tetrad frequencies:

    α and *ga*: 153 PD; 172 NPD; 244 T (43 per cent T).

    α and *ur*: 239 PD; 209 NPD; 383 T (46 per cent T).

    ga and *ur*: 237 PD; 241 NPD; 243 T (34 per cent T).

We may note that the low frequencies of tetratypes in all three types of cross confirms that the three loci are on different chromosomes. Applying the formula to *ga*, we obtain

$$2/3 \left\{ 1 \pm \sqrt{\left(\frac{4 - 2 \cdot 58 - 2 \cdot 04 + 1 \cdot 32}{4 - 2 \cdot 76} \right)} \right\} = 2/3(1 \pm \sqrt{\cdot 56})$$

$$= 2/3(1 \pm 0 \cdot 75)$$

Of the two solution, 1·17 and 0·167, the former is obviously unreal, and we may conclude that *ga* shows about 16·7 per cent second division segregation, and is thus approximately 8·8 map units from its centromere. Knowing the frequency of second division segregation of *ga* we can obtain the frequencies for α and *ur* from the simpler formulae **on the previous page**. They work out to be 35 per cent and 23 per cent for α and *ur* respectively.

This method only has reasonable precision if at least two of the three loci used are fairly near to their respective centromeres. Clearly if two or three of the loci show nearly two-thirds second division segregation, all the tetratype frequencies will be close to two-thirds, which gives little or no information beyond the fact that at least two of the loci must be far from their centromeres. The most useful situation is to find a locus which segregates nearly always at the first division, and which can thereafter be used as a centromere marker.

Whitehouse [576] has pointed out that the same principle can be applied to a group of three loci two of which are linked only if it is justifiable to assume that the sum or difference of the second division segregation frequencies of the two linked loci is equal to their tetratype frequencies. This will only be so if one of these loci is close to the centromere. Other cases are dealt with in Whitehouse's publication.

THE ESTIMATION OF MAP DISTANCE

The map unit

Although linkage data can tell us the order of the genes in a linkage group, it cannot by itself give any certain information about their physical distance apart. It is possible, however, to give some idea of the degree of separation of two markers in terms of the frequency of exchange occurring between them. Since linkage studies were first made with organisms in which only randomized meiotic products could be scored, the measure of exchange frequency which has been adopted is percentage of recombinants among total meiotic products, rather than percentage of tetrads with an exchange in the marked region. These two frequencies are related by a factor of two. Each tetrad with an exchange is a tetratype with two recombinant products out of four. A 10 per cent frequency of exchange per tetrad corresponds to a 5 per cent frequency of recombinants among total products, at least to a first approximation (see below), and the marked region is said to be five map units long. Second division segregation frequencies correspond to frequencies of exchange between marker and centromere per tetrad and thus have to be divided by two to make them equivalent to recombination frequencies among randomized products, and thus to map units.

In principle, if map distances are to be additive, they should be proportional to the mean *total* frequency of exchange rather than to the probability of at least one exchange occurring. Distance in map units is fifty times the mean number of exchanges per tetrad in the interval in question; it will be recalled that *one* exchange gives 50 per cent recombination in each tetrad in which it occurs. This definition raises no difficulty so long as the possibility of more than one exchange can be neglected. In this case true map distances can be obtained simply and directly as percentage recombination or half the percentage of second division segregation. In any but very short regions, however, one must reckon with the possibility of multiple exchange and the relationship between map distance and recombination frequency will depend on whether or not different exchanges can be considered as independent events. In fact, various kinds of 'interference', or non-independent relationships between adjacent exchanges, are possible.

Chromatid interference

The consequences of multiple crossing-over depend on the strand relationships between successive exchanges. As has been pointed out in Chapter 1, if the strand composition of one exchange has no effect on that of the next, one expects two-, three- and four-strand relationships to occur with the relative frequencies 1:2:1. Any significant deviation from this ratio may be termed chromatid interference, either positive or negative depending on whether four- or two-strand relationships are favoured. The data of Table 13 show that, although all investigators agree that all three types of strand relationship always occur, their relative frequencies are sometimes in accord with a 1:2:1 ratio, and sometimes not. Most of the significant deviations from this ratio are in the direction of an excess of two-strand relationships. In *Neurospora crassa* Lindegren's data indicate a marked excess of two-strand relationships among doubles straddling the centromere; among doubles within the same arm there was an excess, of somewhat dubious significance, of four-strand relationships [319]. It has been pointed out by Perkins [399] that asci which, on the face of it, indicate two-strand double crossing-over in adjacent regions separated by the centromere, could also be explained either by division of the centromere at the first, instead of the second division of meiosis, or by an exchange of positions of the two inner nuclei at the four-nucleate stage of ascus development (nuclear passing). Howe [244], in his study of crossing-over in the first linkage group of *N. crassa*, included in his cross a marker, *riboflavin*, which is very close to the centromere of another chromosome. Nuclear passing will give apparent second division segregation for this marker, and this is a rather uncommon event; thus any ascus showing an apparent two-strand double across a centromere and, at the same time, apparent second division segregation for riboflavin, may well be attributed to nuclear passing. Howe did, in fact, record two asci of this type. If nuclear passing occurs, the significance of Lindegren's data is reduced; nevertheless not all of his excess of two-strand doubles can be explained in this way (see Table 13). Subsequently Strickland

TABLE 13. Data bearing on chromatid interference in fungi

Organism	Chromosome region	Numbers of doubles			Reference
		2-str.	3-str.	4-str.	
Neurospora crassa	Linkage group I				
	(a) across the centromere	38 (18)*	25 (23)	21 (21)†	Lindegren & Lindegren[319]
		88	168	77	Bole-Gowda et al.[41]
	(b) within arms	1 (1)*	11 (4)	8 (8)	Lindegren & Lindegren[319]
		73	132	51	Bole-Gowda et al.[41]
	(c) mostly within long arm	191	371	153	Perkins[402]
	Linkage group VI across the centromere	43	54‡	53	Stadler[511]
	Linkage group V region 20 units long, far from centromere	33	63	22§	Strickland[523]
		9	1	7‖	Strickland[521]
Aspergillus nidulans	Mostly within one arm	54	93	37	
Saccharomyces	Linkage group III	12	24	7	Hawthorne & Mortimer[220]
	VII	7	4	12	
	IX	8	11	8	
Coprinus lagopus	Within short region near to, and probably including, the centromere	12	1	0	Day & Swiezynski (unpublished)
Sordaria fimicola	Mostly within one arm	20	61	21	Perkins, El-Ani, Olive & Kitani[403]

Notes: * Figures in parentheses are obtained by discounting all asci which could possibly have been due to nuclear passing or centromere misdivision.
† There were also 7 asci which could have been either 2- or 4-strand doubles.
‡ Only one of the two kinds of 3-strand double was detectable in the scoring system used; the best estimate of total 3-strand doubles would be 108.
§ Data from three crosses combined.
‖ Data from a separate cross.

[523], Perkins [402] and Bole-Gowda *et al.* [41] have all obtained a small but consistent excess of two-strand over four-strand doubles, the two together approximately equalling the three-strand doubles (see Table 13). Day and Swiezynski (unpublished) have shown a very marked excess of two-strand relationships near the centromere of one of the *Coprinus lagopus* linkage groups.

To summarize, many data are not greatly at variance with the assumption of no chromatid interference, but there is an indisputable tendency for two-strand doubles to be in excess. Possible reasons for an excess of two-strand relationships are discussed in Chapter 7.

Chiasma interference

Chiasma interference is the situation where the occurrence of one cross-over affects the probability of occurrence of other cross-overs. If there is no such interference the probability of n exchanges occurring in a given interval, will be given by the appropriate term of the Poisson distribution $(m^n/n!) \times e^{-m}$ where m is the mean number of exchanges within the interval.

If we accept the strong evidence from higher organisms that crossing-over is due to chiasma formation, it is fairly clear that chromosome interference must occur. With rare exceptions, each bivalent at meiosis forms at least one chiasma; indeed, chiasma formation seems to be mechanically necessary to hold the chromosome pairs together in their proper orientation at metaphase I. With no interference, a probability of nearly one for the first exchange implies a high frequency of multiple exchanges. If the unusually high frequency of 5 per cent of bivalents failed to form any chiasma (i.e. $e^{-m} = 0.05$), we would expect, on the basis of the Poisson distribution, the remaining 95 per cent to consist of 15 per cent with one, 22·5 per cent with two, 22·5 per cent with three, 12·7 per cent with four; 10·0 per cent with five, 5·0 per cent with six and 2·1 per cent with seven chiasmata, the mean chiasma frequency being three. Yet in organisms where chiasmata can be counted, as they usually cannot be in fungi, it is commonly found that, though every bivalent has at least one chiasma, many have no more than one and very few have more than three or four. This indicates positive chiasma interference, as if the occurrence of one chiasma tended to inhibit the formation of a second, and so on.

One can arrive at the same conclusion by analysis of genetic data in *Neurospora crassa*. Perkins [402] analysed crosses in which one arm of chromosome 1 was marked with the mutant alleles *cr*, *thi-1*, *nit*, *aur*, *nic-1*, *os*, in that order (cf. Fig. 21). Among 1262 asci, 8 per cent had no detectable cross-over, 44 per cent had a single cross-over, 40 per cent had a double and 7 per cent a triple. Fewer than one per cent had quadruples or multiple exchanges of higher order. Had the same total number of exchanges been distributed according to the Poisson distribution, one would have expected 23 per cent with no detectable exchange, 34 per cent singles, 24 per cent doubles, 12 per cent triples, 5 per cent quadruples and 2 per cent of higher rank. Clearly the formation of each cross-over decreases the probability of finding a further one in the same tetrad, so that single exchanges are favoured at the expense of multiples. On the other hand, Strickland [521] analysed a cross in *Aspergillus nidulans* in one chromosome pair marked with *ribo*, *thi*, *ad-14*, *pro-1*, *paba-1 y* and *bi-1*, and observed, out of 264 tetrads, 67 with no observable cross-over, 89 with one, 63 with two, 39 with three and 6 with four or more. These frequencies fit very closely the Poisson distribution expected on the hypothesis of no interference. Thus the intensity of chiasma interference may vary widely from one species to another.

One can obtain a measure of chiasma interference by calculating the *coefficient of coincidence*. This is the observed frequency of simultaneous crossing-over in two linked regions divided by the value expected with no interference given the frequencies of crossing-over in the regions individually. Linkage data in *Neurospora crassa* have almost always shown positive interference between crossovers in the same chromosome arm though the available data (e.g. [41]) suggest that interference does not operate across the centromere. The overall coefficient of coincidence for a whole chromosome arm has been estimated by Perkins [402] to vary between 0·3 and 0·8 for adjacent inter-marker intervals in the long arm of linkage group I, but to be close to 1·0 for non-adjacent intervals; that is, interference is clearly positive but decreases with distance. Strickland, on the other hand, has shown values of about 0·2 to 0·3 for adjacent regions each between five and ten map units long [521].

Limiting recombination values

As the mean number of exchanges in a marked chromosome segment increases, a further increase comes to have less and less effect on the amount of recombination between the ends of the segment. The recombination fraction will, in fact, approach 0·5 and not rise above this value. If chromatid interference is absent, double crossing-over within a region will, on average, lead to 50 per cent recombination between its ends. This is because the three-strand doubles will, like singles, give two out of four recombinant strands, while the two-strand doubles, which give no recombination, will tend to be balanced by the four-strand doubles, which will each give four out of four recombinant products. The same applies to triples and all higher multiple exchanges (see Table 14). With the addition of each further exchange, those tetrads which otherwise would have been ditypes, are all converted to tetratypes, while those which would have been tetratypes are converted to parental and non-parental ditypes with equal probability; thus PD's will tend to balance the NPD's among tetrads with any number of exchanges greater than zero. As the frequency of tetrads with no exchange in the marked region approaches zero, percentage recombination approaches fifty.

TABLE 14. Mean frequencies of tetrad types resulting from different numbers of exchanges between linked markers, assuming absence of chromatid interference

	Frequency of tetrad type		
Number of exchanges	Parental ditype	Tetratype	Non-parental ditype
0	1	0	0
1	0	1	0
2	$\frac{1}{4}$	$\frac{1}{2}$	$\frac{1}{4}$
3	$\frac{1}{8}$	$\frac{3}{4}$	$\frac{1}{8}$
4	$\frac{3}{16}$	$\frac{5}{8}$	$\frac{3}{16}$
5	$\frac{5}{32}$	$\frac{11}{16}$	$\frac{5}{32}$
n	$\frac{1}{6}+\frac{1}{3}(-\frac{1}{2})^n$	$\frac{2}{3}-\frac{2}{3}(-\frac{1}{2})^n$	$\frac{1}{6}+\frac{1}{3}(-\frac{1}{2})^n$
Many	$\frac{1}{6}$	$\frac{2}{3}$	$\frac{1}{6}$

As shown in Table 14, as the number of exchanges within the marked interval increases, the fraction of tetratype tetrads approaches the limiting value of two-thirds. This is particularly relevant when the interval under study is delimited at one end by the centromere rather than by a marker gene. Patterns of exchanges which, had they occurred between two gene markers, would have given ditype tetrads (whether parental or non-parental) will give first division segregation for a marker when they occur between it and its centromere. Patterns which would have given tetratype tetrads, on the other hand, will give second division segregation. Thus the limiting value of 67 per cent for tetratype tetrads applies also to second division segregation frequencies.

Mapping functions

If one were able to assume the absence of both chromatid and chiasma interference, it would be easy to convert frequencies of recombination, or of tetrad types, to map units. If x is the length of the marked interval in map units (i.e. $2x$ is the mean exchange frequency per tetrad), the probability p_n of n exchanges is given, in the absence of interference, by the appropriate term of the Poisson distribution:

$$p_n = \frac{(2x)^n}{n!} e^{-2x}$$

The frequency of tetratype tetrads (cf. Table 14) will be given by the expression:

$$\sum_{n=1}^{n=\infty} \frac{(2x)^n}{n!} e^{-2x} \cdot \tfrac{2}{3}[1 - (-\tfrac{1}{2})^n]$$

which simplifies to $\tfrac{2}{3}(1 - e^{-3x})$. As x becomes very large this frequency of tetratypes approaches $\tfrac{2}{3}$; the same theoretical limit applies to second division segregation frequencies. The frequencies of PD and NPD tetrads are given by analogous expressions. Since any number of exchanges greater than zero will, on average, give 50 per cent recombination, the recombination fraction among random spores is given by the expression:

$$\sum_{n=1}^{n=\infty} \frac{\tfrac{1}{2}(2x)^n}{n!} e^{-2x}, \text{ or more simply, } \tfrac{1}{2}(1 - e^{-2x})$$

Several attempts have been made to give comparable mathematical expression to situations in which there are various degrees of interference. Barratt et al [19] for example have proposed a model which takes into account a coefficient of coincidence k. The effect of interference is to reduce the number of doubles relative to the singles by the fraction 1-k, the number of triples by the factor $1-k^2$, the number of quadruples by $1-k^3$, and so on. For positive interference k is less than one. Some of the curves relating map distance to tetratype frequency on the basis of this model are shown in Fig. 18, together with some experimental points derived from Perkins' data for one arm of linkage group I in Neurospora [<02]. It will be seen that the model agrees with the data in showing that the frequency of tetratypes, or of second division segregations, can pass through a maximum of more than 67 per cent as the map distance is increased. Such a maximum is easily explained in a qualitative way by saying that chromosome interference results in many tetratype tetrads with only a single cross-over in the marked region, and that it is only as the map length becomes very great that enough multiples are formed to bring the frequency down to near 67 per cent. Very high second division segregation frequencies are uncommon in Neurospora; 75 per cent seems to be about the upper limit in this fungus. Much higher second division segregation frequencies have been reported for *Venturia inaequalis* [279] and *Podospora anserina* [451]. The locus determining mating type in the latter species shows nearly 100 per cent second division segregation (see Chapter 9).

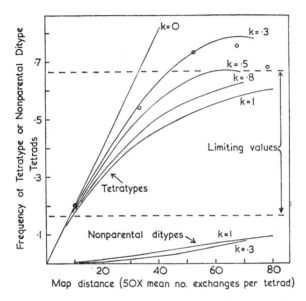

FIG. 18. The relation of tetratype and non-parental ditype ascus frequencies to true map distance between markers, given different degrees of chiasma interference. *k* is defined in the text; *k*=0 corresponds to complete interference, and *k*=1 to no interference. The tetratype curves apply also to second division segregation frequencies, in the case where the interval under consideration is bounded by a marker at one end and the centromere at the other. The limiting values referred to are those which will be approached as the map distance becomes indefinitely great. The experimental points were obtained by Perkins[402] for the right arm of linkage group I of *Neurospora crassa*.

In contrast to the model of Barratt *et al*, Shult and Lindegren [491] have derived expressions relating the frequencies of the three tetrad types to cross-over frequency on the assumption of no chiasma interference but with various kinds of chromatid interference. It would be possible, though complicated, to take both kinds of interference into account. In the absence of any information as to how interference actually operates, however, mathematical descriptions of it must be purely empirical, and the data necessary for a comprehensive description do not yet exist. Even if a complete description of interference in one chromosome of one organism were obtained, there is no guarantee that it would apply to other organisms or even to other chromosomes. Moreover, no matter what justification there is for the assumptions one makes about interference, recombination or second division segregation frequencies must be extremely insensitive measures of map distance as they approach their limiting values.

Estimation of map distances by summation of short segments

The only sound way of measuring the map length of a long interval is to break it down into shorter sections through the use of intervening markers. Provided that the component segments are reasonably short, one can assume without great error that double crossing-over does not occur within them, and that the map length of each is equal to the recombination fraction shown by the markers defining it.

It is reasonable to ask at this point how we can be sure that multiple exchanges do not, in fact, occur within 'short' intervals more often than one would expect from the frequency of apparent single exchanges. If there were a pronounced negative interference operating over short distances, or, in other words, a tendency for exchanges to occur in tight clusters, one might get a high frequency of multiples *within* intervals, even though the coefficient of coincidence calculated on doubles *between* intervals was one or less. There is only one way in which we can detect double exchanges within an interval, and this is through observing non-parental ditype segregations of adjacent markers. Such segregations are most simply explained as the result of four-strand double exchanges within the interval (Table 11).

The expected frequency of four-strand doubles in relation to the number of singles can be calculated from the Poisson distribution, provided we assume the absence of both chiasma and chromatid interference. If the mean exchange frequency within the interval is fairly small, Strickland [521] has shown that the following relationship holds, where N is the frequency of non-parental ditype tetrads, and T the frequency of tetratype tetrads:

$$N = 1/8 \ T^2(1 + 3T/2) \text{ (approximately, assuming no interference).}$$

In Strickland's own data, from *Aspergillus nidulans*, there was, in fact, a moderate but consistent excess of non-parental ditypes over what would have been expected on the basis of the above formula. On the other hand, Perkins [401] has collated all the available data from organisms in which tetrad analysis is possible, and concluded that the result obtained with Aspergillus is quite exceptional. All other organisms for which information is available tend to show many fewer non-parental ditypes than the formula would predict; in other words, they show *positive* interference within intervals. Particularly impressive is Perkins' compilation of data from *Neurospora crassa*. Among, 58,068 ascus classifications with respect to intervals of 15 map units or less, there were only 37 non-parental ditypes; 202 would have been predicted, given the frequencies of tetratypes and assuming no interference.

Thus in Neurospora, and probably in most fungi, it is very unlikely that there is any tendency towards clustering of exchanges unless there is something to prevent each cluster from involving more than two (or at most three) strands. Tight clusters of exchanges, involving two strands in each cluster, can not be ruled out, and as we shall argue in Chapter 7 there is some reason to suggest that they may occur. However, unless one has very closely placed markers one will not expect to detect them; an odd number of exchanges in a cluster involving two strands will usually be indistinguishable from a single exchange, while an even number will usually not be detectable at all.

The lack of constancy of map distance

In the two fungal species *Neurospora crassa* and *Aspergillus nidulans*, whose linkage maps have been studied fairly intensively, linkage groups tend to be rather longer than in well studied higher organisms such as *Zea mays* and *Drosophila melanogaster*. The first linkage group of *N. crassa*, for example, is approaching 150 units in length [400], while the second linkage group of *A. nidulans*, though at the time of writing it holds only ten markers, is already known to be nearly 250 units long [272]. Thus present information suggests that crossing-over occurs rather freely in fungi. Unfortunately it is also, in Neurospora at least, rather variable as between different strains. One of the present authors showed that, although the first linkage group of *N. crassa* was largely or wholly homologous with the corresponding group of the closely related species *N. sitophila*, the map distances in the region of the centromere were greatly reduced in the latter species as compared with the former [150]. Several workers, including Frost [182], have demonstrated heterogeneity in determination of map lengths of the same intervals made in different crosses within *N. crassa*. This may, as Frost suggests, be due to the rather hybrid origin of most laboratory breeding stocks of this species. In view of this, no great reliance can be placed on the reproducibility of map distances. Gene *order* should be invariant, but the frequency or distribution of exchanges may, and often does, vary from one cross to another. A case in point is *tryp*-2, the most distal marker in the right arm of linkage group VI. Stadler's data on this linkage group [510] placed *tryp*-2 some thirty units from the centromere; the much more extensive data of Case and Giles [75], while fully bearing out Stadler's order, place *tryp*-2 only nine units from the centromere. It is quite possible that other fungi, especially if they are somewhat inbred, may prove to give more reproducible map distances, as do Drosophila and maize.

ABNORMAL LINKAGE RELATIONSHIPS

Segmental interchange

Situations exist in which genetic markers tend to remain in their parental combinations without being associated with the same or homologous chromosomes. The best understood of these is the result of breakage in non-homologous chromosomes followed by interchange of segments. If a haploid carrying such an interchange is crossed with another with the normal chromosome complement, a cross-shaped association of four chromosomes will probably result from pachytene pairing at the subsequent meiosis. The situation is outlined in Fig. 19. Depending on the orientation of the four centromeres at anaphase I, and on whether or not exchanges occur between the centromeres and the interchange points, various kinds of spore abortion can occur following the formation of post-meiotic nuclei deficient for one chromosome segment and with another in duplicate. In *N. crassa* McClintock [333] has demonstrated the occurrence of various patterns of spore abortion in asci hybrid for a segmental interchange, which could plausibly be explained on the basis outlined in Fig. 19. Among viable meiotic products, the centromeres of the two non-interchange chromosomes will appear completely linked, as will those of the interchange pair. Markers in the chromosome arms not involved in the interchange, and those distal to the interchange point, will appear linked to a degree depending, in the first case, on distance from the centromere, and, in the second, on distance from the interchange point. Such markers, will in fact behave, so far as analysis of viable meiotic products is concerned, as if linked in a four-armed linkage group, the arms joining at a point corresponding to the four centromeres. Markers between centromeres and the interchange point will generally appear closely linked, even if on different chromosomes, since only crossing-over within these regions, combined with a special sort of disjunction at anaphase I, can result in their recombination (see Fig. 19). This kind of situation is well understood from cytogenetic analysis in higher plants such as maize, and cases, which are almost certainly of the same kind,

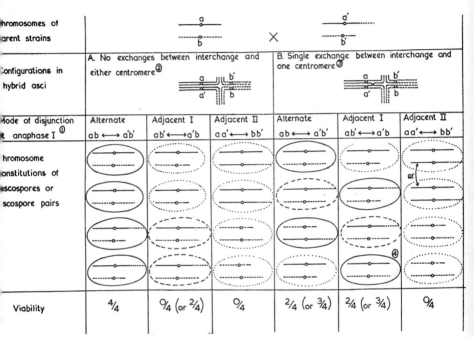

FIG. 19. Genetic consequences of heterozygosity with respect to a chromosomal segmental interchange. Notes: (1) It is assumed that all anaphase I disjunctions will be regularly two-to-two; this will not necessarily always be the case. (2) Exchanges in arms not involved in the interchange, or in arms distal to the exchange point, are not represented; they will not affect viability except insofar as they alter the relative frequencies of different kinds of disjunction. (3) The case of exchange between only one exchange point and its centromere is considered. A single exchange between the other break point and its centromere will have similar results; simultaneous exchange in both regions is likely to be infrequent and is not considered. (4) This is the only case where a product showing crossing-over in the centromere-break point interval is likely to be viable.

Solid circles indicate viable meiotic products, dotted circles inviable products and dashed circles products which may possibly be viable, though almost certainly abnormal, if one interchanged segment is extremely short.

are found not infrequently in studies of radiation-induced Neurospora mutants[404]. McClintock [333], and more recently Barry (Fig. 20), have been able to confirm the interpretation by direct observation of pachytene chromosomes in a few instances.

It is worth noting that if a segmental interchange is very unequal, so that one of the segments involved is minute, meiotic products carrying a duplication with the minute deficiency may be viable (see Fig. 19). A strain of *Aspergillus nidulans* carrying a part of one chromosome in dupli- cate has been used by Pritchard [428] in studies of pairing competition between chromosomes, and very probably had its origin in a very unequal segmental interchange.

Affinity

Another possible reason for non-random assortment of markers on different chromosomes is that the chromosomes concerned are not, in fact, disjoining at random with respect to one another at the first ana- phase of meiosis. This will be the case if one of the centromeres in each metaphase I bivalent has a tendency to become oriented towards one pole of the spindle rather than to the other. Such a polarization is known to occur in megaspore formation in certain lines of maize, and it seems a possibility in any organism in which meiosis occurs in a cell possessing some polarity, such as, for example, an Ascomycete where there is a dis- tinction between the top and the bottom of the ascus. However, the few reported cases of polarized segregation in Ascomycetes show never more than a slight bias, and could be due to differential ripening of asci sub- sequent to meiosis (Mathieson [349]). Alternatively, non-random assort- ment of unlinked markers might occur if there were some sort of *affinity* between non-homologous chromosomes causing them to tend to pass to the same pole of the spindle at anaphase I, or anaphase II. This kind of affinity has been inferred from genetic studies on mice [558] and yeast. Shult, Desborough and Lindegren [115,490] cite, among other examples, a cross between one haploid marked with a gene *cu*, conferring copper resistance, and an auxotrophic mutant (*ch*) requiring choline. The asci obtained from this cross consisted of 6 PD, 48 NPD and 56 T. On intercrossing a *cu ch* and a *cu+ch+* strain, grown from ascospores isolated from the same NPD ascus of the first cross, the relation- ship between the two kinds of ditype asci was found to be reversed; 13 PD, 1 NPD and 13 T asci were obtained. The authors interpreted the 'reverse linkage' shown by the first cross as due to affinity between the centromeres of chromosomes contributed by different parents. Having been brought together in the same haploid as a result of the first cross, they tended to remain together at the meiosis following the second. There was some reason for attributing the property of affinity to centromeres since *ch* was known to be close to its centromere, and other markers elsewhere on the same chromosome showed a lower affinity for *cu*.

While the interpretation given to these data may well be correct, one should always consider segmental interchange as a possible explanation for non-independent behaviour of non-homologous chromosomes. In the case just discussed, for segmental interchange to give the effect of *reverse* linkage it would be necessary for it to occur between non-homologous chromosomes contributed by the *different* haploid parents. This seems rather unlikely. Surzycki and Paszewski [525] have subsequently obtained tetrad ratios in Ascobolus which are very difficult to explain except as due to affinity, either between non-homologous centromeres or other segments of non-homologous chromosomes.

We may tentatively conclude that affinity may be a real phenomenon in certain organisms, although the very extensive Neurospora data have provided no evidence for it. There is a real need for a more systematic investigation of affinity where it does appear to occur; studies on yeasts should produce information far more quickly than those on mice.

Intra-chromosomal rearrangements

Apart from these inter-chromosomal effects, linkage relationships within chromosomes may be affected by various sorts of structural alteration. Inversions of chromosome segments will tend drastically to reduce effective crossing-over between the segment concerned and the homologous segment of a normal chromosome. Thus a diploid heterozygous for an inversion will generally give greatly reduced map distances in the corresponding part of the genetic map, while one homozygous for an inversion may be expected to give a normal map length with an inverted gene order. No well documented cases of this kind are yet available in fungi. If one chromosome of an homologous pair has a segment deleted, crossing-over in that region will obviously be suppressed completely. Minute deletions having this property may occur among radiation-induced mutants of Neurospora (see Suyama *et al.* [533]). Longer deletions have not been found, and are expected to be inviable in a haploid organism.

THE CORRELATION OF LINKAGE GROUPS AND CHROMOSOMES

In any organism which has been well studied both genetically and cytologically it should be possible to show, firstly, that the number of linkage groups is the same as the haploid number of chromosomes, and second-

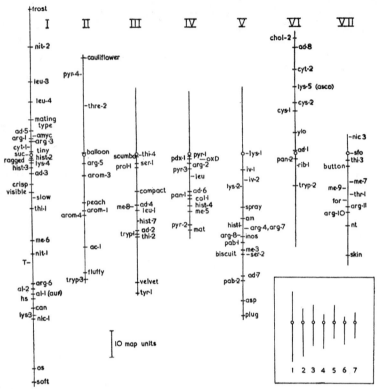

FIG. 21. The seven linkage groups of *Neurospora crassa* with (lower right) the relative sizes and centromere positions of the seven microscopically observable chromosomes according to Singleton [495]. Centromeres are shown as small open circles. Numbering of chromosomes is independent of that of linkage groups and only a few provisional correlations between the two series have been made. Where there are two arms to a linkage group the arbitrarily designated left arm is drawn above, and the right arm below. The order of genes marked by short lines crossing the maps is in all cases fairly well established; genes shown as not touching the maps are approximately in the regions indicated, but their exact positions in relation to linked genes are not established. Linkage maps are those compiled by Barratt & Strickland (ref. 20, with a few subsequent revisions communicated personally); a few unmapped markers are omitted here for simplicity. The map distances can only be taken as a rough guide to what can be expected, since they tend to vary from one cross to another; they are based on the data of Perkins, Maling, Strickland *et al* [338, 400, 405, 524] and Stadler[511]. The meanings of the gene symbols are listed in Appendix II, except for some morphological mutants, the names of which, generally roughly indicative of appearance, are given here in full.

Note added in second edition: A recent paper [404] establishes the position of a few more markers and somewhat alters the best estimates of map distances in some of the linkage groups.

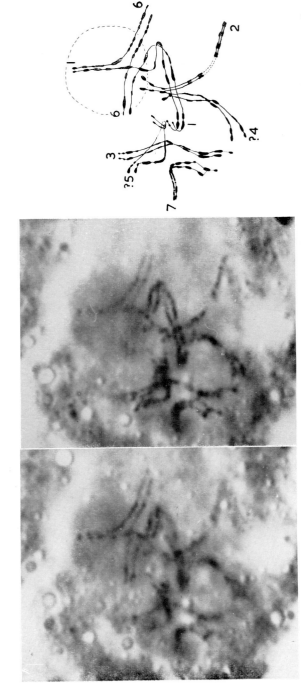
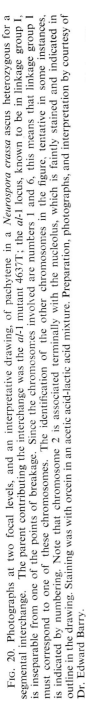

FIG. 20. Photographs at two focal levels, and an interpretative drawing, of pachytene in a *Neurospora crassa* ascus heterozygous for a segmental interchange. The parent contributing the interchange was the *al-1* mutant 4637T; the *al-1* locus, known to be in linkage group I, is inseparable from one of the points of breakage. Since the chromosomes involved are numbers 1 and 6, this means that linkage group I must correspond to one of these chromosomes. The identification of the other chromosomes in the figure, tentative in some instances, is indicated by numbering. Note that chromosome 2 is associated terminally with the nucleolus, which is faintly stained and indicated in outline in the drawing. Staining was with orcein in an acetic acid-lactic acid mixture. Preparation, photographs, and interpretation by courtesy of Dr. Edward Barry.

[facing p. 102

ly, that each linkage group corresponds to one identifiable chromosome.

The first objective has been achieved with quite satisfying completeness in *Neurospora crassa*. The genetic map for this species is shown in Fig. 21. There are seven linkage groups, each with at least a dozen markers. Although not all mutants have been assigned to groups, the number of unmapped ones is relatively small and it seems extremely unlikely that any more linkage groups remain to be discovered. The haploid number of seven chromosomes is well established [495]. In *Aspergillus nidulans* Käfer [272] has established eight linkage groups. Some of these are as yet very sparsely supplied with markers and the genetical data do not argue strongly against the existence of more than eight groups. However, cytological investigations by Elliott [137], show very clearly that the haploid chromosome number is eight, so one will be surprised if a ninth linkage group is discovered. In *Saccharomyces*, Hawthorne and Mortimer [220] have shown that there are at least ten linkage groups; the cytological situation in yeast, however, is still very confused. In *Venturia inaequalis* the haploid number is seven [112], and six linkage groups have been claimed, though not yet published in detail [280]. So far as other fungi are concerned, although linkage mapping has been started in *Sordaria fimicola* [133], *Glomerella cingulata* [572], *Ustilago maydis* [232], *Schizophyllum commune* [442] and *Coprinus lagopus* [111], the maps are still too fragmentary for any useful comparison with chromosome number, even where this is known.

The second object of identifying particular linkage groups with particular chromosomes has been attempted only in *N. crassa*. The general method is to identify the chromosomes participating in associations of four at pachytene in crosses heterozygous for segmental interchanges involving known linkage groups (cf. Fig. 20). Numerous segmental interchanges are known in Neurospora, and the linkage groups involved have been identified in several cases [404]. Singleton has shown that all the chromosomes are microscopically distinguishable, some more easily than others. If information from only one interchange stock is available one will probably not be able to tell which of the two chromosomes involved in the interchange corresponds to which of the two affected linkage groups. However, the study of a second interchange involving one of the same two chromosomes and one different one should serve to identify the chromosome, and the linkage group, common to both interchanges. Work of this kind has been done with *N. crassa*, but has not been published in detail.

8

spora tetrasperma and *Podospora anserina* are special cases which will be described in Chapter 9. In Saccharomyces many or most asci are oval in shape with no definite ordering of the spores. Some, however, are elongated with spores in a linear series; in such asci it seems probable (Hawthorne [218]) that the second division spindles regularly overlap, so that first division segregation usually leads to an alternating arrangement of spores $(+ - + -)$ or $(- + - +)$, while second division segregation will, with equal probability, give $(+ + - -)$, $(- - + +)$, $(+ - - +)$ or $(- + + -)$.

As was explained in Chapter 1, the segregation of a marker at the second, rather than the first division of meiosis is interpreted as resulting from an exchange between that marker and the centromere of the chromosome with which it is associated. Thus analysis of ordered tetrads segregating for a number of linked markers often permits the placing of the centromere in its correct position in the linkage order. The kind of analysis involved is that applied in Table 4 (Chapter 1), the main principle being that the centromere should be so placed as to minimize the number of exchanges necessary to explain the data. Given ordered tetrads, the centromere is almost, but not quite, equivalent to an ordinary genetic marker. A pair of alleles provide more information when multiple exchanges occur with an important frequency. Suppose we have three points in a linkage group, one being the centromere. If the centromere forms one end of the sequence, the simultaneous crossing-over in the two intervals can give only two results, first or second division segregation of the distal marker with, in either case, second division segregation of the proximal marker. Second division segregation of the distal marker indicates a three-strand double exchange (leaving aside the possibility of triple exchange), but first division segregation of this marker is equally consistent with either a two-strand or a four-strand double. If an allelic difference were substituted for the centromere in this example, all kinds of double exchange would be distinguishable. If the centromere is the median of the three-points it is more nearly equivalent to a gene difference; in this situation, two-, three- and four-strand double exchanges between intervals are distinguishable and one fails only to distinguish the two *kinds* of three-strand double distinguishable with three linked marker genes (cf. Table 4, Chapter 1).

Whether the information obtainable about centromere position is worth the extra labour involved in isolating ordered tetrads is a question which becomes increasingly pertinent in Neurospora studies following Strickland's [522] publication of his very rapid method for the isolation

GENETIC CONSEQUENCES OF CHANGES IN CHROMOSOME NUMBER—DIPLOIDY, ANEUPLOIDY, POLYPLOIDY

With the exception of the yeasts, and members of the Blastocladiales, fungi are normally haploid except for the diploid cell in which meiosis occurs. Deviations from the normal haploid state can occur, however, with genetic consequences which must be understood if erroneous conclusions are to be avoided. In some situations the occurrence of aberrant chromosome numbers can be turned to good account in genetic analysis, as demonstrated particularly by Pontecorvo and Roper in their studies on *Aspergillus nidulans*.

CHROMOSOME MAPPING BASED ON VEGETATIVE SEGREGATION IN DIPLOIDS

Until the last decade, the only way of making maps of chromosomes was through the analysis of the products of meiosis. Although Stern had demonstrated the occasional occurrence of crossing-over between homologous mitotic chromatids in *Drosophila*, the phenomenon had remained more of a curiosity than a practical aid to genetic analysis. So far as fungi are concerned, this situation has been entirely changed through the work of Pontecorvo and other members of the Glasgow group. They have demonstrated that the analysis of vegetative segregation may often be a more convenient way of mapping the chromosomes than conventional linkage studies. The method is probably applicable to any normally haploid fungus in which diploid strains can be obtained.

Selection of diploid strains in habitually haploid fungi

In the habitually haploid filamentous fungi the selection of diploids is the essential first step in the analysis. In 1952 Roper [40] obtained diploid strains in *Aspergillus nidulans* by the use of two different methods of selection. Both methods depend on the fact that Aspergillus conidia are always uninucleate, and the first method depends also on the control of the pigmentation of each conidium by its own genotype. If one makes a heterocaryon between two non-allelic *A. nidulans* mutants *white* and *yel-*

low, each mutation being recessive to its wild type allele, all the haploid conidia will be either *white* (genotype $w\ y^+$) or *yellow* (genotype w^+y). Occasional patches or sectors of wild type green conidia proved to be due to the formation, by vegetative nuclear fusion, of diploid nuclei of genotype $\dfrac{w}{w^+}\dfrac{y}{y^+}$. The diploid conidia were distinguishable from haploid conidia by being larger (about twice the volume), and their diploid character was confirmed by their capacity to form sectors of the original component mutant types, as will be discussed below.

Roper's second method for obtaining diploids was rather more efficient in that it used a technique of automatic selection. A prototrophic forced heterocaryon was formed on minimal medium from two auxotrophic mutant strains with different nutritional deficiencies. Since all conidia are uninucleate, all the haploid ones formed by the heterocaryon will be of either one auxotrophic type or the other, and will not grow on minimal medium. A diploid conidium resulting from fusion of unlike nuclei will, however, be prototrophic, just as was the heterocaryotic mycelium. Plating of conidia from such a heterocaryon on minimal medium yielded one diploid prototrophic colony for each 10^6 or 10^7 spores.

Pritchard [426] has found that about one in every hundred ascospores in *Aspergillus nidulans* is diploid. The task of selecting these is a little more difficult than that of selecting diploid conidia; if the two parents of a cross are auxotrophic because of mutations in different loci, haploid prototrophic ascospores will, of course, be formed as a result of recombination during meiosis, and these will usually be numerous in comparison with prototrophic diploids. However, if the mutations in the parent strains are very closely linked, while still being in functionally distinct genes, prototrophic haploids will be rare, and prototrophic diploids carrying the two mutations in different homologous chromosomes will be reasonably frequent in comparison. Under these conditions, diploid ascospores will be selected with fair efficiency by plating on minimal medium.

Roper's original technique of selection of prototrophic diploid conidia is applicable to any mould with monocaryotic conidia, and has, in fact, been successfully applied to *A. niger* [423], *A. oryzae* [251], *Penicillium chrysogenum* [425] and the plant pathogens *Fusarium oxysporum* [65] and *Cochliobolus sativus* [550]. It is not so easy to apply to Neurospora since the macroconidia of this genus are mostly multinucleate, and even the microconidia (which are produced to the exclusion of macroconidia by certain genotypes) are not entirely reliable in their uninucleate character. At all events, attempts to isolate diploid Neurospora strains have failed up to the present; whether this failure is due to technical difficulties or to a real absence of vegetative nuclear fusion is not known.

Holliday [233] has recently used a principle which is basically similar to Roper's second method in the isolation of diploid strains of the maize smut fungus, *Ustilago maydis*. As was described in Chapter 2, this fungus has a life history consisting of an alternation of the yeast-like sporidial free-living phase, in which the cells are uninucleate, and the mycelial obligatorily parasitic phase, in which each cell contains two haploid nuclei of different mating type. The parasitic dicaryon cannot grow vigorously on any artificial medium yet tested. Holliday inoculated maize plants with two different auxotrophs and found that pieces of infected gall tissue, taken from the maize plant before the formation of the brandspores and placed on minimal agar medium, tended to become surrounded by a sporidial outgrowth. These sporidial colonies, which are sporadic in their appearance as if their origin depends on a chance event, are shown to be diploid by a number of criteria. Firstly, they behave as if they carry within them two mating types, in that each cell is capable of initiating an infection of the host plant (*i.e.* they are *solopathogenic*, see p. 267). Secondly, they show vegetative segregation of alleles in which the components of the dicaryon differed. Finally, their cells contain twice the normal amount of deoxyribonucleic acid in their nuclei. It is interesting to note that the diploid strains combine the abilities of the two phases of the normal haploid fungus; they can grow *in vitro* like the sporidial phase and also parasitically like the dicaryotic phase.

The method of synthesizing diploid lines of *Schizosaccharomyces pombe* has been described on p. 29.

The isolation of vegetative segregants from heterozygous diploids

Diploid strains of normally haploid fungi tend to produce sectors showing segregation of alleles originally present in heterozygous condition. It is this property which gives such diploids their usefulness. Segregant cells are sometimes haploid, and sometimes are still diploid but homozygous with respect to loci originally heterozygous. The analysis of vegetative segregation depends on methods for picking out the segregant cells from the mass of mycelium of unchanged genotype. In *Aspergillus nidulans*, a variety of such methods is available.

The simplest technique depends on selection of segregant sectors, or colonies grown from single conidia, by eye. This is possible where the original diploid is heterozygous for recessive colour mutations such as *white* or *yellow*. White or yellow-spored sectors or colonies are easy to see against a background of the dark green wild type.

In addition to visual selection, there are other methods which will select segregants automatically by differential growth. Alleles which permit growth on a given medium are most usually dominant to those which do not; in such cases one cannot easily devise a situation in which the haploid or homozygous diploid segregant has a selective advantage over the heterozygous diploid. There are, however, a few examples of the opposite situation, with a homozygote able to grow where the heterozygote will not. Pontecorvo and Käfer [421] made considerable use of a recessive mutation suppressing the adenine requirement caused by the mutant allele *ad-20*. A diploid homozygous for *ad-20* and heterozygous for the suppressor (*su*) and its dominant wild type allele (*su⁺*) will not grow on adenine-free medium, but segregant conidia carrying *su* in homozygous condition can do so. A second example of growth depending on homozygosis for a recessive gene is provided by Roper and Käfer's [462] work on acriflavine-resistant mutants. One of these mutants (*Acr-1*) is partially dominant in the sense that *Acr-1/Acr-1⁺* is more resistant than *Acr-1⁺/Acr-1⁺*, but less resistant than *Acr-1/Acr-1*. On media containing a relatively high acriflavine concentration, homozygous mutant segregants can be efficiently selected over the heterozygous type. The use of both these automatic selective procedures is illustrated by Table 16.

Another method of selection which permits the isolation of certain kinds of auxotrophic segregant, is Pontecorvo's technique of *starvation selection*, which was outlined in Chapter 3.

Replica plating, in organisms which lend themselves to the method, provides a quicker and more certain method of identifying and isolating auxotrophic segregant colonies. This was the method used in *Ustilago maydis* by Holliday. Diploid sporidia carrying several auxotrophic mutations in heterozygous condition were plated on complete medium, and the resulting colonies were replicated on to minimal; segregant colonies were identified through their failure to appear on the latter medium.

Finally, one can sometimes select for genetic recombination rather than for homozygosis. The situation in which this is possible is explained below (see *Recovery of reciprocal products of mitotic exchange*).

The causes of vegetative segregation

Among the segregants from diploid Aspergillus strains, haploids can be distinguished from diploids by the size of their conidia. There are also genetic differences. While haploids must, of necessity, show segregation of one or other allele at each and every locus originally heterozygous, diploids usually show segregation only for one or a few linked loci, remaining heterozygous for the others.

The principles involved are best illustrated by reference to some actual data. Table 15 shows part of the results of an experiment of Pontecorvo and Käfer [421]. In this experiment segregants were selected visually; the original diploid was heterozygous for both white (*w-2*) and yellow (*y*). These two loci are in different linkage groups, and the *y* group was also marked by the auxotrophic mutations *ad-14*, *pro-1*, *paba-1* and *bi-1*.

Considering first the haploid segregants, we see that among those selected as being white, exactly one-half carried one chromosome of the second pair (marked with *pro-1 bi-1*) and the other half carried the other (marked with *ad-14 paba-1*). This illustrates the not surprising principle, confirmed by a great many other experiments, that haploidization results in a random assortment of different chromosome pairs. It also shows that haploidization is seldom (only once in Table 15) accompanied by recombination within linkage groups. Haploidization is almost certainly a consequence of non-disjunction, which is discussed later in this chapter.

The white diploid segregants, on the other hand, showed no segregation of any of the auxotrophic markers of the second chromosome pair. Neither did any of the diploids segregating for these second chromosome markers show segregation of *white*. Thus homozygosis affects only one chromosome pair at a time, with infrequent exceptions not represented in the data under review.

Turning to the yellow diploid segregants, we see in each case that only some of the marked second chromosome loci have become homozygous. This clearly demands an explanation in terms of intra-chromosomal recombination, or crossing-over. The model for explaining these observations had, in fact, already been derived by Stern from his studies on *Drosophila* [518], and is illustrated in Fig. 22. It is supposed that occasional strictly equal exchanges can occur between the chromatids of homologous chromosomes. Following such an exchange, the two reciprocal exchange chromatids may pass to opposite poles at the subsequent anaphase, or, perhaps equally likely, to the same pole. In the former case all loci *distal* to the exchange point will become homozygous if they were not so before. Homozygosis for one locus will, leaving aside the possibility of double exchanges, necessarily be accompanied by homozygosis for all loci distal to it in the same arm, but not necessarily for more proximal loci. Homozygosis for one arm of a chromosome should occur independently of homozygosis for the other arm.

The results of Table 15 are entirely consistent with the mitotic crossing-over hypothesis. The interpretation was made the more sure by the fact that the genes *ad-14 pro-1 paba-1 y bi-1* were already known, from conventional meiotic analysis, to be linked in that order. Among the diploid segregants, homozygosis for *y* was invariably accompanied by homozygosis for *bi-1+*, which consequently can be placed distal to *y* in the same arm. Homozygosis for *paba* was frequent but not invariable in diploids segregating *y*, while homozygosis for *pro* was less frequent. From this one would conclude that *paba* and *pro* were proximal to *y* in the same arm, with *pro* the nearer to the centromere. None of the yellow diploid

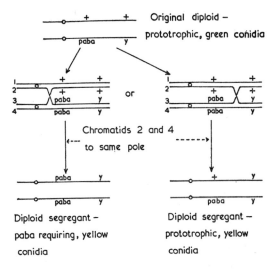

Fig. 22. Segregation patterns resulting from crossing-over at mitosis in a heterozygous diploid (cf Table 15).

segregants were homozygous for *ad-1*. This indicates that the latter locus is probably in the other arm of the chromosome; its linkage to *paba* is clearly shown by the joint segregation of the two markers in the haploids.

The appearance of colonies which are apparently diploid, and yet show simultaneous segregation for all markers in *both* arms of a chromosome is a fairly infrequent but persistently occurring anomaly in *Aspergillus nidulans* diploids. Such exceptions, of which there are two in the data of Table 15, are attributed to chromosome non-disjunction, which is discussed below.

TABLE 15. Vegetative segregation in an *Aspergillus nidulans* diploid
(Data of Pontecorvo and Käfer[421])

Original diploid:

$$\frac{w\text{-}3 \quad + \quad pro\text{-}1 \quad + \quad + \quad bi\text{-}1}{+ \quad ad\text{-}14 \quad + \quad paba\text{-}1 \quad y \quad +}$$
$$\text{I} \quad \text{II} \quad \text{III} \quad \text{IV} \quad \text{V} \quad \text{VI}$$

Colour of segregant selected	Ploidy*	Nutritional phenotype	Genotype	Number of occurrences	Region in which cross-over occured or other interpretation
	2n	prototroph	$\dfrac{w}{+}\;\dfrac{pro \;+\; y \;+}{ad \;+\; paba \; y \;+}$	32	V
	2n	paba	$\dfrac{w}{+}\;\dfrac{+ \; pro \; paba \; y \;+}{ad \;+\; paba \; y \;+}$	78	IV
			$\dfrac{w}{+}\;\dfrac{+ \;+\; paba \; y \;+}{ad \;+\; paba \; y \;+}$	7	III
yellow	2n	ad paba	$\dfrac{w}{+}\;\dfrac{ad \;+\; paba \; y \;+}{?}$	1	Non-disjunction (2n–1 ?)
	2n	pro paba	$\dfrac{w}{+}\;\dfrac{+ \; pro \; paba \; y \;+}{?}$	1	IV, plus non-disjunction (2n–1 ?)

n	ad paba	$\dfrac{+}{\ }\quad ad\ +\ paba\ y\ +$	19	Haploidization without crossing-over
n	paba	$\dfrac{+}{\ }\quad +\ +\ paba\ y\ +$	1	II or III followed by haploidization
2n	prototroph	$\dfrac{w}{w}\quad \dfrac{+\ pro\ +\ +\ bi\dagger}{ad\ +\ paba\ y\ +}$	56	I
n	pro bi	$\dfrac{w}{\ }\quad +\ pro\ +\ +\ bi$ $\quad ad\ +\ paba\ y\ +$	25§	Haploidization without crossing-over
n	ad paba (y)‡	$\dfrac{w}{\ }\quad ad\ +\ paba\ y\ +$	25§	Ditto

white

Note: * Based on conidial size.

† Presumed constitution for second chromosome.

‡ *y* not observable in phenotype, presumed present.

§ Note independent segregation of non-homologous chromosomes.

Recovery of reciprocal products of mitotic exchange

Clear proof of the correctness of the explanation of diploid segregants in terms of mitotic crossing-over requires the isolation of the reciprocal products of one cross-over event. The usual techniques for obtaining cross-over products, depending as they do on selection for homozygosis, ensure that one obtains only one cross-over chromosome, together with a non-cross-over homologue. In certain situations, however, one can select for a cross-over as such without placing any limitation on the nature of the homologous chromosome which accompanies it. This was demonstrated by Roper and Pritchard [427, 463] in the following way. They obtained a diploid of the following constitution with respect to the second chromosome:

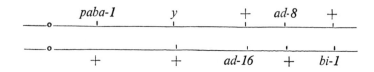

As will be explained more fully in Chapter 7, *ad-8* and *ad-16* are mutations at different sites within the same functional gene. The original diploid requires adenine since this gene is defective in each chromosome. By crossing-over between the two sites, however, a wild type gene can be regenerated, and will confer the ability to grow without adenine on any spore carrying it. The analysis in this case is made easier by the fact that *ad-8* and *ad-16* are distinguishable; *ad-16* strains grow very slightly on adenine-free medium, while *ad-8* strains will not do so at all. Plating conidia from the diploid on to medium supplemented only with *p*-aminobenzoic acid and biotin gave adenine-independent colonies at the rate of about one per 10^7 conidia plated. The chromosomal constitutions of these *ad*+ recombinant colonies were determined by the isolation of haploid segregants from them. In nine out of forty-one cases analysed the genotype was:

$$
\begin{array}{ccccc}
paba & y & + & + & bi \\
\text{───o───┼────────┼────────┼────────┼───────} \\
\text{───o───┼────────┼────────┼────────┼───────} \\
+ & + & ad\text{-}16 & ad\text{-}8 & +
\end{array}
$$

i.e. with the two reciprocal recombinant chromosomes resulting from a single cross-over. The double mutant *ad-16 ad-8* chromosome was identified as such by outcrossing haploid segregants carrying it to wild type, and, by a selective technique which need not be described here, isolating single mutant *ad-16* recombinants distinguished by their slight 'leakiness' on minimal medium.

Holliday [233] has demonstrated the reciprocal nature of mitotic crossing over in *Ustilago maydis* in a rather direct way. In diploid strains heterozygous for a number of genes, segregants homozygous for *ad* are distinguishable on close inspection by a brownish pigmentation. Not infrequently half a sporidial colony will be brown, as if the segregation occurred following crossing-over in the cell from which the colony originated. In such cases, the non-pigmented half of the colony is found to have become homozygous for markers on the other homologue (i.e. the chromosome carrying *ad+*).

Although there is no reason to think that mitotic crossing-over is basically different in mechanism from meiotic crossing-over, there is nothing to show exactly when it occurs. Presumably it could take place any time between the division of the chromosomes into chromatids (probably before prophase) and the following metaphase.

Special features of mapping by vegetative analysis

Mitotic crossing-over is a comparatively rare event; Pontecorvo [419] estimates that it occurs in about one mitosis in a hundred in *Aspergillus nidulans;* it is considerably less common than this in *Ustilago maydis* though its frequency can be greatly increased by ultraviolet light [233] or treatment with certain drugs, such as mitomycin C [235] or fluorodeoxyuridine [130],. which inhibit DNA synthesis. Consequently one would expect the occurrence of more than one cross-over per nucleus to be very uncommon. Although Pritchard's [427] work on intra-genic mitotic crossing-over has indicated that intense negative interference (*i.e.* clustering of exchanges) probably operates over very short intervals, the extensive work of Käfer and Pontecorvo on mitotic reassortment of more widely spaced markers has revealed little or no tendency to clustering. This apparent paradox is discussed in Chapter 7. In ordinary mapping work, as opposed to studies of fine structure, one can, to a first approximation, neglect the possibility of double or multiple crossing-over in mitotic analysis.

The rarity of detectable mitotic double cross-overs makes the ordering of markers within a chromosome arm more straightforward than in meiotic analysis. Thus a single case in which markers *a* and *b* become homozygous simultaneously is a strong indication that they are in the same arm, while the discovery of a second segregant in which *a* becomes homozygous without *b* shows with near certainty that *a* is distal to *b*.

Obviously ordering of markers within an arm involves less labour if the recessive marker alleles are in coupling on the same chromosome.

Unless one can obtain haploid segregants as well as diploid products of mitotic crossing-over, one cannot, by mitotic analysis alone, show any connexion between the two arms of one chromosome. This is, in fact, the present situation in *Ustilago maydis* (but see Rowell [465]). Where haploid segregants can be isolated, as in Aspergillus, the independent segregation of *whole* chromosomes shown by these segregants enables all segregating markers to be assigned to the same or different chromosomes without any ambiguity. Haploid analysis in Aspergillus has recently been made easier by the discovery that haploidization can be induced by treatment of diploids with p-fluorophenylalanine [313].

We have seen that mitotic analysis can provide evidence for linkage and for gene order. Can it also give an indication of the relative distances between linked markers? If one selects for homozygosity for the most distal marker of a sequence within one arm, the distribution of mitotic cross-overs between the marked intervals of the arm will be shown by the frequencies of homozygosis of the more proximal markers. Thus one can space the markers on the map on the basis of a map unit proportional to relative mitotic cross-over frequency; values for absolute cross-over frequencies will not usually be available. The principle is illustrated in Table 16. As will be seen from the comparison, given in the Table, between the spacing of the markers on the mitotic map and the spacing on the corresponding conventional (meiotic) linkage map, the distributions of mitotic and meiotic cross-overs are not necessarily the same. Since there is no reason to suppose that the probability of either kind of crossing-over is constant per unit length over the whole chromosome complement, neither kind of map can be taken as representing physical distances, though the two taken together can give a broad indication of relative distance, especially where very large or very small cross-over frequencies are concerned.

Vegetative recombination in diploid yeast

In yeast, as one might expect for a habitually diploid organism, spontaneous homozygosis with respect to originally heterozygous markers is a rarity. However, as Wilkie and Lewis [581] have shown, treatment with ultraviolet light will induce crossing-over in practically every chromosome pair in a certain 'competent' fraction of the population, estimated as 9 per cent. The very frequency of the induced crossing-over, with multiple exchanges common, tended to obscure the mechanism whereby the subsequent segregation of markers took place. The authors suggested that a process akin to meiosis but without the second division reduction of centromeres, rather than mitotic crossing-over as it is usually understood, would best explain the results. An unequivocal test of this interpretation is still needed, however.

Recombination within genes, detected by the formation of prototrophic cells in heteroallelic auxotrophic diploid yeast cultures, can also be very greatly increased by radiation treatment. Manney and Mortimer [341] have shown that, for a given pair of alleles, the increase is proportional to X-ray dose. Furthermore, the increment of recombinants per unit dose of X-rays seems to be a good measure of the distance between the two sites being recombined (cf. p. 140).

THE PARASEXUAL CYCLE AS AN ALTERNATIVE TO SEX IN IMPERFECT FUNGI

The sequence of events which runs: fusion of unlike haploid nuclei in a heterocaryon ⟶ mitotic crossing-over ⟶ haploidization, has been termed the *parasexual* cycle by Pontecorvo [419]. It has the same effect as a regular sexual cycle in accomplishing genetic recombination and thus increasing the available genetic variation. The relatively low frequency of each of the three stages in the cycle means that it is almost certainly of small effect in comparison with the normal sexual cycle in a homothallic sexually reproducing mould like *Aspergillus nidulans*. However, it may well be of importance in imperfect fungi or in heterothallic fungi in which populations consisting of only one mating type are common. In *Aspergillus niger* [423] and *Penicillium chrysogenum* [425], in which all the stages of the parasexual cycle occur, it seems [419] that the frequencies both of vegetative nuclear fusion and of diploid vegetative segregation may be greater than in *A. nidulans*; such a higher parasexual activity may have an adaptive significance. Since both of these species of imperfect fungi are of importance in industry (for citric acid and penicillin production respectively), the parasexual cycle may be of use in the deliberate breeding of improved strains.

NON-DISJUNCTION AND ANEUPLOIDY

At mitotic anaphase the two chromatids of each chromosome normally pass to opposite poles of the spindle. Occasionally, by mistake, as it

TABLE 16. Mitotic mapping of two chromosomes in *Aspergillus nidulans*

Three selection procedures were used in three different experiments, so numbers are not comparable from one to another. *su* is a recessive suppressor of the effect of *ad-20*. Three markers, irrelevant to the present argument, are omitted for simplicity.

Diploid Y of Pontecorvo and Käfer[329]

Map units from meiotic analysis:

```
         39      19              27    20     8     16    0·2     25           21
   su   ribo-1  thi-1                 pro-1  paba-1  y           Acr-1   w-2
   +     +       +      III°  IV       +   V  +   VI  y   ad-20    +   VII  +    VIII°
   I            II                                       ad-20
```

Selection for	Other markers segregating	Crossing-over in region	Frequency	Derived relative distances between markers (marked region within each arm arbitrarily set at 100)
su/*su* (adenine independence)	none	I	58	
	Acr-1	I and VII	1	

```
   su     ribo-1    thi-1
   |        |         |
   ‾‾‾‾‾‾‾‾‾ ‾‾‾‾‾‾‾‾‾
    22·6      8·0     69·4
```

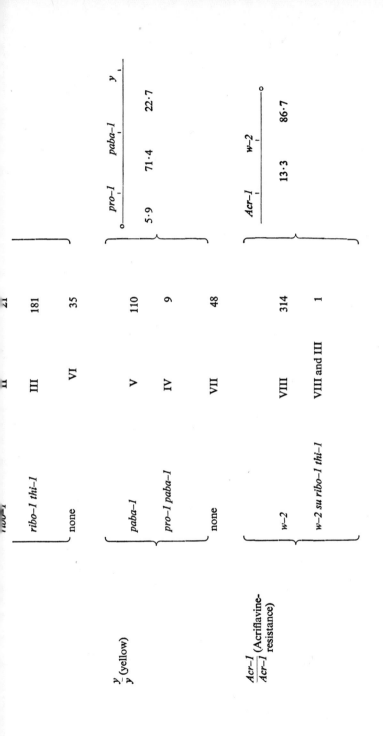

were, both may pass to the same pole. This is called *non-disjunction*, and it is a cause of *aneuploidy*, which is the name given to the presence of a chromosome number other than a multiple of the haploid set.

Käfer [273] has made a detailed study of the origin and genetic behaviour of aneuploids in *Aspergillus nidulans*. Her most striking discovery is the high frequency (1 per cent or more) with which aneuploid nuclei are formed in diploid strains. It seems, indeed, that aneuploid conidia are comparable in frequency to the segregant conidia of the diploid and haploid types considered above. The reason why the aneuploid types have not been prominent in most experiments is that they are usually poor in vigour as compared with euploids, and may either be suppressed altogether by competition for nutrients (particularly at high plating densities) or overgrown by more vigorous haploid or diploid sectors which may be formed at an early stage of colony development. Using a diploid marked at 17 loci spread over several different chromosomes, Käfer selected aneuploids in two ways. Transfers of conidia from colonies or sectors segregant for recessive colour markers yielded many colonies with comparatively feeble growth which tended to be rapidly overgrown by more vigorous *haploid* sectors. Segregation in these sectors of other markers on different chromosome pairs showed that the original feeble colony must have been hyperhaploid (n + 1, n + 2, n + 3, etc.). Of 115 colour segregants analysed, 14 were haploid, 42 were diploid (presumably the result of mitotic crossing-over) while the rest were haploid with respect to the chromosome carrying the colour mutation, but diploid with respect to one, two or three other marked chromosomes in 40, 14 and 5 cases respectively. Since not all eight chromosome pairs were marked, only a minimum estimate of the degree of hyperhaploidy was possible.

The second way in which Käfer obtained aneuploids consisted in plating conidia on complete medium and picking out colonies with low growth rate or poor conidiation. The frequency of these varied from 0·6 per cent to 2·5 per cent depending on the density of plating. The slowest growing colonies all turned out to be hyperhaploids, giving sectors of more vigorous haploid growth with accompanying segregation of markers. The more vigorous abnormal types of colony mostly gave *diploid* sectors showing segregation of whole chromosomes; it is segregation of this type which presumably accounts for the 'non-disjunctional' types of diploid segregant mentioned above in the discussion of mitotic mapping. From the number of linkage groups showing segregation in the sectors of each colony it was possible to assign minimum chromosome numbers of 2n + 1, 2n + 2, etc. to these unstable types.

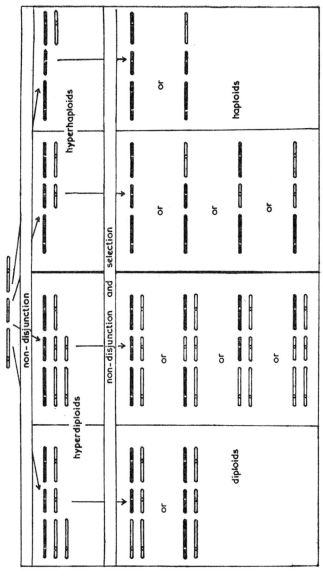

FIG. 23. Non-disjunction as the cause and cure of aneuploidy in diploid strains of *Aspergillus nidulans* showing how the various aneuploid types may be differentiated by the kinds of genetic segregants which they produce in sectors. For simplicity, only three chromosome pairs are represented and are distinguished from each other by size and centromere position. Homologous chromosomes are supposed to differ at marker loci and are distinguished by solid versus outline drawing. Based on Käfer [273].

It seems clear that this analysis of aneuploid formation from diploids provides an explanation of the mechanism of haploidization. Hyperhaploid nuclei probably arising from an initial loss of one or a few chromosomes by non-disjunction, or perhaps by lagging of chromosomes at anaphase, are supplanted by the more viable haploid products of further non-disjunction. Non-disjunction is thus the cure, as well as the cause, of aneuploidy. The process of haploidization may occur in more than two steps; a series of losses of chromosomes, each one giving an improvement in viability as the haploid number is more nearly approached, seems probable, but evidence for such a multi-step process is lacking so far.

These various shifts in chromosome number, and their genetic consequences, are illustrated in Fig. 23. Käfer has published a fuller account of her work in a later paper [274]. Losses of chromosomes leading to haploidization can be induced in Aspergillus by treatment with p-fluorophenylalanine [313].

Pseudo-wild types in Neurospora crassa

Although aneuploidy in the nuclei of vegetative mycelium of Neurospora has not been shown to occur (perhaps because the technical means for demonstrating it have not been to hand), the occurrence of $n + 1$ (or *disomic*) ascospores is well established. The method of selecting disomic ascospores is the same as that used in the selection of diploid ascospores in Aspergillus. A cross is made between two auxotrophic strains mutant in distinct and complementary but closely linked loci. In this situation two kinds of ascospores which can grow on minimal medium may be found. Firstly, of course, there will be haploid prototrophic recombinants formed with a frequency dependent on the amount of crossing-over between the mutant sites. Secondly, any ascospore which carries *both* homologous mutant chromosomes will be expected to be prototrophic if the mutations are in complementary (*i.e.* functionally distinct) loci.

Ascospores which apparently belong in the second category give so-called *pseudo-wild* type cultures (Pittenger [408]). Although such cultures are often indistinguishable phenotypically from wild type, they behave genetically as if they are haploid heterocaryons containing, generally, a 50:50 ratio of nuclei of the two original mutant types. That is to say, on crossing to wild type, about half the asci formed show 4:4 segregation of wild versus one mutant, and the other half show 4:4 for wild versus the other mutant. The heterocaryotic nature of pseudo-wild cultures can be further confirmed by the isolation of single conidia; the two original mutant types are easily recovered from the heterocaryon in this way.

These facts by themselves might suggest that pseudo-wild ascospores are formed already containing two unlike nuclei. The evidence against their being heterocaryotic from the outset consists in the failure to demonstrate that they ever carry both members of more than one chromosome pair. Thus in one experiment, Pittenger [408] crossed a strain carrying the mutant allele *inos*[t], in the fifth linkage group, together with the 'colonial' mutant *col* in linkage group 4, with a second strain carrying another *inos* allele and *col*.+ The *inos* and *inos*[t] alleles were distinguishable, the latter causing an inositol requirement at higher temperatures only, and were also complementary, being able to form a heterocaryon prototrophic at all temperatures. Fifteen prototrophic pseudo-wild types were recovered from the cross, four of which were colonial in growth habit and apparently homocaryotic for *col*. The other eleven did not form any conidia homocaryotic for *col*. Thus, although in all these cultures there were nuclei of both *inos* and *inos*[t] types, they each contained either *col* or *col*+ but never both. It seems simplest to suppose that the original ascospore nucleus in each of these cases was diploid with respect to linkage group 5 but haploid for linkage group 4. Although this experiment does not rule out the possibility that the chromosome number may have been n + 2 or n + 3, rather than n + 1, if such higher numbers occurred at all frequently one would expect sometimes to find heterozygosity for the two marked chromosome pairs simultaneously. Since neither Pittenger, nor any of the several workers who have since observed pseudo-wild types, have reported such a find, it seems likely that most, if not all pseudo-wilds are disomics in origin, giving two kinds of haploid nuclei by non-disjunction (cf. Fig. 23).

Pittenger showed that ascospores disomic for any one of at least five of the seven chromosomes could be selected by the use of appropriate complementary linked mutations. With respect to any one chromosome, the disomic frequency is of the order of $0 \cdot 1$ per cent in most crosses, although this value is subject to variation. One interesting feature of Pittenger's results was the failure to obtain any pseudo-wild types carrying both mating type alleles. It appeared that even where ascospores were heterozygous for other group 1 markers, they were always homozygous for mating type; this situation would, of course, arise as a result of crossing-over between the selected heterozygous loci and mating type, with the inclusion of one parental and one recombinant type chromosome in the same ascospore nucleus. Subsequently, Martin [345] has obtained apparently self-fertile ascospores which probably were disomic and heterozygous for mating type, so this situation may be possible at least in certain strains.

At what stage does the resolution of a presumably disomic pseudowild ascospore into a heterocaryon take place? The failure ever to isolate stable disomic clones, and the usually rather exact one-to-one nuclear ratio of the derived heterocaryons, suggest that the reduction of chromosome number occurs very early, perhaps at the first post-meiotic mitosis in the ascospore. However, Pittenger and Coyle [412], and also Case and Giles [78], have shown that pseudowild cultures occasionally contain a minority of nuclei which seem to have arisen by somatic recombination between the two homologous chromosomes initially present. The fact that the recombinants are in a minority suggests that the disomic condition may persist during several mitotic divisions, with recombination occurring at only a few of them. The discovery of apparent mitotic exchange in Neurospora disomics supports the view that it is a failure to form vegetative diploid nuclei, rather than an inability of homologous chromosomes to recombine during mitosis, which is responsible for the general lack of detectable recombination in ordinary Neurospora heterocaryons.

If pseudowild ascospores were due to non-disjunction at one of the divisions of meiosis, one would expect them to occur in asci containing either two or four ascospores with n-1 chromosomes, depending on whether the non-disjunction took place at the second or the first meiotic division. Such ascospores would certainly be inviable, and so one might expect to find pseudowild ascospores mainly or entirely in asci with two or four aborted spores. Coyle and Pittenger [89] have in fact found pseudowild spore pairs in asci with four aborted spores. However, Case and Giles [78] found 9 fully viable asci (out of a total of 1457) which each contained one pseudowild spore. The other 7 spores in each ascus were homocaryotic and viable and constituted, together with one of the components of the pseudowild culture, a perfectly regular tetrad showing 4:4 segregation for all markers. The second component of the pseudowild culture was supernumerary to the regular tetrad. It was as if one of the eight primary ascospore nuclei had received a gratuitous extra chromosome which was, in every case except one, of one or other of the two original parental types. Exactly how this extra chromosome is introduced, whether before meiosis (in which case it must remain non-dividing during meiosis), or by an extra replication during meiosis, is not clear.

Threlkeld [549] has reported a somewhat different kind of aneuploidy in Neurospora ascospores, with an extra chromosome fragment rather than an extra whole chromosome.

The occurrence of pseudo-wild ascospores may be useful in two ways. In the first place the isolation of the two kinds of non-disjunctional haploid from pseudo-wild mycelium gives two strains identical with respect to six of the seven chromosomes. It is thus possible to study the effects of the difference in the seventh chromosome against a constant genetic background. Secondly, non-disjunctional segregants isolated from a pseudo-wild are always heterocaryon-compatible (see Chapter 6), and the isolation of pseudo-wilds may be a more convenient way than conventional inbreeding for getting two linked mutations into compatible genetic backgrounds.

Is non-disjunction a directed process?

We have seen that in both *N. crassa* and *A. nidulans*, aneuploid nuclei tend to be supplanted rather rapidly by haploid or diploid nuclei during vegetative growth. The suggestion is bound to arise that non-disjunction (or the often indistinguishable alternative of loss of chromosomes by lagging at anaphase) which must be the cause of the loss of aneuploidy, may be especially apt to occur in the division of aneuploid nuclei. One might suppose that such a tendency would have an adaptive value, both in providing for a rapid correction of mistakes in mitosis, and in enabling rapid haploidization to occur as part of the parasexual cycle. However, in Aspergillus there seems to be no good evidence that non-disjunction is any more frequent in aneuploids than in diploids. We known from Käfer's work that in heterozygous diploids non-disjunction is surprisingly frequent, resulting in several per cent aneuploid conidia. There is no indication of any similar frequency in haploids but it is not certain that occasional n + 1 or n + 2 conidia, with the extra chromosomes necessarily carrying no distinguishing markers, could have been noticed even if they occurred. It seems that euploid nuclei have such a strong selective advantage over aneuploids that a frequency of non-disjunction of the order observed in diploids could easily account for the transient nature of aneuploidy in Aspergillus.

Whether this argument holds for Neurospora pseudo-wilds is more doubtful. The development of cultures from pseudo-wild ascospores is fairly regular and reproducible, and it does not appear as if good growth has to wait for a chance non-disjunction to occur. It is possible here that the disomic number is reduced to haploid very early in development and with high probability.

POLYPLOIDY

Triploidy (*i.e.* three sets of chromosomes) and tetraploidy (four sets) are known in fungi from genetic evidence in the yeasts, in *Aspergillus oryzae* [251] and Ustilago (triploids only) [232]. In addition, Emerson and Wilson [141] have shown by cytological investigations the existence of polyploid series in Allomyces species. Different strains of *A. arbuscula* showed 8, 16, 24 and 32 chromosomes in the gametophyte generation and *A. javanicus* strains showed 14, 28 and 50+ (probably 56) chromosomes. To clarify the further discussion, some explanation of the cytological and genetical behaviour of polyploids in general is necessary.

When a polyploid cell arises by failure of anaphase separation of chromosomes, or by vegetative nuclear fusion within a single strain, the

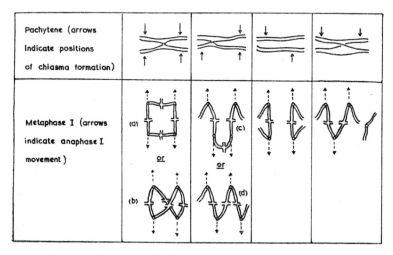

FIG. 24. Some of the more important kinds of behaviour shown by four homologous chromosomes at the first division of meiosis in an autotetraploid

several chromosome sets present are all homologous and capable of pairing with each other in meiosis. This kind of polyploidy is called *autopolyploidy*, and may be distinguished from *allopolyploidy* in which there are chromosome sets of different origin which are, at least partly, non-homologous with each other. Allopolyploidy, though of great importance in higher plants, has been demonstrated in fungi only in artificial hybrids between polyploid strains of *Allomyces javanicus* and *A. arbuscula* [141], and is not known to occur in nature.

Our knowledge of the cytological behaviour of autopolyploids rests

TABLE 17. Tetrad ratios resulting from single factor segregation in autotetraploids

	Type of Tetraploid								
	Duplex *a a a' a'*					Simplex/Triplex† *a a a' a*			
Whether chiasma formed in locus-centromere region between chromosomes carrying unlike alleles	No		Yes*			No	Yes		
Frequencies	ca. ⅓	ca. ⅔	x		1 − x		x		1 − x
Chromosomal constitutions of first telophase nuclei	(a-a+a-a) (a'-a'+a'-a')	(a-a+a'-a') (a-a+a'-a')	(a-a+a'-a') (a-a'+a-a')		(a-a+a'-a') (a'-a'+a'-a')	(a-a+a-a) (a-a+a-a')	(a-a+a-a) (a-a'+a-a')		(a-a+a-a') (a-a+a-a')
Frequencies / Chromosomal constitutions of second telophase nuclei	(aa) (aa) (a'a') (a'a')	(aa') (aa') (aa') (aa')	½ (aa') (aa') / (aa) (aa)	½ (aa') (aa') / (a'a') (a'a')	(aa) (aa') (aa') (a'a')	(aa) (aa) (aa') (aa')	½ (aa) (aa) / (aa) (aa)	½ (aa) (aa') / (aa') (aa')	(aa) (aa) (aa') (aa')
Phenotypic ratio : If *a* dominant	2 : 2	4 : 0	3 : 1	4 : 0	3 : 1	4 : 0	3 : 1	4 : 0	4 : 0
a' dominant	2 : 2	0 : 4	1 : 3	0 : 4	1 : 3	2 : 2	3 : 1	2 : 2	2 : 2

* If the chance of a chiasma in the critical region is not small, the probability of two such chiasmata in the same cell becomes appreciable; this case is ignored here as it leads to no qualitatively different kinds of tetrad.

† Simplex if *a'* dominant; triplex if *a* dominant.

x is the probability of two chromosomes, joined by a chiasma, going to the same pole. This will be nearly zero if only bivalents are formed and ⅓ if quadrivalents are always formed and disjoin two-by-two at random.

entirely on observations of meiosis in vascular plants, and we have only indirect genetic evidence for similar events in polyploid yeast where the cytology is quite obscure. What is observed in higher plant autopolyploids is that *all* the chromosomes of a kind tend to associate at meiotic first prophase, and that though pairing at any point is restricted to two chromosomes, exchanges of pairing partner may occur at points along the length of a chromosome so that all four homologous chromosomes (in a tetraploid) may be associated in one figure. According to where chiasmata happen to be formed, the association of four at pachytene may resolve itself into two bivalents at metaphase I, or a quadrivalent with all four chromosomes joined together, or occasionally, a trivalent plus a univalent. Trivalents and quadrivalents may become orientated in different ways at metaphase; in some plants the chromosomes of a quadrivalent tend to alternate in their orientation, giving a zig-zag or figure-of-eight configuration (Fig. 24 (d) and (b)), while in other species configurations with adjacent chromosomes directed towards the same pole are more common (Fig. 24 (a) and (c)).

The possible genetic situations in an autotetraploid are more numerous and complex than in a diploid. If, to take the simplest example, we are concerned with a recessive mutant allele a and its wild type homologue a^+, we have not one but three kinds of heterozygote, namely *simplex* $a^+/a/a/a$, *duplex* $a^+/a^+/a/a$, and *triplex* $a^+/a^+/a^+/a$. The argument is simplified if we make the assumption, which is true to a first approximation, that all the viable products of meiosis in a tetraploid will be diploid; that is to say, we ignore the possibility of viable aneuploids being formed by random disjunction of univalents and trivalents, and occasional unequal disjunction of quadrivalents. We can then easily calculate the expected frequencies of genotypes among the meiotic products given certain kinds of multivalent formation and orientation. The argument is presented schematically in Table 17. It will be noted that in the case where one allele is dominant over the other, the duplex heterozygote gives both 4:0 and 2:2 phenotypic ratios among tetrads, but that 3:1 tetrads are only possible as a result of crossing-over between the locus of the alleles and the centromere. From the simplex heterozygote one expects predominantly 2:2 tetrads with 3:1 ratios requiring both crossing-over between locus and centromere and the passing of the two cross-over products to the same pole at anaphase I, this latter condition in turn depends on multivalent formation and metaphase orientation as in (a) or (c) in Fig. 24. In the triplex the same situation obtains as in the simplex, with the difference that 4:0 ratios replace the 2:2 ratios; that is to say the recessive phenotype cannot occur at all except as a result of a particular kind of crossing-over and multivalent orientation.

Now let us see how the yeast tetraploid strains fit into this theoretical picture. To take one example, Leupold [305] dissected thirty-three asci from a diploid of genotype a/α, $ad/+$, $pyr/+$, $me/+$, where a, α are the conventional designations of the two mating-type alleles, and ad, pyr and me are auxotrophic mutations of the types implied by the symbols. One of the asci was exceptional in giving ascospores with the following phenotypes:

(a) $ad^+ pyr\ me^+$;   (b) $ad^+ pyr^+ me^+$;

(c) $ad\ pyr^+ me^+$;   (d) $ad^+ pyr^+ me^+$.

In addition to the aberrant ratios for the auxotrophic markers, all four ascospores gave cultures which exhibited neither a nor α mating type reaction but were self-fertile, being able to sporulate without mating. Analysis of the asci formed in these cultures showed the types of segregation expected from diploids of the following genotypes:

(a) $a/\alpha\ ad/+ pyr/pyr\ me/+$;   (b) $a/\alpha\ ad/+ +/+ me/+$;

(c) $a/\alpha\ ad/ad\ pyr/+ me/+$;   (d) $a/\alpha +/+ pyr/+ me/+$.

It is thus quite clear that the original exceptional ascus must have been tetraploid and duplex for all four loci.

A second exceptional ascus from another cross segregated 2:2 for mating type, although 3:1 and 4:0 ratios for other segregating differences indicated its probable tetraploid nature. Further analysis established that spores in this ascus were diploid but each was homozygous for one or other mating type allele. Cultures derived from these spores showed normal mating reactions and could be used to make crosses in which *all* the asci were tetraploid. Some of these asci contained spores of mating type constitution a/a, a/α, a/α, α/α indicating crossing-over between the mating type locus and its centromere.

In some of the diploid × diploid crosses one or more of the loci was in the triplex condition. In such cases most of the asci showed 4:0 ratios for the marker in question, but a few 3:1 ratios did usually occur. This can be taken as indicating that multivalent, rather than exclusively bivalent, associations, do occur in tetraploid yeast meiosis, since otherwise the recessive phenotype could not be segregated from triplex (see Table 17 and Fig. 24).

Other authors, notably the Lindegrens [321] and Roman and his colleagues [459], have shown very much the same sort of evidence for tetraploidy in Saccharomyces. The latter authors, and also Leupold [306] have derived formulae relating the frequencies of different ratios (4:0, 3:1, 2:2) in triplex, duplex and simplex asci, to the frequency of crossing-over between the locus and its centromere. Since such calculations depend on knowledge of the kinds and frequencies of multivalent formation and orientation, and these are difficult or impossible to estimate in the absence of the possibility of observation with the microscope, the method is of very limited usefulness.

Aneuploids from polyploids

If multivalents are formed at all in autotetraploid meiosis in yeast (and some multivalent formation is shown, as we have seen, by the segregation of phenotypically recessive diploid ascospores from triplex asci), one would expect occasional unequal distributions of chromosomes at first anaphase. This is because quadrivalents are not expected always to disjoin two-and-two, and the trivalents-plus-univalents which will almost certainly also occur, will only do so by chance. The most common abnormality resulting from unequal disjunction at anaphase I will be a tetrad with two trisomic $(2n + 1)$ and two monosomic $(2n - 1)$ ascospores.

TABLE 18. Tetrad ratios resulting from single factor segregation in a triploid or trisomic $a\ a\ a'$

Chiasma formation in locus-centromere region between chromosomes carrying unlike alleles	No		Yes		
First telophase	ca. $\frac{1}{3}$ $(a{-}a+a{-}a)$ $(a'{-}a')$	ca. $\frac{2}{3}$ $(a{-}a+a'{-}a')$ $(a{-}a)$	x $(a{-}a)$ $(a{-}a'+a{-}a')$		$1{-}x$ $(a{-}a')$ $(a{-}a'+a{-}a)$
			$\frac{1}{2}$	$\frac{1}{2}$	
Second telophase	$(aa)\ (aa)$ $(a')\ (a')$	$(aa')\ (aa')$ $(a)\ (a)$	$(a)\ (a)$ $(aa)\ (a'a')$	$(a)\ (a)$ $(aa')\ (aa')$	$(a)\ (a')$ $(aa)\ (aa')$
Phenotypic ratio:					
If a' dominant	2:2	2:2	3:1	2:2	2:2
a dominant	2:2	4:0	3:1	4:0	3:1

Notes: (a) No account is taken of the possibility of chromosome loss, though this may happen especially following bivalent + univalent formation.

       (b) If no trivalents and only bivalents are formed, x will be zero; if trivalents are always formed and separate two-to-one at random, x will be $\frac{1}{4}$.

One such ascus was found by Roman *et al* [459] in a tetraploid strain which was simplex for ability to ferment galactose versus inability to do so (*i.e.* $g^+/g/g/g$). In this ascus all four spores were disomic and heterozygous with respect to mating type (a/α), and each of the derived cultures could be induced to sporulate. One of the four, while showing regular 2:2 ratios for all other markers, gave 3:1 as well as 2:2 ratios for galactose-positive versus negative. It was inferred that this culture was trisomic for the chromosome carrying the g locus, with genotype $g^+/g^+/g$. The theoretical possibilities in such a case are set out in Table 18. It will be seen that 4:0 tetrads should also be found in a sufficiently large sample. Two of the other ascospores from the exceptional ascus gave galactose-negative cultures and, though self-fertile, gave only one or two-spored asci, at least two meiotic products being inviable in every case. It was reasonable to assume that these two ascospores were monosomic with just one chromosome carrying g. The fourth ascospore, which behaved as if diploid and homozygous for g in that it gave complete asci with all the products galactose-negative, was in all probability trisomic $g/g/g$.

Given the existence of diploids homozygous for mating type, and these are readily obtained from tetraploid segregation, it is a simple matter to construct triploids by crossing diploid a/a × haploid α, or diploid α/α × haploid a. Ascus formation in triploids would be expected to give a high proportion of inviable ascospores with grossly unbalanced chromosome numbers, but among the viable ones there should be many trisomics. This is, in fact, probably the best method of obtaining trisomics. Little work on yeast trisomics has been published, though, potentially, they provide valuable tools for genetic analysis, particularly for the assignment of markers to linkage groups. If a trisomic is heterozygous for a number of markers one needs only a few asci to establish which markers are associated with the chromosome which is represented three times, and which are not.

CHAPTER 6

THE GENE AS A FUNCTIONAL UNIT

In preceding chapters we have explained how the sites of mutations can be represented by linear maps corresponding to chromosomes. One may ask whether there is any obvious order or pattern in the sites so mapped. Experience shows that a pattern does exist in that the sites of mutations giving the same kind of phenotypic effect are not scattered about the chromosome complement in a random fashion but tend to be concentrated in one or a few short segments. Generally, as we shall see in Chapter 8, the more precisely we can define the phenotypic effect, the more specific does the relationship between a given effect and a given chromosome segment appear. Thus there are at least eleven short regions in *N. crassa* which can mutate to give a growth requirement for arginine. There are a number of different ways in which the synthesis of arginine can be interfered with, and if we look only at those mutants which are blocked in the terminal step in the biosynthetic pathway (the splitting of argininosuccinic acid) we find that they are *all* due to mutation within the same very short region of linkage group VII. Such a short segment within which mutation has a specific type of effect is often called a *locus*, but it corresponds to what most geneticists mean by the term *gene*. Our present knowledge of the specific biochemical functions of loci (genes) will be discussed in Chapter 8. In this Chapter we shall be concerned with the problem of how to define a gene without necessarily undertaking a full biochemical analysis of its function.

COMPLEMENTATION TESTS

The general method for determining whether two mutations are in the same gene may be called the complementation test, and is applicable to all cases where the mutants in question have some clear functional deficiency as compared with the wild type. If the chromosome sets of two defective mutants are brought together into the same cell, whether by heterocaryon formation or by formation of a diploid nucleus, a normal phenotype will result if each mutant can supply what is deficient in the other, but a mutant phenotype, probably resembling the less extreme of the two mutants where they are different, will be obtained if both chromosome sets are deficient in the same function. The first alternative, known as *complementation*, is evidence (though not, as we shall see, conclusive evidence) that the two mutations are non-allelic, i.e. in different genes. The second alternative, absence of complementation, is near proof of allelism. The two situations are contrasted in Table 19.

TABLE 19. The principle of the complementation test

	Case A: mutations in different functional loci	Case B: mutations in the same functional locus
Haploid mutant genotypes	$a \qquad\qquad b^+$ _____ _ _ _ _____ $+$ _____ _ _ _ _____ $a^+ \qquad\qquad b$ N.B. a and b may or may not be linked.	a' _____ $+$ _____ a''
Phenotype	Wild, since both a^+ and b^+ functions are supplied	Mutant, since a^+ not present

Complementation tests by heterocaryosis

In filamentous fungi the simplest method of carrying out the complementation test is to form a heterocaryon. Beadle and Coonradt [24] showed in 1944 that mixed inocula of mutant strains of *N. crassa* of different auxotrophic types would grow readily on minimal medium as a result of hyphal fusion and heterocaryon formation. This sort of test was afterwards used very widely by Neurospora workers for the rapid classification of mutants having the same growth requirement into different categories, corresponding to different genes. It was plain from the outset that, in *N. crassa*, strains would only fuse readily to form stable heterocaryons if they were of the same mating type. Later, both Garnjobst [189, 190] and Holloway [237] (see p. 233) showed that the component strains had to be similar at several other loci as well. Unfortunately, by the time this latter discovery was made, laboratory stocks of Neurospora had been cross-bred to the point where heterocaryon-incompatibility due to heterogeneity at these other loci was of common occurrence. In more recent years, efforts have been made in many laboratories to base work on a single wild type stock from which all mutants are derived, or with which they are highly inbred. Provided one has stocks with a fairly closely similar 'background' genotype, and of the same mating type, the Neurospora heterocaryon test will usually give a clear yes-or-no answer in twenty-four hours. The usual procedure in the case of auxotrophic mutants is to superimpose inocula of conidia or hyphal strands on minimal agar medium. The test is not restricted to auxotrophic mutants, though these give the most clear-cut results. Any morphological variant which is significantly slower or more compact in growth than wild type can be tested against other slow growing or auxotrophic mutants.

While a positive heterocaryon test is good *prima facie* evidence for the physiological distinctness of the mutants concerned, a negative test may merely mean that the strains being used will not form a heterocaryon at all. For this reason, negative results can be greatly strengthened by the use of the *forced heterocaryon* technique. As an example we may take two of the mutants of the *am* series in Neurospora which grow poorly on minimal and respond to alanine [154]. A heterocaryon was 'forced' by inoculating together on minimal medium plus alanine the two double mutants am^2 arg-1 and am^3 arg-10, arg-1 and arg-10 being non-allelic mutants with blocks in arginine synthesis at different steps. A culture resulted which must have been heterocaryotic since it was arginine-independent, but on transferring to minimal medium it grew no better than the am^2 or am^3 homocaryons. It could be concluded from this that am^2 and am^3 were truly non-complementary.

In *Aspergillus nidulans* heterocaryon formation is not so easily accomplished as it is in Neurospora. Nevertheless, forced heterocaryons can be made, and have been widely used in complementation tests as well as in the selection of diploids (see pp. 30 & 104).

Some caution is necessary in basing conclusions on quantitative characteristics of heterocaryons. The ratio of the two kinds of nuclei in even a forced heterocaryon can vary between wide limits, and the growth rate given by a heterocaryon formed from complementary auxotrophic mutants will be sensitive to changes in nuclear ratio, at least when the ratio is far from equality. As the experiments of Pittenger and Atwood [410, 413] have shown, it cannot be assumed that an optimal nuclear ratio will become selected during growth of the heterocaryon. Indeed, their results show that once a heterocaryon has been established and started growth, its nuclear ratio may be remarkably stable even if it is outside the range giving optimal growth. Apparently the selection of hyphal tips with significantly deviant ratios is difficult, either because of the large numbers of nuclei in the growing tips, or because of rather free nuclear migration. The nuclear ratio actually achieved depends primarily on the ratio of conidia of the two component types present in the mixed inoculum, but also on the extent to which each type can grow before the whole inoculum becomes fused into a heterocaryotic mass. Ideally any conclusion that two mutants show quantitatively imperfect complementation should be checked by making and testing heterocaryons with a range of nuclear ratios. However, tests adequate for determining whether any complementation at all occurs can be, and generally have been made without any control of nuclear ratio.

Complementation tests through diploidy

In yeasts the phenotype, prototrophic or auxotrophic, of the diploid clone resulting from the copulation of two auxotrophic haploids, provides a simple test of whether the two mutant chromosome sets are complementary or not (e.g. Roman [458]). A particularly neat technique for the classification of *ad-1* and *ad-2* mutants in *Saccharomyces cerevisiae* has been devised by Bevan and Woods [36]. These classes of adenine-requiring mutants accumulate an intracellular red pigment, evidently a derivative of an adenine precursor. If an *ad-1* haploid is mated with an *ad-2* strain the resulting diploid cells no longer accumulate the precursor and are white. By transferring cells by the replica plating technique from a number of pink haploid colonies, all of one mating type, to another plate carrying a lawn of haploid cells of another pink strain and of the other mating type, complementation is shown by the development of patches of white diploid growth against the pink haploid background. Not only was any *ad-1* mutant complementary with any *ad-2* mutant, but certain combinations of *ad-2* mutants gave white or pale pink diploids. This sporadic intra-class complementation could be represented by a complementation map, of the kind discussed below.

Diploid complementation can sometimes be shown in filamentous fungi which do not have a persistent diploid stage as a regular part of their life cycle. Pritchard [427], for example, found that among large numbers of ascospores, formed in crosses between different mutants of the *ad-8* series, fewer than one in a thousand were capable of forming colonies in the absence of adenine. This showed, not only that the mutant sites were closely linked, but also that they must have been non-complementary, since about 1 per cent of *A. nidulans* ascospores are known to be diploid and heterozygous with respect to any differences between the parents [426]. The opposite result, i.e. the occurrence of prototrophic ascospores with a frequency of the order of 1 per cent would, of course, have been ambiguous, since it could have been due to haploid recombination by crossing-over rather than to diploid complementation.

Pseudo-wild types as a proof of complementation

In Neurospora, the occurrence of pseudo-wild ascospores among the progeny of a cross between closely linked auxotrophic mutants is proof of complementation between such mutants. This principle has been used by Pateman and Fincham [397], Case and Giles [77] and de Serres [117] in demonstrating complementation or absence of complementation between alleles. If closely linked mutants are complementary, the frequency of disomic ascospores is apparently such that some pseudo-wilds should always be readily detectable among the progeny of the cross between them.

The cis-trans test

Absence of complementation in the tests outlined above is not completely satisfactory evidence of allelism, though it can be accepted as such with a fair degree of safety. The reason for having some reservations here is as follows. The situation in a diploid or heterocaryon formed from two haploids which are actually mutant at different loci may be represented as $\frac{a\ +}{+\ b}$; in a diploid the two genomes are necessarily present in equal 'dose' while in a heterocaryon there is the possibility of having variation in nuclear ratio. We compare the resulting phenotype with that of the wild type $\frac{+\ +}{+\ +}$. Clearly, we can only expect the two to be the same if each mutant individually is recessive to the corresponding wild type allele, i.e. if $\frac{+\ +}{a\ +}$ and $\frac{+\ b}{+\ +}$ are each similar to $\frac{+\ +}{+\ +}$. Recessiveness is, indeed, the general rule for auxotrophs and most other kinds of mutant, at least when growth rate and gross morphology are used as criteria. Yet even if complete recessiveness of the individual mutations can be demonstrated, there is still the possibility, however remote, that $\frac{a}{+}$ together with $\frac{b}{+}$ will give a mutant phenotype through some sort of cumulative effect, even though neither does so by itself. The only really fair comparison is of the *trans* heterozygote or heterocaryon, $\frac{a\ +}{+\ b}$, with the corresponding '*cis*' arrangement $\frac{a\ b}{+\ +}$. If the two are the same, the two sites of mutation are complementary; if the former is mutant and the latter wild in phenotype the case for non-complementation (allelism) is made, and the two mutations fall into the same 'cistron', as Benzer [28] defines the term.

In practice the rigorous *cis-trans* test is very seldom carried through, owing to the difficulty of obtaining and identifying the double mutant where the two mutations are closely linked and similar in effect. Since such a double mutant will usually be similar in phenotype to the single mutants, it is usually only practicable to look for it by tetrad or mitotic analysis, either of which techniques give one a chance of isolating the double mutant formed at the same as the readily identifiable wild type. Case and Giles obtained double mutants at the *N. crassa pan-2* locus (see Table 20, Chapter 7) by tetrad analysis, while Pritchard [427] used mitotic recombination to recover an *Aspergillus nidulans* strain doubly mutant at the *ad-8* locus. However, the difficulty in many cases is increased by the fact that, as shown in the next chapter, recombination at the intragenic level is very often not exactly reciprocal, so that the formation of a wild type from an inter-mutant cross may not be accompanied by the formation of the reciprocal double mutant. Generally speaking the investigator feels justified in assigning two mutants to the same gene or cistron on the basis (i) of their individual recessiveness to wild type, and (ii) their failure to complement each other in heterocaryon or diploid. Nevertheless, the strict *cis-trans* comparison should be made in doubtful or critical cases if at all possible.

One might expect there to be cases where two chromosomal units can complement each other when in the same nucleus in a diploid, but not when in different nuclei in a heterocaryon. The possibility of comparing the two kinds of test is, so far, restricted to *Aspergillus nidulans* and related species in which both heterocaryons and diploids may fairly readily be made. It seems to be exceptional for the two kinds of complementation test to give different results, but at least a few examples have been reported [456, 420].

PARTIAL COMPLEMENTATION
AND COMPLEMENTATION MAPPING

A combination of linkage mapping and complementation tests leads to the important generalization that non-complementing mutations are always at very closely linked sites. This implies that each physiologically active unit or cistron is confined within a short chromosome segment. However, the converse statement, that complementation between mutant sites implies their location in distinct and separate segments, is not always true. Many workers in recent years (particularly those working on *N. crassa* such as Catcheside [81], Fincham [154], Woodward [590]

10

and Giles [77]) have found the following situation. (a) Among a group of auxotrophic mutants, all phenotypically similar and all mappable within the same short chromosome region, *most* pair-wise combinations are non-complementary; indeed it is often found that most of the mutants are non-complementary with *all* the others. (b) Some pair-wise combinations show complementation, though the resulting phenotype is often more or less sub-wild as regards growth rate and, in cases so far analysed, always sub-wild as regards the level of the relevant enzyme (see Chapter 8). (c) The complementation relationship within the group of mutants can be represented by a 'complementation map' in which non-complementary pairs of mutants are shown as overlapping segments and complementing pairs are shown as non-overlapping segments. An example of a relatively simple complementation map, and the data on which it was based, is given in Fig. 25. (d) The mutants which appear functionally to overlap other mutants are not deletions, since they are often capable of back-mutation to something resembling wild type, and, moreover, behave in linkage mapping experiments as point mutations (Fig. 38, Chapter 8). (e) Where an enzymic analysis of the basis of the auxotrophy is available, all the mutants, whether mutually complementary or not, appear to be deficient in the same enzyme (see Chapter 8). In this kind of situation it seems entirely reasonable to consider the whole series of mutations as being in the same gene.

A comparison of the complementation map of the Neurospora *pan-2* locus and the recombinational map of the same region is shown in Fig. 41. It will be seen that there is a fair degree of correspondence between the two maps, as if the complementation map does, in fact, tend to give a picture of the physical structure of the locus. Such complementation maps have now been made for some dozen *N. crassa* loci. So far, exceptions to the rule that intra-locus complementation relationships can be represented by a linear map have been uncommon, though they do occur. Bernstein and Miller [33] found four *N. crassa iv-3* mutants, out of 382 which were complementary in some combinations, which did not fit a linear complementation map; some very complex maps, notably those for *pan-2* [77] and *ad-4* [590] in *N. crassa* and *ad-2* in *Saccharomyces cerevisiae* [36], show no such exceptions in spite of a large number of possible ways in which mutants could fail to fit. The significance of this rule will be discussed in Chapter 8. There seems to be a fairly sharp distinction between loci showing complementation and loci which do not (i.e. all mutants equivalent to class A of Fig. 25). Catcheside [81] lists a number of loci within which he found no complementation, and a study by De Serres

[117] on the *ad-3A* locus has failed to detect any complementation.

Does the existence of complementation between different mutants within a gene mean that we must regard it as consisting of several units with distinct physiological functions? The answer seems to be no. In cases where one can make a complementation map it is difficult to consider each segment of the map (e.g. the four segments of the *arg-1* map, shown in Fig. 25) as an independent cistron or unit of function. The con-

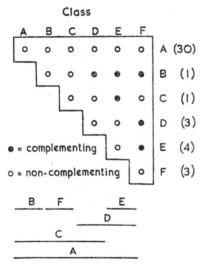

Fig. 25. Patterns of complementation of six classes of *N. crassa arg-1* mutant, and (below) the representation of the results by a complementation map (Catcheside & Overton[81]). Figures in parentheses are the numbers of independently occurring mutants in each class. Note that the order of *B* and *F* in the complementation map is arbitrary.

tinuous series of overlaps seems, as it were, to tie the whole segment together, and the fact that many point mutations knock out the function of the entire locus makes it particularly difficult to ascribe any functional autonomy to the segments of the map. In this kind of situation the cistron concept really loses its usefulness unless we modify the definition of the cistron to mean a region within which different mutations may be either non-complementing or *imperfectly-complementing*, and this definition, though it is probably the best available for the physiological genetic unit, may sometimes be difficult to apply without careful biochemical analysis.

The justification for considering such loci as *arg-1* and *pan-2* as single genes in the sense of units of function is discussed further in Chapter 8.

Without entering here into a discussion of the biochemistry of gene action, the point can be made clearer by contrasting these genes with the *ad-3* region of the first linkage group of *N. crassa*. Mutations within this region all tend to give the same phenotypic effect, namely a growth requirement for adenine combined with an accumulation in the growth medium of a purple pigment, apparently a side product of an accumulated adenine precursor. All the 'adenine-purple' mutants are quite closely linked, and were at first attributed to a single gene. De Serres [116], however, showed by heterocaryon tests that they fell clearly into two groups, *ad-3A* and *ad-3B*. The *ad-3A* mutants were non-complementing (or at least, very poorly complementing) among themselves, as were the *ad-3B* mutants, but every member of each group complemented with *any one* of the other group. Furthermore the inter-group complementation was always complete, as judged by the wild type growth rate of the heterocaryons. In this case it is clearly reasonable to consider *ad-3A* and *ad-3B* as distinct genes, a view which is strengthened by the demonstrations, referred to in the next chapter, that they appear as two non-overlapping segments by recombinational analysis.

We seem, then, to have adequate criteria based on complementation tests for defining what may be called the gene, the locus or the functional genetic unit. Recessive mutants belonging to different genes are always fully complementary. A pair of mutant alleles of the same gene are usually non-complementing; if they are complementary, there are always other alleles of the same series which are non-complementary with both, and the complementation is often incomplete as judged by morphology or growth rate and probably always incomplete as judged by the quantity and quality of the gene product, where this can be determined (see Chapter 8). So far as is known, each gene so defined is concerned in the formation of a single kind of polypeptide chain, and different genes with the formation of different ones. The evidence for this generalization which, if valid, will eventually provide the best and most concise definition of the gene, will be considered in Chapter 8.

CHAPTER 7

THE FINE STRUCTURE OF GENES AND THE MECHANISM OF GENETIC EXCHANGE

INTRA-GENE MAPPING AND 'NEGATIVE INTERFERENCE'

In the previous Chapter we have discussed the problem of defining the functional genetic unit on the basis of complementation tests. The conclusion reached was that it is possible to assign mutations (with the exception of those which appear to be deletions of extended chromosome segments) to functional genes, or loci. Each gene is a piece of chromosome which is very small in the sense that different mutations within one gene appear, to a first approximation, to be completely linked.

However, independently occurring mutations in the same gene seem seldom to give the same mutant allele. Their non-identity can be established by one or more of the following criteria: (a) they may show widely differing rates of reversion to a state phenotypically similar to wild type, as shown by Giles [194] for *inos* mutants in *N. crassa,* and in many other cases since; (b) the *kinds* of reversion which may occur are often characteristic of a particular mutant allele (Giles *et al* [197] and Pateman [394]; see discussion in the next Chapter); (c) different alleles may show different complementation relationships, as discussed in the preceding Chapter; (d) it is very commonly found that crosses between allelic mutant strains of independent origin (*heteroallelic* crosses) yield low frequencies (10^{-3} to 10^{-6}) of non-mutant progeny, while inter-crosses of strains carrying the same mutation (*homoallelic* crosses) yield either none, or several orders of magnitude fewer. Differences in these four respects may exist even though there are no detectable phenotypic differences between the alleles in question; on the other hand, minute scrutiny at the biochemical level may often reveal small phenotypic differences as well, as shown in the following Chapter.

Such differences between mutant alleles assigned to the same gene are a strong indication that a gene is a complex structure capable of many kinds of mutational change, at many different sites. This conception of different sites within a single gene is further strengthened by the pos-

sibility, now demonstrated in many cases, of actually mapping the sites
in a linear order by recombination studies.

The first and simplest way of constructing a map of sites within a gene
may be illustrated by reference to the detailed data of Case and Giles
[75] on *pan-2* in *N. crassa*. It is based on the assumption, the validity of
which will be discussed presently, that the occasional wild types resulting
from crosses between allelic mutants are the result of some kind of
exchange occurring between homologous chromosomes in the region
between the mutant sites. If this is so, then prototroph frequency should
increase with the inter-site distance, in the same way as recombinant
frequency varies with map distance in ordinary linkage studies. By
determining the prototroph frequencies from inter-crosses in all combinations of a series of alleles, it should be possible to establish a map of
the sites of mutation. Case and Giles found that, although prototroph
frequencies were not precisely additive, they were so to an encouraging
degree. Figure 26 shows a part of the relevant data. Comparable results
have been obtained by Pritchard [429] for *ad-8*, and Siddiqi [492] for
paba-1, in *Aspergillus nidulans*, by Leupold [307] for *ad-7* in *Schizosaccharomyces pombe*, by Ishikawa [248] for *ad-8* in Neurospora, and in
many other cases.

Inspection of Fig. 26 will show that the order of some of the sites is
subject to a good deal of uncertainty, and it might even be maintained
that such agreement as the data do show with a linear map is no more
than could have occurred by chance. On the other hand, when one looks
at other sets of data of this type for other genes, the general tendency
towards linearity is convincing. In Ishikawa's data, which are some of
the most detailed, the linearity is very clear and the additivity of map
distances is relatively precise [248]. One can easily explain away depar-
tures from linearity in intra-gene maps, where they occur, as due to
some variation in exchange frequency from one cross to another.

Mention should be made here of a novel method for intra-genic
mapping in yeast due to Manney and Mortimer [341]. Instead of using
the frequency of prototrophic meiotic products from an interallelic
cross as a measure of genetic distance, these workers used the frequency
of X-ray induced reversion to prototrophy in a heteroallelic diploid.
Presumably such reversion is due to some kind of genetic recombination
occurring between homologous chromosomes in vegetative cells (cf.
p. 114).

There is a second method available for intra-genic mapping. If
intragenic recombinants are due to genetic cross-overs of the same sort

as are encountered in ordinary linkage mapping it should be possible to obtain an unequivocal indication of the order of sites within a gene by reference to genetic markers close to the gene and on each side of it. If we have two auxotrophic mutants, m' and m'', within one gene, and outside markers p and q spanning the gene, with the overall order $p-m'-m''-q$, then, from a cross $p\,m'+q \times +\,+m''+$, prototrophic recombinants should all be of the constitution $+\,+\,+q$, assuming that multiple crossing-over within the short p-q interval is negligible. Reversing the outside markers relative to the intragenic mutants, so that the cross becomes $+m'+\,+ \times p+m''q$, will result in all the prototrophic recombinants being of constitution $p+\,+\,+$.

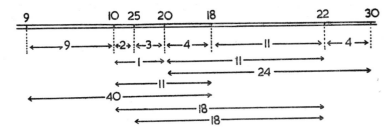

Fig. 26. Map of *pan-2* in *Neurospora crassa*, after Case & Giles[75]. Figures above the map refer to mutant isolation numbers. Figures below refer to prototroph frequencies, expressed as numbers per 10^4 viable ascospores, from crosses between mutants.

This principle was first used for intragenic mapping in a fungus by Pritchard, who ordered a number of sites within the *ad-8* gene of *Aspergillus nidulans*, selecting *ad-8*[+] recombinant ascospores by plating on minimal medium. Some of his results are shown in Fig. 27. It will be seen that while the order of the sites is established quite unequivocally, there were substantial numbers of *ad-8*[+] recombinants which, on the simple cross-over hypothesis, would have to be explained by double or even triple cross-overs. The frequency of these apparent multiple cross-overs is far higher than would have been expected from the fairly close spacing of the markers concerned. This situation was described as *negative interference*, in other words, as a tendency for cross-overs to occur in clusters. Pritchard proposed that recombination between linked markers may be limited not so much by the frequency of exchanges within paired chromosome regions as by the number and lengths of the regions which are *effectively paired*. Even though chromosomes may appear to be fully paired at pachytene it is possible that exchanges are limited to short regions of especially close contact. Within each of these regions more than one exchange may tend to occur, though in the far longer regions which do not happen to be so closely paired no exchanges occur at all. On this basis one would expect clustering of exchanges within short intervals. From Pritchard's data the mean number of exchanges within each closely paired region was calculated to be about 0·6.

In spite of this appearance of negative interference, one outside marker recombinant constitution was so much more common than the other among *ad-8+* recombinants that there was no doubt about the order of

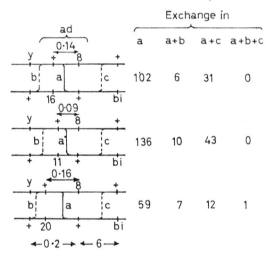

FIG. 27. Some of Pritchard's [427] data from crosses between allelic *ad* mutants in *Aspergillus nidulans*. The crosses illustrated were *yellow (y) ad8* × *ad16 biotin-requiring (bi)*, *y ad8* × *ad11 bi* and *y ad8* × *ad20 bi* from top to bottom respectively. Adenine-independent recombinants were selected and classified with respect to the outside markers *y* and *bi*. The figure shows the patterns of exchanges which will most simply account for the observed genotypes among the recombinants. Estimated map distances between the markers are shown in terms of per cent recombinants.

the *ad-8* sites. A similar unambiguous ordering of sites was shown by Siddiqi [492] within *paba* in Aspergillus.

Case and Giles [75] used the same principle to confirm the order of mutant sites within *pan-2* of Neurospora. The outside markers used in this case were *yellow (ylo)*, 3·6 map units to the left of *pan-2*, and *tryp-2*, 7·7 map units to the right. Some of their data are given in Table 20.

TABLE 20. Crosses indicating the order of *pan–2* mutant sites in relation to outside markers (data of Case and Giles[75]). Ascospores from the crosses were plated on sorbose medium containing tryptophan but devoid of pantothenic acid.

Cross	Frequency of pan+ prototrophs (%)	Number of pan+	Percent of pan+ in the four classes				Indicated order of sites
			ylo +	+ tryp	+ +	ylo tryp	
ylo pan⁹ + × + + pan¹⁰ tryp	0·09	270	22·0	32·3	26·1	19·4	}ylo − pan⁹ − pan¹⁰ − tryp
ylo pan¹⁰ + × + pan⁹ tryp	0·09	266	29·5	34·7	9·3	26·4	
ylo pan¹⁰ + × + + pan²⁵ tryp	0·03	66	21·6	19·6	33·3	25·7	}ylo − pan¹⁰ − pan²⁵ − tryp
ylo pan²⁵ + × + pan¹⁰ tryp	0·03	60	18·3	36·6	18·3	26·6	
ylo pan²⁵ + × + + pan²⁰ tryp	0·02	90	15·8	20·0	54·4	10·0	}ylo − pan²⁵ − pan²⁰ − tryp
ylo pan²⁰ + × + pan²⁵ tryp	0·04	120	8·0	27·0	13·0	52·0	
ylo pan²⁰ + × + + pan¹⁸ tryp	0·04	59	8·4	5·0	79·6	6·7	}ylo − pan²⁰ − pan¹⁸ − tryp
ylo pan¹⁸ + × + pan²⁰ tryp	0·04	76	3·1	5·2	10·5	71·0	
ylo pan¹⁸ + × + + pan²² tryp	0·12	187	12·0	10·0	74·2	3·0	}ylo − pan¹⁸ − pan²² − tryp
ylo pan²² + × + pan¹⁸ tryp	0·10	272	10·0	13·0	6·0	71·0	
ylo pan³⁰ + × + + pan²² tryp	0·03	112	7·7	6·4	29·8	55·8	}ylo − pan²² − pan³⁰ − tryp
ylo pan²² + × + pan³⁰ tryp	0·05	249	5·0	14·0	51·5	29·0	

Again we see that most pairs of *pan-2* sites can be placed in order with respect to the outside markers. In these data, however, the proportion of prototrophic recombinants which have to be explained as due to double or triple cross-overs is much higher than in the Aspergillus data. In several crosses the apparent doubles (parental combinations of outside markers) approached the apparent singles in frequency, while in the $pan^{10} \times pan^{25}$ cross the apparent triples (*i.e.* the 'wrong' outside marker recombinant genotype) were not convincingly less frequent than the presumed singles. On the basis of a hypothesis of reciprocal crossing-over the data indicate a much higher degree of negative interference (*i.e.* a higher mean number of exchanges per cluster) than do the Aspergillus data. This same tendency has been found over and over again in Neurospora. Indeed, the *pan-2* results are rather unusual in the nearness to which they approach the comparatively clear-cut Aspergillus situation. The only fine structure mapping work in Neurospora which lends support to a model of reciprocal crossing-over without extremely high negative interference involves recombination between different though closely linked genes (*e.g. ad-3A* and *ad-3B* [116]). Intra-genic recombination at some Neurospora loci has not given any clear and meaningful departure from random reassortment of outside markers in the selected recombinants [172, 396, 471].

All the results considered so far have been derived by the analysis of randomly mixed products of meiosis. One can find out much more about the mode of origin of intra-genic recombinants by analysing *tetrads* from intra-genic crosses. This is a laborious type of investigation because of the comparative rarity of intra-genic recombination, but it has now been undertaken on a large scale in several cases. The results show unequivocally that the hypothesis of reciprocal crossing-over with extreme clustering of exchanges is *not* an adequate one, in as much as the great majority of intra-genic recombinants arise through some kind of *non*-reciprocal process. The evidence is reviewed in the following section.

'GENE CONVERSION' OR NON-RECIPROCAL GENETIC EXCHANGE

In 1955 M. B. Mitchell [357, 358] published data which showed that *pdx⁺* recombinants from inter-crosses of Neurospora *pdx* mutants did not always arise by reciprocal crossing-over. The essential finding was that some asci containing a *pdx⁺* recombinant ascospore pair did *not* contain the expected reciprocal recombinant of double mutant type. This is to say, with respect to one of the sites of mutation, the ascus showed a 3:1 ratio as if the site had been converted non-reciprocally from mutant to wild type on one of the four strands at meiotic first prophase. Mitchell's data forced geneticists to take more seriously the phenomenon of 'gene conversion' which had long been claimed by Lindegren [e.g. 316] to be fairly frequent in yeasts. Her main results have since been repeated by several other workers on different Neurospora loci, and by Lissouba and Rizet in some remarkable studies of loci determining ascospore colour in *Ascobolus immersus* [325], which are discussed in more detail below. 'Gene conversion' might be explained away as spontaneous mutation were it not for the fact that it is several orders of magnitude more frequent than mutation of the same alleles in vegetative nuclei or in crosses between strains carrying the same mutation. Furthermore it seems to be a directed process; the mutant allele changed by 'conversion' seems always to be converted to the type of allele present in the homologous chromosome.

Recombination of the conversion type is not confined to Neurospora, nor does it appear to be restricted to certain genetic loci. The notable studies of Rizet, Lissouba *et al.* on *Ascobolus immersis* [325] show clearly that, in this species as well as in Neurospora, recombination of closely linked sites is more often than not a consequence of non-reciprocal conversion type events. A similar conclusion can be drawn from the data of Roman [458] and of Kakar [275] on the origin of prototrophs, apparently by mitotic recombination, during vegetative growth of heteroallelic diploids of yeast. Nor is the phenomenon of conversion confined to cases where one is selecting for intragenic recombination. It seems very probable from the work of Strickland on both Neurospora and Aspergillus [522, 520], and that of Kitani and his colleagues on Sordaria [284, 285] (see further discussion below), that analysis of a large number of tetrads will reveal occasional 3:1 ratios with respect to practically any genetic marker.

How can this violation of elementary genetic principles be explained? Most speculations have been based on the concept of 'copy-choice'. Let us suppose, as first proposed by Belling and later by Lederberg [297], that exchange between homologous chromosomes occurs by a process in which new genetic strands are laid down alongside old ones by a process of copying, with the possibility that copying may switch from one parental strand to its homologue where the two are closely paired. Let us further suppose, to simplify the argument, that chroma-

tids are single stranded. Then, if we imagine that the copying of two
new strands from two old ones is not exactly synchronized, it is easy to
imagine a situation in which a section of one strand is copied twice and
the corresponding section of the homologue not copied at all. This
would depend, of course, on each newly copied strand being released
from its model fairly promptly so as to make room for a second copying.
The idea is illustrated in Fig. 28. In order to explain why 3:1 ratios are
practically never found to affect more than one gene or part of a gene
one must suppose that there is a strong tendency for the 3:1 distribution

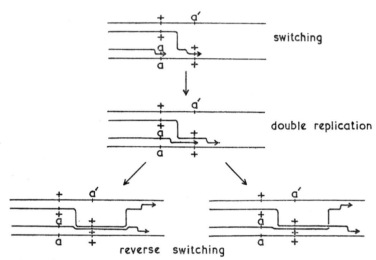

Fig. 28. Non-reciprocal switching as a consequence of non-synchronized copy-
choice.
    Note: Switching back (a) of same strand; (b) of sister strand. a and a′
represent mutant sites within a gene (see text).

of strands to be corrected within a very short distance, either by switching
back of the original strand, giving the effect of conversion without
crossing over, or by a compensating switch on the part of the other
new strand, giving the effect of a short region showing a 3:1 ratio with
reciprocal recombination of outside markers spanning the region. This
switch hypothesis was, in its essentials, first proposed by Freese [172].
    Crossing-over has, in the past, been envisaged as an exactly reciprocal
process. On this view, conversion would have to be a different kind of
process superimposed on crossing-over. The most pleasing aspect of
the switch hypothesis is that it makes it unnecessary to postulate two

different mechanisms of recombination. It suggests that every exchange involves a short region in which the strands are in a 3:1 ratio. Whether or not such a region is detected will depend on whether any markers fall within it. In ordinary chromosome mapping, with the markers relatively widely spaced, the short region of non-reciprocity will go undetected in the overwhelming majority of tetrads, but as the markers between which an exchange is being sought become very close together the region of non-reciprocity will become more and more likely to include one of them.

If these speculations have any substance it would be expected that a 3:1 segregation for any single marker would tend, perhaps in 50 per cent of cases taking Fig. 28 literally, to be associated with orthodox crossing over of markers spanning the one showing the aberrant ratio. Such an association has, in fact, been demonstrated by Stadler and Towe in Neurospora [514] and by Kitani *et al.* in *Sordaria fimicola* [284, 285] (see pp. 158–159).

Now let us see how the switch hypothesis fits some of the data on gene conversion and its association with crossing-over from some of the more detailed studies in Neurospora. Extensive ascus analysis, with outside markers, has been made of inter-allelic recombination at two loci. Case and Giles have published three sets of data on *pan-2* [75, 76, 78] while Stadler and Towe have carried out much the same sort of study on *cys*, both genes being in linkage group VI.

The most striking finding from Case and Giles' earlier investigations [75] was the same as that of Mitchell, namely that the appearance of a prototrophic recombinant spore pair in a cross between two allelic auxotrophic mutants was not, in most cases, accompanied by the appearance of a double mutant spore pair in the same ascus. In a minority of asci containing a *pan-2*+ recombinant spore pair these authors did find a spore pair doubly mutant within *pan-2*, and most of their later studies consisted of further genetic experiments on double *pan-2* mutants so obtained. Double mutants could be distinguished by their inability to give recombinants when crossed to *either* parental *pan-2* mutant. The data summarized in Fig. 29 are from a cross of a double *pan-2* mutant to a strain carrying a single *pan-2* mutation at a third site mapping between the other two. The three single mutants corresponding to these sites showed complementation in all combinations (cf. Fig. 41) and the three kinds of mutation could be distinguished singly or in combination by simple complementation tests. Among 1457 asci which were completely analyzed 11 showed 3:1

segregation with respect to one or more of the *pan-2* sites. The constitutions of these asci are indicated in Fig. 29. In Fig. 30 are shown the constitutions of 14 asci, out of 1651 dissected by Stadler and Towe [514], which were aberrant with respect to one or both of two mutant sites within the *cys* gene. These two sets of data will be discussed together since they lead to many of the same generalizations.

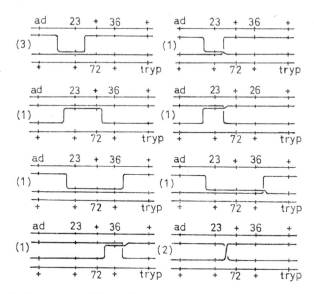

FIG. 29. Constitutions of 11 asci showing aberrant ratios of *pan-2* mutations out of 1457 asci from the cross *ad-1 pan-2*[23,36] *tryp-2*[+] × *ad-1*[+] *pan-2*[72] *tryp-2*. The numerals 23, 36 and 72 refer to three mutant sites within *pan-2*. The constitutions of the four meiotic products (spore pairs) in each case are indicated by strands switching as necessary between the two parental genotypes to give the result actually observed. This can be regarded as representing the actual mechanism of recombination or, alternatively, merely as a conventional representation of the results. The figures in parenthesis indicate the numbers of asci observed in each class. Two asci showing 5:3 ratios with respect to certain of the sites are omitted for the sake of simplicity. After Case and Giles [78].

The first point to note is that, though the asci show 3:1 ratios with respect to parts of the *pan-2* or *cys* gene, the outside markers show regular 2:2 segregations in every case. In other words, the segment affected by the conversion event is always a very short one, never extending to cover either of the outside markers. Secondly, with few exceptions, only one or two of the four strands emerging from meiosis have participated in the recombinational events; the other two strands tend to be of the parental types, at least so far as the chromosome segments covered by the markers are concerned. Both these features would be expected on the basis of the type of switch hypothesis outlined in Fig. 28. In Figs. 29 and 30 the tetrad types are represented in terms of the simplest pattern of switches which would lead to the observed result; this is convenient whether we regard it as indicating the mechanism of recombination or merely as a formal representation of segregation patterns. It should be noted that if an inexactly reciprocal double switch is regarded as one event there is no real indication of negative interference in these tetrad data; only two asci (in Fig. 30) require for their

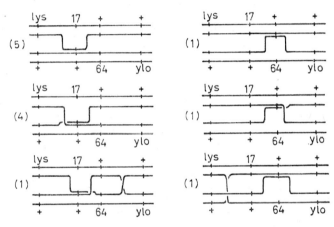

FIG. 30. Constitutions of 13 asci containing cys^+ recombinant spore pairs from 1651 asci analysed from the cross lys-5 cys^{17} $ylo^+ \times lys$-5^+ cys^{64} ylo. Representation of the data as in Fig. 29. One ascus showing a 5:3 ratio with respect to the cys^{64} site has been omitted for the sake of simplicity. After Stadler and Towe [514].

explanation a second reciprocal exchange between the gene under study and an outside marker, and this is scarcely more than might have been expected on the basis of the known map distances without any negative interference. Finally it may be pointed out that the data tell against any idea that conversion is a kind of directed mutation *with different sites mutating independently*. It is very clear from Figs. 29 and 30 that there is a strong tendency for two or even three sites within the same gene to 'mutate' coincidentally, and this is much more suggestive of a mechanism involving switching of strands than of mutation of sites. However, as will be argued later (pp. 164–170), some kinds of directed mutation theory are not ruled out.

Where does this leave our interpretation of the random spore data, for example those of Fig. 27 and Table 20? It is clear that the types of tetrad shown in Fig. 29 would be quite adequate to account for the Aspergillus data. The occurrence of a minority of tetrads showing gene conversion without recombination of outside markers will have the same consequences, so far as random spore data are concerned, as the reciprocal intragenic crossing-over with negative interference which Pritchard postulated. Many data from Neurospora, however, of which those of Table 20 can be taken as representative, are of somewhat greater complexity in that many interallelic crosses show high frequencies of the 'wrong' recombinant outside marker genotype among the selected prototrophs. These must, adopting the diagrammatic convention of Figs. 29 and 30, be represented as arising from a *triple* switch of one strand with a compensating triple or single switch of the other. In some sets of data [134, 396, 471] recombinants of this sort are almost or quite as common as those attributable to single switches. This implies a tendency for several switches of the same strand to occur in a short interval, which brings us back to something like the idea of negative interference which we first considered, with the difference that the clusters of events are non-reciprocal switches rather than reciprocal cross-overs.

POLARITY IN INTRAGENIC RECOMBINATION

Before any further discussion of possible mechanisms for intragenic recombination it is necessary to review evidence, which has been accumulating during the last few years, that the pattern of genetic exchange depends on the precise relative positions of the marker mutations. In most crosses between allelic mutants, as we have seen, most wild type recombinants appear to arise by conversion at one or other of the two mutant sites. It might be expected that the chance of conversion occurring at any mutant site would depend only on the intrinsic nature of that site and not on its position. A number of instances are known, however, in which a linear sequence of mutant sites within a gene shows a gradation of conversion frequency from one end to the other. Effects of this kind have been shown, in different fungi, by both tetrad and random spore data. Tetrad data in Neurospora are not sufficiently numerous to demonstrate the point, which is, however, very strikingly illustrated by the very extensive ascus analysis carried out in the Discomycete species *Ascobolus immersus* by Rizet, Lissouba and their colleagues in France [324, 325, 326, 452].

Ascobolus immersus normally has dark brown ascospores, but patient searches among large populations of spores has resulted in the discovery of a large number of mutants with pale spores. Crossing any one of these mutants to wild type gives asci which, with very rare exceptions, show 4:4 segregation of dark versus light spore colour. Crossing of the various mutants with each other has led to their classification into a number of series, with members of the same series showing close linkage, and members of different series showing no linkage or only loose linkage when crossed together. Presumably each series corresponds to a gene (or cistron in the qualified sense of Chapter 6) or perhaps in some cases to a small number of functionally related genes, but this is not really known because of the difficulty of carrying out complementation tests with mutants which affect only the colour of the normally haploid and homocaryotic ascospore. The important results of Rizet *et al* concern the relationships between mutants within several different series.

When two independently isolated mutants of the same series are intercrossed, the overwhelming majority of the asci formed have eight pale spores. A few, however, generally show two dark spores, which when isolated and tested prove to be genetically wild type, and six pale ones. The analysis of large numbers of asci is greatly facilitated by making use of the fact that the ascospores are discharged from ripe asci in coherent groups of eight. If discharged spores are collected on some suitable surface, the groups of eight can be observed, and, if necessary, isolated, without the necessity of ascus dissection. On the basis of our knowledge of the results of inter-allelic crosses in Neurospora, discussed above, we would expect these 6:2 asci to be the result either of apparently reciprocal crossing-over between different closely linked mutant sites, or of 'conversion' affecting one of the sites. In fact in the Ascobolus data as a whole both kinds of 6:2 asci were found. The 'cross-over' asci each contained, among their six pale spores, one pair of one parental mutant type, one pair of the other parental mutant type, and one pair carrying both mutations. The double mutants were distinguished by their failure to give any 6:2 asci when crossed to either parental type. The 'conversion' asci contained no double mutant spore pair, but, instead, two spore pairs of one of the parental mutant types. The parental type which occurred twice in a 'conversion' ascus was referred to as the *majority parent*, and the other as the *minority parent*.

Perhaps the best example of the type of result obtained is provided by the ascospore mutant series '46'. Within this series the mutants can, with very little ambiguity, be arranged in a linear order on the basis of the frequencies of 6:2 asci given by inter-crosses. The resulting map is shown in the upper part of Fig. 31, and is no different in principle from the map of *pan-2* in Neurospora, discussed above (Fig. 26), and many other cases in Neurospora, Aspergillus and yeast. In series 46 *all* the 6:2 asci (leaving out of account mutant 277 for the moment) are of the conversion type and never contain a double mutant spore pair; in this respect the data resemble those relating to the *cys* locus of Neurospora [514]. A more novel feature was that in the 6:2 asci from any one cross the majority parent is always the same. The mutants of the series can (again leaving aside 277) be arranged in a linear order such that, in any inter-mutant cross, the majority parent is the one to the left. The order based on the majority-minority relationships in 6:2 asci is shown in the lower part of Fig. 31, and is the same as that based on the frequencies of wild type recombinant spores. The mutant 277 conflicts with this otherwise remarkably consistent pattern in that it, alone out of all the mutants of the series, gives a proportion of 6:2 asci apparently due to reciprocal crossing-over in crosses with the other mutants. The explanation given to the very striking agreement between the two methods of ordering the mutants, and to the exceptional behaviour of mutant 277 is considered below. For reasons to be explained, 277 is placed to the right of the other mutants in the map. A broadly similar situation was found in a second series of mutants in a different chromosome region.

These observations suggested a switch hypothesis modified by the addition of two simple postulates. The first of these is that replication is polarized, always proceeding from left to right, 'left' and 'right' being defined by the majority-minority relationships in 6:2 asci, as in Fig. 31. The second is that the points at which switching-back takes place are predetermined. It follows from the second of these postulates that the switch which gives the intragenic recombinant must always be the *first* one, except in the rather special case where the two mutant sites span a switch-back point. Reference to Fig. 28 will show that, if this model applies, and replication is from left to right, the mutant site to the right will be the one to show conversion, that is double replication of one homologue and no replication of the other. Actually, all the consequences will be the same if the direction of replication is considered to be reversed with the *first* switch predetermined as to position, but with a low probability, and the switch-back variable in position. Lissouba and Rizet have called a region of polarized replication, within which there is no reciprocal crossing-over, a *polaron*. In their view, the chromosome can be regarded as sub-divided into polarons connected by switch-back points or *linkers* (*structures de liaison*).

As well as accounting for the correlation between the map of mutant sites based on recombination frequencies and the majority-minority relationships of mutants in 6:2 asci, the polaron hypothesis predicts that mutant sites rather close together but spanning a linker will tend to show reversed polarity relative to that shown in the array of mutants as a whole. This is because most exchanges between such sites will be switch-

Fig. 31. Two ways of ordering the pale-ascospore *Ascobolus immersus* mutants of series 46. Figures in the map, represented as a double line, refer to mutant isolation number. Distances between sites, shown *above* the map, are given in terms of frequencies per 1,000 of 6 : 2 asci (6 mutant spores : two wild type). The arrows *below* the map indicate relationships of mutants within 6 : 2 asci. Single-ended arrows indicate conversion-type asci and point from the majority to the minority parent. Double-ended arrows indicate reciprocal cross-over asci. The numbers on the arrows are the numbers of asci analysed giving the indicated result. For further discussion, see text. From data of Lissouba *et al.*[325].

backs following a primary switch which occurred further to the left. This is the interpretation given by Rizet to the relationships of the mutants 137 and 277 (Fig. 31); placing 277 to the right of 137 is plausible if a linker is supposed to be situated between them. Actually, the primary reason for postulating a linker in this position is the fact, mentioned above, that 277 shows some reciprocal crossing-over with 137, and with mutants further to the left. If one postulates that switching back is restricted to specific points, the linkers, and this is the simplest explanation of the polaron effect, then the occurrence of reciprocal exchange must be taken as demonstrating the presence of a linker in the region concerned.

Attractive as it is, the polaron hypothesis is not free from difficulties even applied to the Ascobolus data. One would expect that, if all switches are reversed at the first available linker, the frequency of recombination between sites spanning a linker, even if they are physically quite close together, should not be less than the *total* frequency of switching in the whole polaron to the left. From this point of view the low frequency of recombination between mutants 137 and 277 (Fig. 31) is anomalous unless we assume that most switches within the series 46 polaron are not corrected at the first opportunity but at some subsequent linker. If chromosomes in general really do consist of polarons connected by linkers the latter should appear in fine structure maps as spaces, representing relatively high exchange frequencies, between clusters of sites.

What evidence is there that the polaron effect occurs in fungi other than Ascobolus? Neurospora random spore data contribute much that is relevant to this question. Some studies, such as those of Case and Giles on *pan-2*, have shown no polarity at all. None have shown the absolute polarity demonstrated by at least some of the Ascobolus data, but a relative polarity has been indicated in several cases. Murray's [370] data on *me-2* are worth considering in some detail.

Murray studied a series of mutants belonging to the *me-2* locus in linkage group IV, approximately 6 map units distal to *tryp-4* and 4 units proximal to *pan-1*. She was able to arrange the mutant sites in an approximate linear order by reference to prototroph frequencies from inter-mutant crosses, and also by reference to the behaviour of the outside markers (cf. Table 20). The mutants tended to fall into four clusters, called α, β, γ and δ, and while the order of sites within each cluster was none too certain the order of the four clusters was not in doubt. Crosses were made between many pairs of *me-2* mutants differing at both outside marker loci, and in many cases the same two mutants were crossed twice with the outside markers reversed in one cross relative to the other. For illustrative purposes the data relating to crosses involving one mutant representative of each cluster are given in Table 21A. One can see at once that the distributions of me-2^+ recombinants among the four outside marker classes show a consistent pattern. In every case one recombinant outside marker genotype is more numerous than the other, the direction of the inequality being always consistent with the order of the mutant sites based on prototroph frequencies. Even more striking is the convincing and consistent inequalities of the two *parental* outside

TABLE 21A. Data showing polarized recombination from Murray [370]. The crosses involved four mutants at the me-2 locus of *Neurospora crassa*, K44, P2, K5 and P24, representative of the regions α, β, γ and δ of the me-2 gene map. The outside markers were *tryp-4*, to the left of α and *pan-1* to the right of δ.

	Number of me-2+ recombinants in outside marker class			
Cross	*tryp*+	+*pan*	*tryp pan*	+ +
tryp α+ $\times$ +β *pan*	26	**59**	16	**56**
tryp β+ $\times$ +α *pan*	**84**	23	**87**	15
tryp α+ $\times$ +γ *pan*	11	**50**	12	**41**
tryp γ+ $\times$ +α *pan*	**94**	17	**82**	18
tryp α+ $\times$ +δ *pan*	29	**139**	31	**53**
tryp δ+ $\times$ +α *pan*	**117**	17	**36**	22
tryp β+ +γ *pan*	55	**91**	51	**89**
tryp γ+ $\times$ +β *pan*	**114**	66	**83**	36
tryp β+ $\times$ +δ *pan*	37	**80**	30	**41**
tryp δ+ $\times$ +β *pan*	**53**	19	**18**	13
tryp γ+ $\times$ + δ *pan*	47	**116**	34	**66**
tryp δ+ + γ *pan*	**67**	20	**24**	13

N.b. For each cross the majority parental and recombinant outside marker classes among the me-2^+ recombinants are shown in bold type to emphasize the pattern of the results.

marker combinations. In each cross there is an appearance of strong linkage between the me-2 site to the left and the *tryp* marker, while recombination between the site on the right and the *pan* marker was much more frequent, approximating to free reassortment in several cases. One can express this situation by saying that, with one obligatory exchange between the *me-2* sites (to generate the *me-2+* recombinant), a second exchange involving the same strand is almost as likely as not to occur between the right-hand site and *pan* but is improbable between the left-hand site and *tryp*. Interpreting the appearance of an *me-2+* recombinant with a parental combination of outside markers as due to *conversion* of the *me-2* allele originally associated with these markers (and this seems justified in the light of tetrad data of the type shown in Figs. 29 and 30), one can say that the *me-2* site to the right (i.e. on the *pan* side) is always the more likely to be converted, that is, always the more likely to be the *minority* mutant type in tetrads.

An increasing number of analyses both in Neurospora and in other Ascomycetes seem to conform, more or less, to the pattern of Murray's results on the *me-2* locus. In Neurospora this is true both of Stadler and Towe's [514] random spore data on *cys* and of Smith's [500] very convincing and consistent results from a similar analysis of *his-5*. There are indications of a similar relative polarity in Pritchard's data on the *ad-8* locus of *Aspergillus nidulans*, while Siddiqi's [492] results with *paba* mutants of the same fungus show the effect very clearly.

It is clear that Rizet and Lissouba's polaron model will not, as it stands, accommodate the range of polarity effects, varying from strong (though incomplete) to barely detectable, which are found in Neurospora and other fungi. Indeed, subsequent studies in Rizet's own laboratory show that the original polaron concept was an oversimplification even for Ascobolus. Rossignol [464], while fully confirming the polaron effect as a strong general *tendency*, found a number of exceptions to the polaron principle (reversals of polarity, occasional reciprocal exchange within supposed polarons) from further crosses between mutants of series 46. Two attempts have been made to formulate a hypothesis of more general application. Stadler [514] suggested a *modified polaron* model in which, though many switches are reversed at the linker at the end of the polaron, some are reversed *within* the polaron only a short distance from the primary switch. On this hypothesis a proportion (though often a minority) of the exchanges between two mutant sites could be return switches following a primary switch to the left. To the extent that this happened it would give the effect of conversion of the site to the left rather than of the site to the right. It might also be suggested that a similar weakening of the polaron effect would be the result if replication was not 100 per cent polarized in one direction but sometimes proceeded in the opposite direction.

Quite a different hypothesis has been proposed by Murray [370] following a rather similar suggestion by Stahl [515]. This is the hypothesis of subdivision of the chromosomes into short *fixed pairing regions*. These regions are akin to the regions of close pairing postulated by Pritchard except that they are supposed to be predetermined segments with their ends more or less fixed. Within any such region clusters of exchanges are envisaged and, in order for the strands to emerge from each end in a 2:2 genetic ratio the number of switches within each region, counting both strands, must be even. If, in a cross between two closely linked mutants, the two mutational sites are within the same fixed pairing region, but asymmetrically placed with respect to the ends of the region with the left hand site (m') closer to the left end than the right hand site (m'') is to the right end, then we would expect the left hand site to show a higher conversion rate than the right hand one. This is because, given one switch between the two sites to generate a wild type strand, it is more likely that the second compensating switch will be to the right of m'' than to the left of m', simply because there is a longer interval to the right in which the second switch can occur.

Such a model as this does not predict a unidirectional polarity such as is suggested by many of the data so far available. Rather it would lead one to expect a relative polarity which becomes reversed somewhere in the middle of each fixed pairing region. A unidirectional polarity could arise as a special case when a series of mutants happens, as a whole, to be relatively very close to one end of a fixed pairing region.

Both the fixed pairing region and the modified polaron hypotheses make many of the same predictions and both are quite flexible since each contains a parameter about which one has no *a priori* expectation and which can be adjusted to fit the data in any particular case. In the fixed pairing region hypothesis the adjustable parameter is the position of the exchange being selected for in relation to the ends of the fixed pairing region. In the modified polaron model it is the proportion of switches which are reversed before the end of the polaron. The common feature of both models is the concept of fixed points or regions of the chromosome with special properties with respect to recombination. Neither hypothesis requires that this segmentation with regard to recombinational properties need bear any special relationship to the segmentation into functional genes. Siddiqi's [492] data are perhaps the only ones which seem to argue strongly against the fixed pairing region hypothesis. He found that in the *paba* gene of Aspergillus the tendency for the second exchange to be to the right rather than to the left seemed to be stronger for interallelic crosses involving mutant sites further towards the right end of the gene. This is the opposite of what the fixed pairing region idea would lead one to expect.

POST-MEIOTIC SEGREGATION

At this point we feel obliged to introduce yet a further complication which has to be taken into account in any attempt to explain the mechanism of genetic recombination at meiosis, and which has arisen particularly from the work of Olive, Kitani and El Ani on *Sordaria fimicola* [284, 285]. This fungus, like Ascobolus, has an ascospore pigment which is subject to dilution or modification as a result of mutation at any one of a number of different loci. Again like Ascobolus, ascospore colour is autonomously determined by the genotype of the spore itself, so that the segregation of alleles affecting spore colour can be observed directly in the ascus. One of the genes controlling spore colour, *gray*, is particularly suitable for study, since it has been mapped in a fairly detailed linkage group with other markers, affecting morphological characters, closely placed on each side of it [133].

In crosses of *gray* (*g*) × wild type, the overwhelming majority of crossed asci (the fungus is homothallic, so some perithecia result from selfing) show a regular mendelian 4:4 segregation for ascospore colour. About one ascus in 2,000 however, shows 6 *g* and 2 *g+*, or 2 *g* and 6 *g+*, the latter ratio being about five times more frequent than the former. These abnormal asci also tend to show recombination of markers spanning the *g* locus, though they often do not do so. Thus far the situation resembles gene conversion as found in Neurospora or Ascobolus, though the different frequencies of the two kinds of 6:2 asci is something which has not been apparent in other data [324].

The really unexpected anomaly in the Sordaria data is the occurrence, with a frequency roughly equalling that of the 6:2 asci, of asci showing 5:3 ratios. A 5:3 ratio means that a pair of spores which ought to be identical, since they derive their nuclei by mitosis of a single product of meiosis, show segregation with respect to the spore colour alleles. One cannot suppose that the anomalous ratio is due to one or more extra mitotic divisions in the ascus, with non-survival of some of the resulting nuclei, since three other segregating markers always showed perfectly regular 4:4 segregations in these asci. One has to conclude that chromosomes at the end of the second division are two-stranded, and that conversion can affect a single strand rather than both strands of a chromosome. Actually, the possibility of microscopically visible exchanges involving half-chromatids has been known in higher plants for some years [294], even though evidence for post-meiotic segregation has been totally lacking in higher organisms. Assuming that this conversion occurs during the first division of meiosis, and not during the second (though the latter is not ruled out), this implies that the four chromatids of each bivalent are each two-stranded. The occurrence of half-chromatid conversions, as we may call them, was shown by Kitani *et al.* to be highly correlated with crossing-over of outside markers; more so, in fact, than were the 6:2 asci, or whole chromatid conversions.

Observations of 5:3 segregations in asci have also been reported for *Ascobolus immersus* by Lissouba *et al.* [325] and for Neurospora by several authors [78, 514]. In these fungi exchanges or conversions involving half chromatids seem infrequent in comparison with the whole chromatid events. Case and Giles [78] also found that some 5:3 ratios in Neurospora were due to one of the eight spores being a pseudowild type (cf. p. 122). The origin of this sort of aneuploidy is obscure and may have nothing to do with the subject matter of this Chapter.

The relative frequencies of 6:2 and 5:3 segregations seem to depend on the particular marker under observation. Kitani *et al.* [285] found that one of their spore colour mutants in Sordaria gave about the usual low frequency of 6:2 asci when crossed to wild type, but no 5:3's at all. Somewhat similar observations were also made by Rizet and Lissouba on Ascobolus spore colour mutants.

DIFFICULTIES OF THEORIES INVOLVING COPY-CHOICE

Although the type of hypothesis involving copy-choice with inexactly reciprocal switching will, with suitable qualifications, explain many of the facts in a reasonably economical way, there are a variety of complications which are not accounted for by any of the forms of the hypothesis which we have considered so far.

In the first place, following Kitani's results on Sordaria we have to take into account recombinational events involving *half* chromatids. If these are to be explained on the basis of copy choice we have to postulate that one half of a chromatid can switch independently of the other half.

Secondly, we know from tetrad analysis that all four strands of a bivalent can be involved in genetic exchanges. It is thus impossible to suppose that two strands are old and two new, with the exchanges confined to the latter, unless we also postulate a recombination of new and old strands by some kind of breaking-rejoining process (sister-strand crossing-over). The fact that some data (see Table 13 of Chapter 4) do show some excess of two-strand double cross-overs could be taken as evidence in favour of a copy-choice mechanism, but if we accept this argument we must also suppose that in most examples sister exchanges are so frequent as virtually to randomize the strand relationships between successive exchanges. If we are going to admit so high a frequency of sister-strand crossing-over, we may as well have non-sister strand crossing-over occurring by the same process, and copy-choice switches are redundant so far as accounting for crossing-over is concerned.

Thirdly, there is considerable evidence, well summarized by Taylor [544], that in higher organisms DNA replication is completed well before pachytene, when the homologous chromosomes are visibly paired, and no further DNA synthesis occurs until after meiosis. Pritchard [428, 429] has countered this argument with the suggestion that homologues might be paired in short regions well before pachytene without such pairing being apparent in the obscurity of the leptotene or pre-leptotene nucleus. Postulating such limited short regions of pairing serves the second purpose of accounting for the phenomenon of negative interference, that is, a tendency for exchanges to be clustered together, with relatively long regions between the clusters with no exchanges. On this view, pachytene pairing is irrelevant for genetic exchange, and diplotene chiasmata are merely the delayed visible manifestations of exchanges which actually occurred much earlier. In most organisms chiasmata already present at pachytene could hardly be distinguished from points where one chromosome was passing over the other because of the normal slight relational coiling. If determinations of the time of DNA synthesis during or preceding meiosis in, for example, an Ascomycete could be made, critical evidence against copy-choice switching during DNA synthesis might be obtained, since homologous chromosomes are not even present in the same nucleus until immediately before meiosis begins.

A final objection to the simple type of switch theory is that it implies a *conservative* type of chromosome replication. That is to say, a mode of replication in which the original chromosomes remain intact and completely new chromosomes are copied from them. Taylor *et al.* [546] have obtained evidence that in mitosis in bean roots, at any rate, this is not the case. By arranging for chromosomal DNA to be isotopically labelled with tritiated thymidine and then allowing the chromosomes to divide in a non-labelled medium, he was able to show that the original chromosomal material was not conserved in the original strand, but was instead equally divided between two daughter chromatids. At the following division, however, labelled DNA was segregated from non-labelled. This result is only interpretable on the basis that each apparently undivided chromosome or chromatid is bipartite, one part being derived from the parental chromosome, and the other newly-synthesized. At the next division the two parts are separated, one going into each daughter chromatid. The scheme bears, in fact, a remarkable resemblance to the mode of replication of DNA molecules in *Escherichia coli* demonstrated by Meselson and Stahl [351] (see Appendix I), If both higher plant chromosomes and DNA molecules show semi-conservative replication it seems most unlikely that fungal chromosomes are replicated by a conservative mechanism. It should be mentioned, however, that there is not universal agreement on the interpretation of Taylor's results [293], and, from the biochemical side, Cavalieri [82] has long maintained, though against heavy odds, that the conserved unit shown by the Meselson-Stahl experiment is more likely to be a duplex molecule than a single strand. Some authors, for example Bernstein [32], attempt to save the copy-choice type of hypothesis by simply assuming that DNA molecules replicate conservatively, at least during meiosis.

The most serious attempt to construct a model for genetic exchange combining copy-choice with the most generally accepted theory of DNA replication has been made by Taylor [545]. As we have already mentioned, Taylor's experiments led to the conclusion that the chromosome replicates semi-conservatively, as if it were one enormous molecule of DNA. The great size and thickness of even a fungal chromosome, in comparison with that of a DNA double-helix, makes it seem quite unlikely that there could be only one DNA molecule per chromosome. To overcome this difficulty Freese [173] proposed that a chromosome consists of a large number of two-stranded DNA molecules placed end to end (though no doubt folded or coiled in quite a complex way) and joined by linkers of non-DNA material attached only to one strand at the end of each molecule. In Appendix I it is explained how a polynucleotide chain is polarized, with the polarities of the two members of a DNA duplex molecule running in opposite directions. The linkers are supposed to be joined only to one pole of each chain. When the chromosome replicates, it is supposed that the strands of each DNA molecule unwind, and that the linkers separate into two parallel linear arrays, alternate linkers going together and disjoining from adjacent linkers.

Then, if a new partner strand is formed alongside each separated strand, and a new linker is formed to connect to one end of each of the newly-formed strands, we have two chromosomes, each one deriving half of its material from the parent chromosome. The idea is illustrated in Fig. 32.

Taylor's model for conversion and associated crossing-over depends on copy choice switching of the newly-forming strands of paired homologous chromosomes. Since, on the basis of the model, these strands are equivalent to half chromatids, such a process might be expected to give half chromatid conversions (Fig. 32 B, D, E). Such a phenomenon had not been observed at the time of Taylor's original proposal, and this fact detracted from the plausibility of the whole scheme. The subsequent observation of such events by Kitani *et al* [284, 285] has provided unexpected support for the model. To explain why conversion usually appeared to affect whole chromatids Taylor was forced to postulate that when one strand switched, the pre-existing complementary strand on which it had been replicating almost always broke at the switch point. Following this break, the complementary strand was supposed to recommence growth from the break, replicating on the switched strand. This reversal of roles, the copy being copied by an extension of the original model, would result in a section of DNA both strands of which were newly formed. The other portion of the broken strand was supposed to be joined by the newly-formed strand of the same polarity, whose own further growth would thus be stopped; this would result in a section of DNA composed entirely of parental material (Fig. 32 C). Thus the supposed events in the region showing the conversion effect amount to the conservative replication of a section of chromosome which, as a whole, is replicating semi-conservatively. This part of the hypothesis is somewhat complicated and unconvincing but it must be admitted that it seems unlikely that anything simpler would fit the known facts, given semi-conservative replication.

The region of double replication, whether it affects a half chromatid or a whole chromatid, is supposed to come to an end when the switched strand is called upon to make attachment to a linker. At this point the switched strand will be in competition with the strand which is legitimately copying from the same model. Depending on which of the two strands makes attachment to which of the two newly-formed linkers, crossing-over may occur (Fig. 32 D, F) or not (Fig. 32 E. G). In any event, the attachment of the two strands to different linkers associated with different chromatids will put an end to the situation of both copying

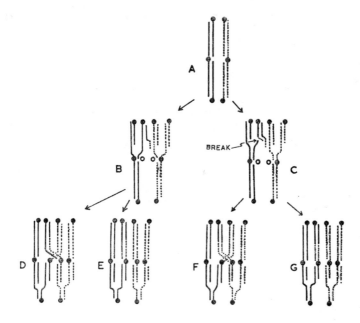

FIG. 32. A model for whole chromatid and half-chromatid conversions correlated with crossing-over, based on semi-conservative replication of chromosomes (after Taylor[545]).

A: Short section of paired homologues before replication. Each chromosome consists of DNA double-helices (here shown as pairs of parallel lines for diagrammatic simplicity). The two strands of each double-helix are of opposite polarity, and one end of each is attached to a linker of non-DNA material, here shown as a solid circle. *B* and *C*: Double-helices begin to unwind and each separated strand has a new complementary strand of opposite polarity laid down alongside it beginning at the top of the diagram; here new strands are shown to the right of the old ones. The second new strand from the left switches and begins to replicate from the homologous chromosome. In *C*, but not in *B*, the old strand *from which* the switch has been made breaks at the switch point. The upper broken end recommences growth, copying from the switched strand, while the lower broken end is joined by the new strand of similar polarity which was being formed to the left. In both *B* and *C* it may be a matter of chance which of the two strands replicating from the same model attaches to the new linker formed in association with the right hand chromosome. In *D* and *F* the switched strand makes this attachment leaving the other strand no option but to cross-over to the new linker of the left hand chromosome. In *E* and *G* the switched strand reverts to the chromosome with which it was originally associated. The four alternative results are: *D*, half chromatid conversion with crossing-over; *E*, half chromatid conversion without crossing-over; *F*, whole chromatid conversion with crossing-over; *G*, whole chromatid conversion without crossing-over.

from the same parental strand. Figure 32 is an attempt to explain the whole process in diagrammatic form.

Since the scheme was proposed the idea of predetermined points at which 6:2 ratios of strands revert to the regular 4:4 ratio has become additionally attractive as a means of explaining the polaron effect of Rizet, Lissouba et al., though additional assumptions would have to be made to accommodate the incomplete polarity revealed by some of the Neurospora data. A serious objection is that, if the chromatid is really constructed according to the model, the occurrence of three-strand double cross-overs is not accounted for, since switching between chromatids will be confined to two mutually exclusive pairs by the polarities of the newly forming DNA strands. Thus, if cross-overs are exclusively the consequence of switching on the Taylor model, only two- and four-strand doubles will be possible. The way out of this difficulty would seem to be at postulate that strictly reciprocal exchanges can occur at linkers between any two chromatids; such exchanges would not need to be associated with DNA replication. In fact there is, as we saw in Chapter 4, room for a hypothesis that will permit a moderate excess of two-strand relationships between successive cross-overs.

MODELS INVOLVING LOCAL REPAIR WITHOUT NET DNA SYNTHESIS

In view of the many complications which become necessary in any of the various kinds of switch hypothesis, it is not surprising that some workers have sought quite a different kind of explanation of genetic recombination and gene conversion. Two authors [234, 236, 577] have recently proposed mechanisms in which short regions of 'hybrid' DNA can be formed during chromosome pairing by unwinding of single homologous DNA chains from different chromatids and their rewinding on each other. Crossing-over and gene conversion then follow as processes restoring normal DNA structure. Both models assume that the chromatid is basically a sequence of two-stranded DNA molecules arranged end to end, and both involve discontinuities in the linear structure at which free ends of DNA chains can become unwound from their partners. Somewhat simplified illustrations of the two models are given in Figs. 33 and 34.

Both models involve unwinding a strand from each chromatid and its rewinding with a complementary strand derived from the other. Holliday [234,236] supposes that this latter strand is an original strand of the homologous chromatid, while Whitehouse [577, 578] suggests that it is a newly synthesized strand. As a corollary of this difference between the two hypotheses, Holliday's scheme states that the two strands which unwind from their respective chromatids are of like polarity while Whitehouse's supposes that they are of opposite polarity. Following the formation of the 'hybrid' DNA, different mechanisms are invoked by the two authors for restoring the separateness of the two chromatids. Holliday suggests that the half-chromatid exchanges are resolved by breakage and reunion which could occur in two ways, either leading to crossing-over of outside markers or not. In Whitehouse's scheme there has been additional DNA synthesis, and it has to be supposed that the supernumerary single stranded DNA which is left over after hybrid formation is broken down enzymically, the free ends then being joined to restore a regular two-stranded structure. This model in its simplest form will *always* give crossing-over as a result of hybrid DNA formation (it was in fact, originally conceived as a model for crossing-over rather than for gene conversion), while in Holliday's model crossing-over will only follow at about half the loci at which hybrid DNA is formed. These points should be made clearer by reference to Figs. 33 and 34.

The novel feature of both models is that they invoke actual conversion, in the sense of removal and replacement, rather than double replication by miscopying. It is supposed that, in hybrid DNA, mismatched base pairs not capable of mutual attachment by hydrogen bonds tend to be removed by enzyme action and replaced by other bases. If the replacement is such as to restore complementarity of bases the new structure will be a stable one. The most likely way in which a mismatched base pair could be corrected would be through replacement of just one member of the pair, though if replacement were at random both might well be replaced before a properly matching combination was hit upon. The result of this latter course of events would presumably be a new kind of mutant rather than conversion of one kind of parental site to the other, but in most experimental systems this result would probably not be recognized even if it occurred. Though the assumption that mis-matched base pairs can be corrected may seem a somewhat arbitrary one it gains some plausibility from the results of Setlow [485] and Howard-Flanders *et al.* [49] who showed that thymine dimers formed as a result of ultraviolet irradiation could be metabolically removed from *Escherichia coli* DNA and replaced by normal thymine. The repair mechanism at work in the *E. coli* system seemed to involve the excision and replacement of short sequences of nucleotides rather than

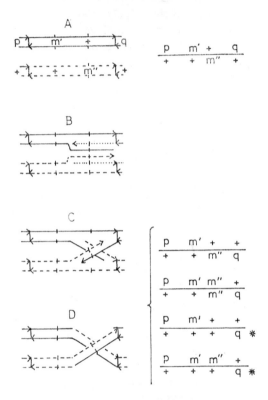

FIG. 33. Whitehouse and Hasting's scheme for crossing-over and gene conversion through hybrid DNA formation [578]. Two of the four chromatids present at first prophase of meiosis are shown; each is supposed to consist of a series of double stranded DNA molecules arranged end to end. Short vertical lines connecting the two strands represent 'linkers', and the opposite polarities of the strands are indicated by arrow heads. The stages are as follows: A. Before recombination. Parental-type chromatids are shown as carrying different mutations m' and m'' within a gene and also as having different outside marker constitutions, pq and p^+q^+. B. Detachment of the ends of homologous DNA chains from their respective linkers, with subsequent unwinding and new synthesis of DNA (indicated by dotted lines) to restore the duplex structure. C. Unwinding of newly synthesized DNA chains and their rewinding with the homologous strands from the opposite chromatids which were unwound at stage B. D. Breakdown of the DNA chains left unpaired at stage C and rejoining of free ends. In this example the hybrid DNA is supposed to have included site m'' but not site m'. Conversion at the hybrid site can occur from m'' to + or from + to m'' in each of the two chromatids, giving the four possibilities shown in the lower right of the Figure. Asterisks indicate chromatids which are now wild type with respect to gene m. This is only one of the possible sequences of events discussed by Whitehouse and Hastings. To

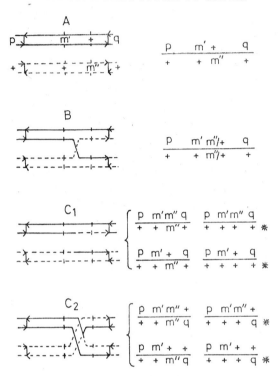

FIG. 34. Holliday's scheme for gene conversion, with or without crossing-over of outside markers [236]. The diagrammatic conventions are the same as in Fig. 33. No new DNA synthesis is involved in this scheme; the DNA chains unwound in B are supposed to rewind on the already existing chains of the opposite chromatids. The half-chromatid chiasma formed in B is supposed to be resolved by breakage and rejoining of DNA chains in either of the two ways shown in C1 and C2. The eight possibilities generated by gene conversion in hybrid DNA are shown to the right of C1 and C2; asterisks indicate chromatids which have become wild type with respect to gene m. The hybrid DNA region can also be envisaged as covering both m' and m''. For further discussion see text.

explain gene conversion in m, without crossing over of outside markers, hybrid DNA formation is supposed sometimes to occur simultaneously on both sides of a linker giving, in effect, a two strand double cross-over within a very short interval. Hybrid DNA covering both m' and m'', or occurring in one chromatid and not the other, can also be considered, and different possibilities are generated in each case.

of single bases, but this could be true of fungal gene conversion too without necessitating much alteration of the models under consideration.

There is little point in choosing between the Whitehouse and Holliday schemes in the present state of our knowledge. Both should be regarded rather as examples of *possible* mechanisms than as unique solutions, and both are obviously subject to considerable modification in detail. Both contain enough arbitrarily variable parameters to make critical testing of all their details impossible. It seems, however, worth while to consider some of the predictions made by both models and comparing them with those made by the various kinds of copy choice hypothesis considered earlier in this Chapter.

Both schemes involve discontinuities in the genetic material, and both predict that conversion frequencies should be related to the positions of the genetic markers in relation to these discontinuities. To this extent the repair models resemble the polaron and fixed pairing region hypotheses, but they differ sharply from the latter in predicting that conversion should be more frequent close to a discontinuity rather than less frequent. The available data are not sufficiently precise to discriminate between these apparently contrasting predictions; the general indications of relative polarity which have been so far obtained seem as easily explained on the one type of hypothesis as on the other.

One important advantage which the repair models have over all copy-choice models is that they predict that crossing-over should occur equally well between any two of the four chromatids during the first prophase of meiosis. Thus they do not require any additional assumptions to explain three and four-strand double cross-overs, though there would still have to be some reason, perhaps relatively trivial, why 2-, 3- and 4-strand doubles do not occur in exactly a 1:2:1 ratio (cf. pp. 90–91). Another advantage shared by both models is that they explain half-chromatid conversions (5:3 ratios) in quite a natural way. It need not be supposed that the mechanism for the correction of mismatched base pairs is 100 per cent efficient, and when the mismatching is corrected in one chromatid and not in the other the result will be a 5:3 ratio. Differences, from one mutant site to another in the same organism, in the relative frequencies of 6:2 and 5:3 ratios, and differences between the frequencies of conversion from wild to mutant and from mutant to wild at a given site (cf. pp. 158–159), seem quite consistent with a repair mechanism which discriminates to some extent between different bases, but are very hard to interpret on the basis of copy-choice.

The weakest point in both the Whitehouse and the Holliday models is that, without further assumptions, they provide no basis for the frequency of intragenic recombination (in so far as it is due to conversion) being related to the distance between sites. Holliday, recognizing this difficulty, suggests that the chance of formation of hybrid DNA, and hence of gene conversion, is drastically decreased when the strands concerned have two or more genetic differences (i.e. mismatching base pairs) which are relatively close together. This would account for low recombination frequencies between closely spaced sites, but there is no apparent reason why this effect should operate in such a way as to lead to a linear map with *additive* recombination frequencies. It is, however, part of Holliday's case that strict additivity in intragenic mapping is not, in fact, generally found. He assembles evidence that a more general rule is a systematic deviation from additivity, with closely spaced sites showing too little recombination relative to widely spaced ones. At present it is true that the additivity shown in most gene maps is very far from perfect, but more precise data are needed to assess the adequacy of Holliday's explanation.

We saw during our discussion of the Neurospora tetrad data (Figs. 29 and 30) that different sites in the same gene tend to show coincident genetic conversion. This was readily explained by some form of switch hypothesis. It is also compatible with either of the repair models, since a region of hybrid DNA might well cover several sites within a gene. Where there are three markers within the gene, as in Case and Giles study (Fig. 29), all three sites might be converted simultaneously if all were included within a region of hybrid DNA. In such a case, however, there is no reason in any model involving independent conversion of single bases why all three should be converted in the same direction. If we represent the hybrid region including the three sites as $\dfrac{a \quad b \quad c}{+ \ + \ +}$, conversion giving the result $\dfrac{+ \ + \ +}{+ \ + \ +}$ should be no more likely than conversion to $\dfrac{+ \ b \ +}{+ \ b \ +}$. On any switch or copy-choice type of hypothesis the latter result would require one switch in each inter-site interval, and this might often be very much less likely than the former result, which would require no special positions for the switches provided that they spanned the marked region. Case and Giles' data, in fact, show no instance of the middle marker being converted independently of the two outside markers, while they do contain two asci with all three sites converted in the same direction. These numbers are not large enough to be meaningful, but further data along the same lines should serve to distinguish between conversion of individual sites and conversion of extended segments of genetic material. The latter alternative would, however, still be compatible either with copy-choice or with removal-and-replacement.

CHROMOSOME STRUCTURE IN RELATION TO RECOMBINATION THEORY

All the ideas about recombination which have been discussed in this Chapter have been based on the assumption that the undivided chromosome or chromatid consists essentially of one double-stranded DNA molecule or a single file of such molecules. Almost the only reason for making this assumption is that it fits the genetic evidence, namely that half-chromatids, or even whole chromatids (cf. p. 74) mutate as units, and that half chromatids or whole chromatids seem to be the minimum units which can participate independently in genetic recombination. This last piece of evidence may be less strong than it appears, since the fact that most fungi do not have more than eight spores in their meiotic tetrads means that genetic exchange involving less than half a chromatid would usually lead to genetic segregation subsequent to spore delimitation, which is something which is not normally looked for. From this point of view more extensive studies on a 16-spored Ascomycete such as Chromocrea might be fruitful.

The only evidence other than from genetics for only one DNA duplex per chromatid is that of Taylor [545] from higher plants, indicating semi-conservative replication of chromosomes. La Cour and Pelc [293], on the basis of similar experiments, concluded that a multistranded chromosome with conservative replication of individual strands would fit the facts better. The cytological evidence that chromosomes of higher organisms are multistranded is quite strong [449, 517], and it is certainly easier to think that the relatively great thickness of a chromosome is accounted for by a number of parallel DNA molecules rather than by an enormously complex folding of a single DNA molecule, as supposed by Freese [173]. Practically nothing is known of the structure of fungal chromosomes which are, of course, very much smaller. How a multiplicity of identical DNA molecules in a chromosome could be consistent with its behaving as a genetic unit is not at all clear, though Holliday's idea [234], that genetic uniformity of the DNA molecules in one chromatid would tend to be maintained by chain reactions of gene conversions, is attractive in some ways. Until we know more in detail of how chromosomes are constructed theories purporting to explain genetic recombination in molecular terms are unlikely to carry total conviction.

THE RESOLVING POWER OF GENETIC ANALYSIS

Throughout this Chapter we have assumed that the genetic material consists of DNA (see Appendix I) and that genetic exchange, at least within genes, involves changes within DNA molecules rather than reassortment of whole molecules. There is, in fact fairly convincing evidence that the minimum units of mutation and recombination must be some orders of magnitude smaller than whole DNA molecules.

TABLE 21B. DNA content of fungal nuclei

Species	Ploidy	DNA per nucleus $\times 10^{14}$ (grams)	Number of base pairs $\times 10^{-7}$	Number of units of M.W. 10^7 (thousands)	Reference
Neurospora crassa	n	4·6	4·3	2·8	Horowitz & McLeod[243]
Aspergillus nidulans	n	4·4	4·1	2·7	Roper *et al* [222, 422]
	2n	9·0			
Aspergillus sojae	n	7·3	6·8	4·4	Ishitani *et al* [251]
	2n	14·6			
Ustilago maydis	n	5·7	5·3	3·4	Esposito & Holliday [130]
	2n	11·5			
Saccharomyces	n	2·4	2·2	1·4	Ogur *et al* [385]
	2n	4·8			
	3n	6·5			
	4n	9·9			

Note: Calculations of numbers of molecules or base pairs made on the basis of $6·02 \times 10^{23}$ molecules per gram molecule (Avogadro number); DNA has a molecular weight of 650 per base pair.

Table 21B shows the available information on the DNA content of the nuclei of several fungal species. It will be seen that, in species where a series of strains of differing ploidy is available, the DNA content is proportional to the number of chromosome sets. In the Table the amounts of DNA are shown in terms of numbers of nucleotide base pairs per nucleus, and also of numbers of units of 10 million molecular weight. This molecular weight is roughly what is found in preparations of genetically transforming DNA from bacteria, and the size of the molecules in intact cells may well be higher; it has been shown that very careful isolation of DNA from bacteriophage T4 can yield molecules of about 100 million molecular weight [64, 26, 68]. In *Neurospora crassa*, if one assumes a total genetic map length of 1,000 units (500 units have been mapped so far, see Fig. 21), there are fewer than three DNA molecules of size 10 million per map unit. If all the DNA is not genetic, and the rather wide variations in DNA content between otherwise apparently comparable organisms suggests that some DNA may be genetically redundant, then there will be even fewer molecules per map unit.

It is quite clear, at least in Aspergillus and Neurospora, that the genetic material is much more finely subdivisible than would be possible if DNA molecules were inherited as indivisible units. This conclusion leads us to the question of whether there is *any* limit to the subdivisibility of a DNA molecule in genetic recombination, other than that imposed by the indivisibility of a single nucleotide base. Pontecorvo and Roper [422] have made a calculation leading to the conclusion that the closest linkage observed in *Aspergillus nidulans* corresponds to a separation by no more than about eight nucleotide base pairs. The calculation assumes that the linear sequences of sites *within* genes are an integral part of the larger sequence comprising the whole linkage group. As we have seen, this seems likely; the evidence from the use of outside markers tells against the possibility of genes being attached as side-chains. It also assumes, more dubiously, that the probability of recombination per unit length of genetic material is the same throughout the genome, and the same within genes as between genes. The relevant evidence, though very meagre, tends to support this last assumption, there being no indication that genes are separated from each other by regions where exchanges occur especially freely. For example, Pritchard's [427] data for *Aspergillus nidulans* shows some *ad-8* sites as being more closely linked to *yellow*, clearly a distinct gene, than to other *ad-8* sites.

Table 22 shows similar calculations for some of the more thoroughly studied genes in *Neurospora crassa*. It will be seen that the result is quite similar to that obtained for Aspergillus. The wide variation between the lengths of genetic map occupied by different genes, especially the difference between *am* and *pan-2*, is very striking. It is difficult to believe that these differences represent proportionate differences in the DNA content of the genes concerned. It is possible that the known sites within *am* all fall, for some reason, within the same small part of the gene, but it is also possible that *pan-2*, which seems to be a quite unusually 'long' gene, may be in a chromosome region with a relatively high recombination frequency. If the latter possibility is actually the case, the calculation on *pan-2* may give too high an estimate of the minimum number of base pairs separating recombinable sites. One should also bear in mind that the tendency, observed in fine structure mapping, towards clusters of recombination events each counting as a single exchange or even being

TABLE 22. Calculated chemical dimensions of some *Neurospora crassa* genes

Gene	Map distance* between		Calculated number base pairs† between		Reference
	Ends of gene map	Closest sites	Ends of gene map	Closest sites	
pan-2	1·1	0·002	$4·7 \times 10^4$	86	Case & Giles[75]
me-2	0·10	0·004	$4·3 \times 10^3$	172	Murray[369]
pyr-3	0·034	0·001	$1·5 \times 10^3$	43	Suyama *et al*[533]
am	0·017	0·0002	$7·5 \times 10^2$	9	Pateman[395], and personal communication

Notes: * Double the percentage frequency of prototrophs from inter-mutant crosses, on the assumption that an equal number of double mutant recombinants are formed.

† On the basis of a total map length of 1,000, and $4·3 \times 10^7$ base pairs in each haploid nucleus (see Table 21).

unobservable in coarse mapping, will tend to make estimates of total map length too low in relation to frequencies of intra-genic recombination. This too will result in overestimation of the amount of DNA between closely linked sites. In spite of the several uncertainties in the calculations, they do suggest that recombination can occur at very many different points within a DNA molecule, and that it can very likely occur between any two adjacent base pairs.

CHAPTER 8

THE BIOCHEMICAL ANALYSIS OF
GENE FUNCTION

We have seen in the previous chapters that the two approaches of re-combinational and complementation analysis combine to give a picture of the gene as a complex linear structure, containing many different and separable sites capable of mutation, but acting as a single unit of function. We now turn to the analysis, in biochemical terms, of the functions of genes. Though it would be quite artificial to restrict discussion of this very general question to fungal examples, most of the currently important ideas can, in fact, be well illustrated by reference to *Neurospora crassa*, and the greater part of the present chapter will be concerned with this fungus.

THE BIOCHEMICAL BASIS OF AUXOTROPHY

Single versus multiple growth requirements

The work on *Neurospora crassa*, which was initiated by Beadle and Tatum in the early nineteen-forties, led almost at once to an important generalization. This was that the great majority of mutants unable to grow on minimal medium but able to grow on 'complete' medium each require only a single substance. On this basis it seemed highly probable that most mutations affected only a single metabolic pathway, and further evidence, which will be considered below, tended to show that only a single step in the pathway is generally 'blocked'. The obvious inference was that single gene mutations each affect the activity of a single enzyme. This 'one gene–one enzyme' hypothesis is, with certain refinements and qualifications, which will emerge from our further discussion, still accepted today.

Much effort was at one time devoted to arguments for or against the generality of the one gene–one enzyme hypothesis. The main question at issue was whether mutations could occur which affected more than one enzyme. There are quite numerous examples of mutants which have multiple growth requirements. Most of these, however, can be adequately explained as representing more or less complex end effects of single metabolic blocks. Some of these cases will be mentioned below. The most serious argument against the general truth of the one-to-one hypothesis is that auxotrophic mutants may be unrepresentative of mutations in general. If mutations with multiple (or pleiotropic) primary effects did occur they might well not be identifiable among auxotrophs, since they might lead to failure to grow on any available medium. We know that many kinds of mutations do have apparently irreparable effects and can only be perpetuated in heterocaryons, where they are supported by the presence of normal nuclei [10]. However, such irreparable effects could be due to single metabolic blocks in the syntheses of substances which are unable to penetrate the hyphae from the outside. The whole argument is inconclusive; one cannot be sure the relatively simple effects on metabolism observed in auxotrophic mutants are representative of the effects of mutation in general, but neither is there any clear evidence to the contrary. It is clear, in any case, that auxotrophic mutants represent one very important kind of effect of mutation.

It now seems that the most significant exceptions to the rule of one gene—one enzyme are to be found among mutants in which only one biosynthetic pathway is blocked. It is well established in bacteria, and suspected to be the case in fungi, that the formation of enzymes catalyzing sequential steps in the same pathway may be subject to a common control. Mutations acting on the control mechanism may affect the production of two or more enzymes simultaneously. This possibility is considered later in this Chapter (see pp. 204–205).

Information from auxotrophs on biosynthetic pathways

Having found that an auxotrophic mutant will grow if given a single supplementary substance in minimal medium, one can test all known or postulated metabolic precursors of this substance for growth-promoting activity. If the biosynthesis is blocked in the mutant only in a single step, and if all intermediates in the pathway are able to penetrate the mycelium, any intermediate occurring *after* the block should be able to substitute for the end-product, while any occurring *before* the block should be inactive in supporting growth. As good an example as any of this kind of reasoning is provided by studies on arginine auxotrophs of *N. crassa*, first described in the classical paper of Srb and Horowitz [509], and extended subsequently by other workers.

A number of different kinds of arginine auxotroph can be distinguished from each other on the basis of complementation tests, and their differing responses to amino acids metabolically related to arginine. The growth responses of the ten *arg* loci, which are clearly distinct as judged by their mutual complementation and different chromosomal locations, are summarized in Table 23; for the sake of completeness a mutant type responding specifically to proline is also included. Fig. 35 shows a scheme for the biosynthesis of arginine and of proline based partly on Table 23 and partly on other experimental findings, mentioned below.

Accumulation of intermediates and absence of enzymes as further evidence for the positions of metabolic blocks

On the basis of the kind of evidence shown in Table 23, tentative conclusions can be drawn as to the metabolic step affected by a mutation. However, arguments based on nutritional requirements alone are not always safe, as Newmeyer and Tatum [378] have shown. These authors found that mutation at the *nt* locus could give a specific growth requirement for nicotinic acid, or an alternative requirement for nicotinic acid or any one of a series of precursor compounds, depending on the state of other genetic loci. These 'modifier' loci perhaps determined, in some unknown way, the ease with which the various precursors could gain access from the growth medium to the nicotinic acid biosynthetic pathway; at all events, they made it difficult to determine the function of the *nt* gene by growth experiments.

Two other kinds of evidence can be used to confirm and extend conclusions based on growth experiments. Firstly, if a metabolic reaction really is inoperative in a mutant, the intermediate formed immediately before the block should often accumulate in the mycelium or the growth medium or both. Secondly, if an enzyme catalysing the reaction in ques-

TABLE 23. Growth responses of different classes of *N. crassa* auxotroph to arginine and related substances[508, 509, 553, 554, 100])

Genetic class	Glutamic acid	Proline	Glutamic γ-semi-aldehyde	Ornithine	Citrulline	Arginine
arg–1, arg–10	–	–	–	–	–	+
arg–2, arg–3 arg–12	–	–	–	–	+	+
arg–4, arg–5 arg–6, arg–7	–	–	–	+	+	+
arg–8, arg–9	–	+	+	+	+	+
prol–1	–	+	–	–	–	–

N.b. *arg-4* and *arg-7* are probably not distinct (Catcheside, personal communication).

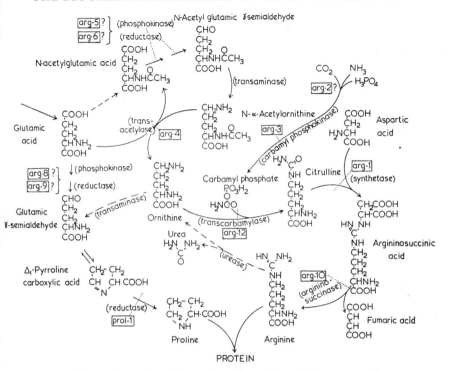

FIG. 35. The pathways of arginine and proline synthesis in Neurospora. The positions of the metabolic blocks in the various classes of mutants are based on the work of Yura [603] for *prol-1*, Davis [98] for *arg-3*, Vogel and Vogel [555] and De Deken [113] for *arg-4* (probably not distinct from *arg-7*), Fincham and Boylen [157] for *arg-10*, Newmeyer [377] for *arg-1*, and Davis and Thwaites, and Woodward and Schwartz [100, 594] for *arg-12*. The blocks assigned to *arg-5*, *6*, *8* and *9* are speculative, but seem likely to be in the positions shown. There is no obvious place in the scheme for a metabolic block in *arg-2*; this mutant may possibly affect the availability of carbon dioxide for carbamyl phosphate synthesis.

tion is present in the wild type but absent, or much lower in activity, in the mutant, one can be fairly confident that one has identified the site of the block correctly.

The work of Ames [3, 4, 5], Webber [565, 566] and Catcheside [80] on histidine-requiring mutants provides some excellent examples of these two approaches to the analysis of metabolic blocks. The system of reactions shown in Fig. 35 also affords some good examples. Two kinds of mutant, the *arg-1arg-10* and classes, grow if given arginine but do not respond to citrulline or any other related compound. Catcheside [81] has isolated of the order of a hundred arginine-specific auxotrophs in *N. crassa*, and all fall into one or other of these two groups; it seems unlikely that any other locus can mutate to give this specific growth requirement. The *arg-1* and *arg-10* loci are on different chromosomes (cf. Fig. 21) and the two mutant classes are strikingly different in their physiology. The difference is clearly seen in a paper chromatographic analysis of the free amino acids present in extracts of mycelium (see Fig. 36). Extracts of *arg-1* mycelium, grown in minimal medium, resemble wild type extract in containing glutamic acid, glutamine and alanine as the most conspicuous ninhydrin-reacting components. An extract of an *arg-10* mutant, on the other hand, shows an entirely new component which has a position on the chromatogram identical with that of argininosuccinic acid and which almost certainly is this compound. The accumulation of argininosuccinic acid in *arg-10* mycelium is greatly increased by the addition of citrulline to the growth medium; when citrulline is provided, citrulline itself and argininosuccinic acid can become the most prominent free amino acids in the mycelium, the more normal amino acids being depleted presumably through the utilization of their amino nitrogen for the citrulline to argininosuccinic acid conversion. Citrulline added to *arg-1* or to wild type cultures does not lead to any accumulation of argininosuccinate though citrulline itself is accumulated to some extent in *arg-1* mycelium. The double mutant type *arg-1 arg-10* accumulates no more than a trace of argininosuccinic acid but does accumulate citrulline (Fig. 36). These facts are quite consistent with the metabolic blocks in the two mutant types being as shown in Fig. 35, with *arg-1* mutants unable to convert citrulline to argininosuccinate and *arg-10* mutants unable to split the latter to give arginine. Enzyme studies on cell-free preparations confirm this hypothesis. *arg-10* mutants lack argininosuccinase, which is almost certainly the enzyme responsible for arginine formation in wild type and which is always demonstrable in wild type extracts [157]. *arg-1* mutants have abnormally low activity of the enzyme responsible for the condensation of citrulline and aspartate to give argininosuccinate [377].

A relatively complete analysis has also been made of the situation in the one known mutant of the *prol-1* class. The mutant is not stimulated by glutamic γ-semialdehyde, even though this substance will substitute for proline in supporting growth of *arg-8* and *arg-9* mutants (see Table 23). Wild type cells contain an enzyme, pyrroline-5-carboxylate reductase, which reduces the cyclized product of glutamic γ-semialdehyde to

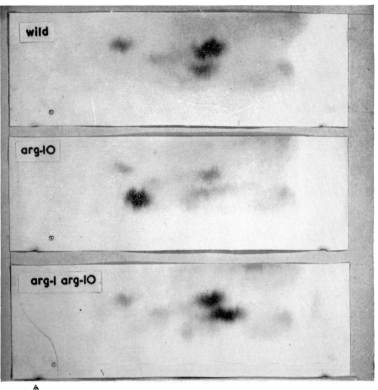

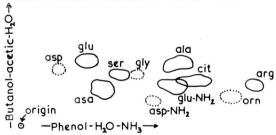

FIG. 36. Ninhydrin stained two-dimensional chromatograms of extracts of *Neurospora crassa* mycelia showing the accumulation of argininosuccinic acid and a little citrulline in an *arg-10* mutant, and the almost complete blockage of this accumulation, and the accumulation of citrulline instead, in an *arg-10 arg-1* double mutant (cf. Fig. 35). Mycelia were grown on minimal medium supplemented with 0·001 M L-arginine, which is a growth-limiting concentration for arginine-requiring mutants ; the accumulation of the precursors is much more pronounced under these conditions than when arginine is supplied in excess. Key: ala = alanine, asa = argininosuccinic acid, asp = aspartic acid, asp-NH₂ = asparagine, arg = arginine, cit = citrulline, glu = glutamic acid, glu-NH₂ = glutamine, gly = glycine, orn = ornithine, ser = serine.

[*facing p.* 178

proline in the presence of reduced triphosphopyridine nucleotide (TPNH).*

The activity of this enzyme is extremely low in the *prol-1* mutant. Yura [603] was able to show that *prol-1* produces an abnormal and catalytically inefficient variety of the reductase; this will be referred to again below. The fact that the mutant has a little of the enzyme activity is, no doubt, the explanation of its ability to grow slightly on minimal medium.

Several other cases could be cited in which a similar satisfying correlation exists between growth requirement, accumulation of an intermediate, and absence or reduced activity of an enzyme, and some of these will be mentioned later in this chapter. One might suppose that all kinds of auxotroph can be explained in this sort of way, and perhaps most of them eventually will be. However, it must be admitted that a proportion of Neurospora auxotrophs have so far resisted analysis, and may ultimately have to be accounted for in other ways. Some of these refractory cases are discussed below.

Apparent metabolic blockage without enzyme deficiency

From time to time auxotrophic mutants have been reported in which all the enzymes necessary for catalyzing the synthesis of the required metabolite are apparently present. For example Kakar and Wagner [276] found three classes of isoleucine-valine mutants in yeast which had fairly clear single enzyme deficiencies, but also a fourth class which seemed to have the complete enzymic equipment for synthesizing the required amino acids. Again, among the Neurospora arginine auxotrophs the *arg-2* class seem to produce all the enzymes which are *known* to be necessary for arginine synthesis (cf. Fig. 35).

Various explanations have been put forward to account for such anomalous situations. Firstly, it is conceivable that there is no real block in the biosynthesis at all; the required metabolite may be required in larger amounts than usual, and in larger amounts than the organism can make, in order to redress a metabolic unbalance having its origin in some other metabolic pathway. Possible examples of this kind of effect have been described by Emerson [143]. The isotopic tracer studies of Bonner *et al.* [44] showed that several kinds of mutant requiring nicotinic acid or a related substance for growth are, in fact, able to synthesize nicotinic acid from the constituents of minimal medium. It is certainly unsafe to assume that a growth requirement signifies a complete metabolic blockage.

* See footnote, p. 186

Secondly, it has been suggested in some instances that an enzyme, though present, is ineffective *in vivo* because of a faulty integration in the cell structure. Such an explanation has been most recently suggested by Wagner *et al.* [556] for two groups of Neurospora isoleucine-valine mutants (shown as *iv-1* and *iv-3* in Table 24) which lack, respectively, dehydrase and condensing enzyme activities in intact mycelium and in the mitochondrial fraction of cell-free preparations, but which show these activities in the soluble protein fraction. Wild type shows good activity of both these enzymes in both the mitochondrial and soluble fractions. Wagner's explanation is certainly a possible one, but it is also possible that the enzyme activities detected in the soluble protein fraction are due to proteins which are unrelated to the enzymes responsible for isoleucine-valine synthesis *in vivo*, and whose own *in vivo* functions are quite different. It may be not that the *iv-1* and *iv-3* mutations are affecting enzymes in a novel way but rather that the assays for the enzymes concerned are not sufficiently specific.

This brings us to a third type of explanation for the anomalous presence of an enzyme in a mutant which was expected to lack it. This is simply that attention is being focused on the wrong enzyme. The history of the study of the arginine pathway in Neurospora illustrates this point. Several investigators [151, 154, 554] have speculated about the significance of the presence of ornithine δ-transaminase in all the mutants blocked between ornithine and citrulline. It now seems extremely likely that this enzyme is not involved in ornithine synthesis at all, and that the immediate precursors of ornithine are acetylated compounds, as has long been known to be the case in *Escherichia coli* (cf. Fig. 35).

Multiple effects of simple metabolic lesions

Although, as we have just seen, the biochemistry of some auxotrophs is still obscure, others in which the situation seemed at first sight quite complex have been very successfully accounted for on the basis of primary effects on single known enzymes. A particularly good example is the *arom-1* type of mutant studied by Gross [208, 210]. *arom-1* mutants need a mixture of phenylalanine, tyrosine, tryptophan and *p*-aminobenzoic acid for growth, and are also peculiar in possessing two enzymes, dehydroshikimic acid dehydrase and protocatechuic acid oxidase, not detectable in extracts of wild type grown under the same conditions. In addition, the *arom-1* mutants lack an enzyme normally present in wild type, dehydroshikimic acid reductase. It is this last observation which gives the clue to the explanation of the whole complex of effects. Dehydroshikimic acid is apparently an essential precursor of all four aromatic compounds required by *arom-1* cultures. When its normal conversion is blocked by the absence of the reductase it tends to accumulate, and in so doing induces the formation of detectable levels of the two 'new' enzymes, both involved in dehydroshikimic acid degradation. Presumably wild type can also make these enzymes, but normally does so only to a very slight extent. The situation is outlined in Fig. 37.

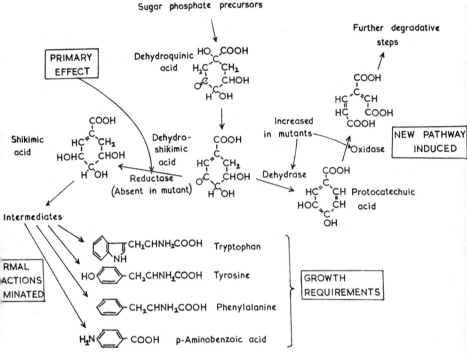

FIG. 37. Multiple effects of the primary enzymic deficiency in *arom-1* mutants of *Neurospora crassa*.

Multiple growth requirements due to a failure in the synthesis of a common precursor are not particularly surprising or unusual; another well known example is the requirement for either homoserine, or *both* threonine and methionine in *N. crassa* mutants unable to synthesize homoserine, which is a precursor of both threonine and methionine [547]. Another kind of explanation for a multiple requirement which applies in at least one case is that a single enzyme has been lost or damaged which is essential in two distinct pathways. This is the explanation

in the case of mutants requiring *both* isoleucine and valine. Although the
earlier steps in the biosynthesis of isoleucine are quite different from
those in the valine pathway, the later reactions are so similar in the two
pathways (differing only in the presence of an extra methyl group in the
isoleucine synthesis) that it is not surprising that the same set of enzymes
can catalyse both. The situation here is shown in Fig. 38.

Quite frequently, auxotrophs are strongly inhibited by substances
which do not affect the growth of the wild type to an appreciable extent.
For example, arginine mutants tend to be inhibited by lysine, and *vice
versa*, and histidine mutants are inhibited by arginine, methionine, and
some other amino acids to such an extent that they scarcely grow on
complete medium. The histidine-requiring mutants have been the most
completely analysed from this point of view [350], and it seems clear that
here the inhibitory amino acids have their effect by inhibiting the *uptake*
of histidine from the medium, probably through competition for some
carrier molecule which provides an essential pathway into the cell. The

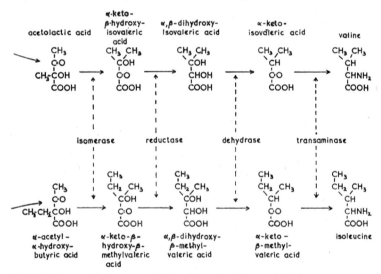

FIG. 38. The terminal stages of isoleucine and valine synthesis in Neurospora
catalysed by the same series of enzymes. The isomerase and reductase
activities are probably functions of a single enzyme, defective in *iv-2* mutants.
The dehydrase is probably a different enzyme, defective in *iv-1* mutants. No
mutants lacking the transaminase are known in Neurospora. [557, 33, 433, 371].
The preceding steps, involving condensation of pyruvate with another molecule
of pyruvate (to give acetolactate) or with α-ketobutyrate (to give α-acetyl-α-
hydroxybutyrate) are also catalysed by a common enzyme, defective in *iv-3*
mutants [556].

wild type has no need to take up histidine and so is immune to such inhibition. This explanation will probably be found to apply to a good many similar cases.

THE NATURE OF GENETICALLY DETERMINED CHANGES IN ENZYME ACTIVITIES

The gist of our argument so far is that many mutations, even some of those whose effects are at first sight quite complex, have their primary effects on single specific enzymes, and that there is no firm evidence for the existence of genes in fungi with pleiotropic primary effects. How, then, do mutations affect enzymes? In particular, do they act by simply altering the *rate* of enzyme formation, or do they affect the *kind* of enzyme formed? Though these need not be mutually exclusive alternatives, many well investigated cases in Neurospora point to a qualitative rather than a quantitative effect. So far as known examples permit us to generalize, genes determine enzyme *structure*.

Protein structure

Since each enzyme is a specific kind of protein, a brief discussion of protein structure seems in order here. Proteins consist of polypeptide chains; either a single chain as in mammalian ribonuclease [229], or a number of probably identical chains as in muscle phosphorylase [336], or two kinds of chain, as in haemoglobin [407]. Single polypeptide chains in proteins each consist of up to twenty different kinds of amino acid residues, the number of occurrences of each amino acid, and the sequence in which the different kinds are linked together, being fixed for each polypeptide. A single polypeptide chain may contain between a hundred and several hundred amino acids: that is, its molecular weight may be anything between about 20,000 and 50,000 or even more. In two cases of relatively small single chain proteins, mammalian ribonuclease [229] and the protein of tobacco mosaic virus, the complete amino acid sequence has now been determined [552]. In principle it should be possible to do this for any protein available in pure form in adequate amounts; in practice the task becomes rapidly more difficult as the length of the chain increases beyond 150 to 200 residues. The amino acid sequence or *primary structure*, is, however, only a part of the story. Proteins do not exist in nature mainly in the form of extended polypeptide chains, though they tend to assume this form on treatment by heat or other denaturing agencies. Rather each chain tends to be held by hydrogen bonding in a helical configuration, the well known α-helix of Pauling and Corey. Such a helical conformation may not prevail along the whole of the chain, but those few soluble proteins whose three-dimensional structure is known from X-ray crystallographic analysis (notably myoglobin and haemoglobin) are largely in this form. This intra-chain coiling, held in place mainly by hydrogen bonding, is known as *secondary structure*. The three-dimensional models of myoglobin and haemoglobin resulting from the X-ray analysis of Kendrew and Perutz [407, 283] show the tubular α-helices to be bent and folded back on themselves at certain points, so that each polypeptide chain comes to have an overall globular shape. This second order folding is called *tertiary structure*, and in some proteins is held in place by disulphide bridges formed between cysteine residues in different parts of the chain, and perhaps sometimes by electrostatic interactions (salt bridges) or other kinds of bonding. Finally, the entire protein molecule may consist of two, four or even more of such elaborately folded polypeptides, fitted together in a very specific way and held together by forces which are still not well understood but which are probably very generally of the kind loosely described as *hydrophobic interactions*; this may be called *quaternary structure*.

In none of the cases of genetically altered enzymes in fungi has the nature of the alteration been fully analysed. However, a good deal of information on kinds of changes in enzyme *function* is now available, and these should ultimately be explicable in structural terms. In the following sections some representative examples are given.

Complete and partial losses of enzyme function

In most cases where a nutritional requirement can be attributed to a change in a particular enzyme, the enzyme activity has apparently been completely lost. Does this mean that the enzyme-forming system has stopped working, or that it is producing an inactive form of the enzyme? There are two ways of seeking to answer this question, both very well exemplified by the work of Bonner, Yanofsky, Suskind and their colleagues on Neurospora tryptophan synthetase (TSase). The first depends on the recognition of an inactive enzyme variety through its capacity to combine with antibodies which are otherwise specific for the enzyme. The second is possible when, as in the case of TSase, the enzyme has several catalytic functions one of which may survive in the mutant form.

Wild type Neurospora TSase will catalyse the following three reactions:

(1) Indoleglycerol phosphate (IGP) $\rightleftharpoons$ Indole + 3-phosphoglyceraldehyde;

(2) Indole + Serine $\rightleftharpoons$ Tryptophan;

(3) Indoleglycerol phosphate $+$ Serine $\rightleftharpoons$ Trytophan $+$ 3-phospho-glyceraldehyde.

The enzyme is a single protein which has been obtained in near-pure form [528]. Reaction (3) is not merely the sum of reactions (1) and (2); in (3) indole is not formed as a free intermediate and apparently is held on the enzyme surface. Reaction (3) is thought to represent the *in vivo* function of the enzyme, while the other two reactions are artefacts obtained by supplying and withholding substrates in test tube systems. In reactions (2) and (3) pyridoxal phosphate is an essential co-factor.

Bonner and co-workers [43] have isolated a large number of trypto-phan-requiring mutants of *N. crassa* all lacking enzymic activity for reaction (3). All these mutations can be mapped at the *tryp-3* (*td*) locus of linkage group II; all fall into the same gene by the complementation criterion, since most of them are non-complementary with all the others. Most of the mutants produce protein which, while unable to catalyse reaction (3) is still sufficiently similar to TSase to be able to precipitate anti-TSase antibodies formed by rabbits after injections of the enzyme [529]. This immunologically cross-reacting material (CRM) is generally formed by mutants in amounts comparable to the amount of enzyme normally formed by wild type, one unit of CRM being defined as that quantity which will precipitate antibody which would otherwise be able to remove one unit of TSase activity.

It seems certain that the CRM's produced by different mutants differ from each other, and that each is an abnormal variety of TSase. Many of the CRM types retain part of the activity of TSase [526, 530]. Among the *td* mutants are several which accumulate indole during growth. All these produce a CRM which is able to catalyse reaction (1), the hydrolysis of indoleglycerol phosphate, but cannot bring about the coupling of serine. Extracts of some of the indole accumulators show this IGP-ase activity to a maximal extent only when both serine and pyridoxal phosphate are present in the reaction mixture; in other cases only pyridoxal phosphate is required, while in yet others neither substance stimulates activity. It seems that some of the CRM types are activated for IGP hydrolysis by combination with pyridoxal phosphate even though the catalytic activity for which this co-factor is normally required has been lost. Among the *td* mutants which do not accumulate indole, at least one can actually utilise indole in lieu of tryptophan. This mutant forms a type of CRM which is active in reaction (2) but unable to promote the reactions involving indoleglycerolphosphate. It thus seems that, as a result of different mutations within the *td* locus, a whole series of altered varieties of TSase can be produced. Some have serological cross-reactivity but no known enzyme activity, others have lost the capacity to activate IGP but retain catalytic activity towards indole and serine, and yet others retain activity towards IGP but are unable to couple serine to indole even though some of them evidently remain able to combine with serine since they are activated by this substance in the presence of pyridoxal phosphate. Some *td* mutants, of which *td*[1] is the best investigated, produce no TSase nor any detectable CRM. It is possible that in these mutants no protein of any kind corresponding to TSase is produced, though it is also possible that some kinds of configurational change in the enzyme might cause a masking of the groupings which combine with antibody.

This whole series of investigations is an excellent example of the variety of functional alterations which can be detected in an enzyme catalysing a rather complex reaction sequence.

Conditional enzyme activity in mutants

We have seen from the cases just discussed that auxotrophic mutants have very often lost either the entire function of an enzyme, or an essential part of it. In absolute auxotrophs this loss is usually complete and unconditional, as we would expect. However, one may occasionally find a situation in which a mutant produces an abnormal enzyme variety which is apparently ineffective in the normal milieu of the cell, but which shows some activity under rather special conditions in a test tube system. Several such cases are known among the *am* mutants of *Neurospora crassa*, investigated by one of us.

The *am* mutants, of which some dozen are now known, all require one of a number of α-amino acids for normal growth. They also have a tendency to accumulate ammonia when supplied with nitrate. These symptoms are adequately explained by their lack of normal triphosphopyridine nucleotide (TPN)*-linked glutamic dehydrogenase activity. The glutamic dehydrogenase reaction is shown in Fig. 39, and is normally the major pathway through which ammonia is utilized for the synthesis of the α-amino groups of amino acids. It may be mentioned in passing that *am* mutants *are* able to grow on minimal medium after a lag but still without producing any detectable TPN-linked glutamic dehydrogenase. The probable reason for this has been shown by Sanwal and Lata [478], who found that Neurospora also produces a diphosphopyridine nucleotide (DPN)-linked glutamic dehydrogenase which is not affected by the *am*

* Now more usually known as nicotinamide-adenine dinucleotide phosphate (NADP).

mutations, and which can presumably function as an auxiliary ammonia utilizing mechanism. Here the effect of a mutation on enzyme constitution is much sharper than its effect on growth.

The TPN-linked glutamic dehydrogenase has now been isolated from the wild type in near-pure form. A similar isolation procedure applied to several of the *am* mutants has resulted in the isolation of proteins which, in their fractionation characteristics, electrophoretic mobility, and in the number and kinds (as judged by electrophoresis and chromatography) of the peptides which they yield on hydrolysis by trypsin, are very similar to wild type enzyme [155]. In the mutant *am*[1] this related protein has no enzyme activity that has been demonstrated. In *am*[2] and *am*[3], neither of which can grow any better on minimal medium than can *am*[1], the related protein does have some glutamic dehydrogenase activity, though of a very peculiar kind. In both the activity is shown only in the reaction mixture for the glutamate oxidizing reaction (i.e. glutamate plus TPN), and it requires abnormally high glutamate concentrations for maximum expression. In the reaction mixture for glutamate synthesis (i.e. α-keto-

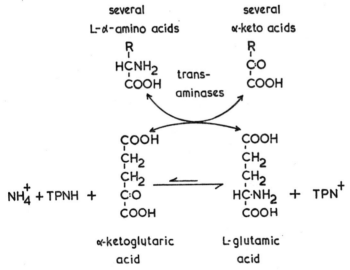

Fig. 39. The reversible reaction catalysed by glutamic acid dehydrogenase, and its relation to the synthesis of other amino acids through the mediation of transaminases.

glutarate, ammonium salt and TPNH) neither the *am*[2] nor the *am*[3] protein shows any activity. The *am*[3] enzyme, but apparently not the *am*[2], is active in glutamate synthesis provided the *products* of the reaction,

TPN and glutamate, with the latter at fairly high concentration, are present in the reaction mixture. It seems that these two mutants each produce an enzyme variety which is normally in an inactive form, but which can be activated, perhaps through a shift in its pattern of folding, through contact with glutamate and TPN. This activation is, at least in the case of the am^2 enzyme, actually antagonized by the substrate mixture necessary for glutamate synthesis. It is understandable that neither of these mutant enzyme varieties can be effective in glutamate synthesis *in vivo*.

Abnormal enzyme activity in partial revertants

Pateman [394] showed that ultraviolet irradiation of conidia would induce, in most of the *am* mutants, occasional reversions to ability to grow normally on minimal medium. Not all *am* mutants yielded revertants equally freely, nor were the kinds of revertants obtained constant from one mutant to another. The mutant am^1 gave only revertants which were indistinguishable from standard wild type. From am^2 and am^3, both of which, as we have seen, produce grossly defective but still somewhat active enzyme, revertants were obtained some of which appeared to be normal wild type, while others produced enzyme varieties which were decidedly improved in activity compared with the original mutant product, but which were still not as active as wild type enzyme under standard assay conditions. Three of these 'partial revertants' have been analysed in detail, and their respective enzyme varieties purified.

Each of the three secondarily derived mutant enzymes proved to be capable of nearly wild type activity in glutamate synthesis under optimum conditions, which were different for each enzyme variety. The mutant am^{2l}, derived from am^2, produced an enzyme which was almost inactive at room temperatures but which was activated to the extent of 50 per cent or more of the wild type level by brief warming to 35°C to 40°C [152]. This activity was slowly lost at room temperature, but could be regained by renewed warming. It seems likely that a shift in the folding configuration of the protein molecule is involved here. Activation could also be achieved by incubating the enzyme in a mixture containing fairly high concentrations of α-ketoglutarate and TPNH at room temperature. The mutant am^{3b}, derived from am^3, also produced an enzyme variety which required activation, and again incubation with an α-ketoglutarate-TPNH mixture was effective [155]. In this case, however, heat treatment alone had no effect. The glutamic dehydrogenase of a second mutant derived from am^3, am^{3a}, showed only slight activation on warming or incubating with substrate mixtures; its main distinguishing feature was an increased requirement for all substrates as compared with wild type enzyme. The effect was particularly pronounced in the case of ammonium ion for which the Michaelis constant (i.e. the concentration sufficient to keep the enzyme half saturated while the reaction is going on) was some thirty-five times as high with the am^{3a} as with the wild type enzyme [156]. In addition to these and other peculiarities, these three enzyme varieties were all less stable to heat than the wild type enzyme; the degree of instability was characteristic of each variety, being highest in am^{2l} and most nearly normal in am^{3b}. This is a common effect, and it seems that almost any mutational change in an enzyme is likely to cause some change in thermal stability.

It has been shown [394, 500] that all the mutations just discussed, both the primary ones and the secondary ones, lie within the same short segment of the fifth linkage group, and may thus reasonably be regarded as falling within the same gene. It is clear that am^1, am^2 and am^3, at least, are mutant at different sites within the gene. One imagines that the am gene determines the amino acid sequence of the polypeptide subunit of the glutamate dehydrogenase protein and, indeed, Barratt [17] has shown a clear difference in one tryptic peptide between am^{3a} and wild type enzyme, although the change in amino acid sequence has not yet been defined. The striking thing about several of the mutants, however, is that they produce varieties of the enzyme which need special conditions for activity and which tend to lapse into inactive forms when these conditions are not provided. The defect in such mutant proteins is evidently not the absence of essential catalytically active groupings, but rather in the conformation of the molecule on which the correct spatial relationship of the active groups and substrates is dependent. While the configuration of the wild type enzyme seems to be stable over a wide range of conditions, several of the mutant varieties appear to be unstable in this respect. Such instability could well be due to alterations in the primary structure (amino acid sequence), but this has not been proved as yet in the present instance.

Altered enzymes in 'leaky' mutants

It is not uncommon to find that an auxotrophic mutant has a slight residual ability to grow on minimal medium. This may be due to utilization of an alternative biosynthetic pathway, as is probably the case with the am mutants mentioned above, or to an incomplete loss of enzyme activity.

The latter explanation applies to the *prol-1* mutant of *N. crassa* investigated by Yura [603]. There a particularly thorough investigation of the properties of pyrroline-5-carboxylate reductase in the mutant showed that the effect of the mutation was to cause an alteration in the enzyme which results in greatly decreased stability to heat and greatly decreased catalytic efficiency (increased activation energy of the enzyme-substrate complex).

Naturally occurring cryptic enzyme differences

Horowitz's studies on Neurospora tyrosinase [241] show that naturally occurring strains may differ in the type of a given enzyme that they produce. One would expect enzyme variants to survive in nature only if the enzyme concerned is not an essential one, or if the variations are of a kind which do not have an important effect on activity. Tyrosinase seems not to be an essential enzyme for vegetative growth though it plays an important role in sexual reproduction, and the genetic differences discovered by Horowitz *et al.* are cryptic, in that they do not affect activity at physiological temperatures.

Two criteria serve to distinguish at least four tyrosinase varieties in strains of *N. crassa* from different parts of the world. First, there are large differences in thermostability between one enzyme type and another. Secondly, the different kinds of enzyme all carry somewhat different electric charges at pH values near neutrality and are thus separable by electrophoresis.

Since these variations in tyrosinase do not affect growth, there are no selective methods available for detecting possible crossing-over between the gene determining one variety and that determining another. Consequently a large scale recombinational analysis is not possible in this case; nevertheless the genetic evidence shows that the determinants of the different varieties are closely linked and are probably alleles of a single gene locus. This postulated single locus is called T. In heterocaryons containing two different T alleles, *both* corresponding enzyme types are produced; in other words, there is no indication of dominance.

THE SPECIFICITY OF GENETIC CONTROL OF PROTEIN STRUCTURE

In recent years the 'one-gene-one-enzyme' hypothesis has come to be understood as meaning that the structure of each kind of polypeptide is determined by one gene, and that each gene has only one such function. We know nothing that would discredit the idea that genetic determination of protein structure is always through the determination of amino acid sequence, and the actual demonstrations of gene-control of amino acid sequence in several proteins in diverse organisms (human haemoglobin, tobacco mosaic virus protein, bacterial alkaline phosphatase) tend to support this simple hypothesis.

There are two questions to be considered in this connexion. The first is whether each gene governs the structure of only one enzyme. So far as we know this is so. Thus although there are a number of mutations in Neurospora which seem at first sight to have multiple effects on metabolism, these, have generally turned out to have primary effects on single enzymes, even though some of these single enzymes have relatively complex functions. It should perhaps be pointed out here that where an enzyme can catalyse more than one reaction its 'singleness' has to be demonstrated by showing that its several activities maintain a constant ratio to each other through all stages of purification. The second question is whether mutations capable of altering the structure of a given enzyme are confined to one gene. This question deserves more detailed consideration.

In almost all cases where an extensive series of mutants deficient in a particular enzyme has been isolated, the most striking genetic fact has been the localization of all the mutational sites within the same very short chromosome region. This has been shown most strikingly in *N. crassa* for series of mutants deficient in, respectively, tryptophan synthetase [43] (*tryp-3*), argininosuccinase [376] (*arg-10*), adenylosuccinase [197] (*ad-4*), glutamic dehydrogenase [154] (*am*), and the enzymes of isoleucine and valine synthesis [557] (*iv-1, iv-2*). In each of these examples a single gene, in the sense of a single region of non-complementation, is concerned. In a number of cases mutations within the gene are known to cause structural changes in the corresponding enzyme. Table 24 lists the *N. crassa* genes which have been fairly certainly implicated in the structural determination of specific enzymes together with others for which such a function seems probable.

There are, however, some enzymes whose genetic control seems more complicated. An example is nitrate reductase, both in Neurospora [504] and in *Aspergillus nidulans* [88a]. In Neurospora there are three genes, *nit-1*, *nit-2* and *nit-3*, which have been found to mutate to give an apparent loss of the enzyme and a consequent failure to grow on nitrate as sole nitrogen source. Two of these classes of mutant (*nit-2* and *nit-3*) also lack cytochrome c reductase activity while *nit-1* strains seem to lack nitrate reductase specifically. It may be that *nit-2* and *nit-3* are deficient in some cofactor necessary for both enzyme activities, or,

TABLE 24. Enzyme-determining genes in *Neurospora crassa*

Gene	Linkage group	Enzyme	Control of enzyme structure	Reference
ad–4	III	adenylosuccinase	+	591
am	V	glutamic dehydrogenase	+	155
arg–1	I	argininosuccinate synthetase	?	377
arg–10	VII	argininosuccinase	?	157
arg–12	?	ornithine transcarbamylase	?	98, 100
arom–1	II	dehydroshikimic acid reductase	?	210
his–1	V	imidazole-glycerolphosphate dehydrase	?	3, 5
his–3	I	histidinol dehydrogenase	?	3, 566
his–4	IV	histidinolphosphate phosphatase	?	3, 4
iv–1	V	dihydroxy acid dehydrase	?	371, 557
iv-2	V	α-keto-β-hydroxyacyl reductoisomerase	?	556
iv-3	IV	pyruvate$+\alpha$-ketoacyl condensing enzyme	?	556
prol–1	III	pyrroline-5-carboxylate reductase	+	603
thr–2	II	threonine synthetase	?	159
tryp–3	II	tryptophan synthetase	+	43
T	I	tyrosinase	+	240, 241

N.B. iv-1, iv-2 and *iv-3* [33] are assumed to correspond to Wagner's [556] group (2+3), group 1 and group 4 mutants respectively. For approximate map locations see Fig. 21.

conceivably, that they control the structures of polypeptide chains common to both enzymes, if they are indeed distinct. In *Aspergillus nidulans* there are at least six genes whose mutation can eliminate effective nitrate reductase activity, but Cove *et al.* [88a, 396a] have evidence that only one of them is concerned with the structure of the enzyme protein and that at least some of the others are necessary for the synthesis of a complex cofactor which is needed both for nitrate reductase and for xanthine dehydrogenase activity. Some enzymes do consist of more than one kind of polypeptide chain, and in these cases structural control by more than one gene is to be expected. The isomerase which functions in the biosynthesis of leucine in Neurospora is probably a case in point; this enzyme activity can be lost by mutation either of *leu-2* or of *leu-3* [209, 211].

Suppressor mutations

Further questions relating to gene-enzyme specificity are raised by the phenomenon of suppressor mutation, in which the effects of one mutation seem to be overcome by a second mutation in a different gene. There are two general ways in which suppression might be explained. Firstly, the suppressor mutation may be bringing about some change in metabolism such that the lesion produced by the primary mutation no longer matters. Secondly, it may be in some way correcting the product of the primary mutant gene so that the metabolic lesion is abolished. Both explanations apply in different cases.

The best examples of the first kind of suppressor action come from studies of the interactions of mutations affecting arginine and pyrimidine synthesis in Neurospora. As was indicated in Fig. 35, *arg-12* mutants are defective in ornithine transcarbamylase and thus have partial or total blocks in their utilization of carbamyl phosphate for citrulline synthesis; *arg-2* and *arg-3* mutants, on the other hand, seem to be defective in their ability to make carbamyl phosphate for this purpose (though the exact nature of the defect in *arg-2* is obscure). In the pyrimidine pathway, some *pyr-3* mutants lack aspartate transcarbamylase (ATC) and thus cannot utilize carbamyl phosphate for pyrimidine synthesis, while other *pyr-3* mutants produce this enzyme but seem to be unable to form the carbamyl phosphate which acts as a substrate for it. The explanation of this apparently dual function of *pyr-3* is unknown; it is difficult to postulate two independently functioning genes since many *pyr-3* mutations eliminate both functions simultaneously. It may be that *pyr-3* controls one enzyme with both functions or, perhaps more likely, that it represents two closely linked genes functionally co-ordinated in an *operon* (cf. p. 203). At all events, it looks as if there are two 'pools' of carbamyl phosphate in Neurospora, one used for arginine and one for pyrimidine synthesis, and that they are supplied by different enzyme systems under separate genetic control. The work of Davis and Woodward [98, 100, 594] and of Reissig [448] has shown that the effects on growth of a blockage of carbamyl phosphate *formation* in one pathway (by mutation at *pyr-3*, *arg-2* or *arg-3*) can be overcome by a blockage of carbamyl phosphate *utilization* in the other (by mutation of *arg-12* or *pyr-3*). In other words, *arg-12* suppresses some *pyr-3* mutants (the aspartic transcarbamylase-positive ones), while other *pyr-3* mutants (ATC-negative) suppress *arg-2* or *arg-3* mutants. Apparently one carbamyl phosphate pool can flow into the other when its own normal outlet is dammed.

Some of the best examples of suppressor action through actual correction of the primary enzymic defect were provided by Bonner, Yanofsky and Suskind through their studies on *tryp-3* mutants of Neurospora. Although, when tryptophan synthetase deficient mutants are isolated from wild type all the mutations turn out to be within the *td* (*tryp-3*) locus, mutations having the effect of restoring enzyme activity to *td* mutants can occur not only at the *td* locus itself but, in many cases, at

other loci also [599]. The distinction between a further mutation at the
td locus and a 'suppressor' mutation at another locus, depends on the
analysis of a cross of each apparent revertant to wild type. The principle
is illustrated in Fig. 40. Once a suppressor allele has been separated
from the *td* allele with which it was originally associated, it can be combined by appropriate crosses with other *td* alleles. When this is done it is
found that each suppressor tends to be specific for one particular *td*
allele; no suppressor was found which could suppress the effects of *td*
mutations in general. This observation clearly rules out any suggestion
that any of the suppressor loci can take over the function of *td* and act
independently of it. It is much more consistent with the following argument. We have already seen that different *td* alleles produce different
varieties of the tryptophan synthetase molecule. Some of these may be

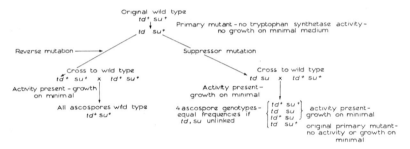

FIG. 40. Procedure for distinguishing between suppressor mutation and
reversion at the original mutant locus.

inactive under the normal intracellular conditions in which the wild type
enzyme is active, but may become active in some degree if the intracellular conditions are somewhat altered by a second mutation. Since the
nature of the defect in the enzyme is probably specific for each *td* allele,
the kind of suppressor mutation which can ameliorate its effects should
also be allele-specific. This idea is attractive partly because it avoids the
necessity of supposing that two or more genes can affect the structure of
the same enzyme.

Suskind [527] has demonstrated that this hypothesis fits the facts in at
least one case. The mutant *td*24 shows no tryptophan synthetase activity
in crude extracts nor growth on minimal medium at 25°C, though it does
show a little growth and activity at 35°C. Purification of inactive crude
extracts revealed the presence of an active synthetase whose presence
had been masked by an inhibitor, which was recovered in a separate
fraction and which was identified as a metal whose inhibitory effects

could be duplicated by zinc. The mutant td^{24} contains no more zinc than wild type, but its tryptophan synthetase is far more sensitive to zinc inhibition than the wild enzyme. A specific suppressor mutation su-24 permits growth of td^{24} at all temperatures and restores activity in crude extracts. The enzyme produced in td^{24} su-24 is, however, of the same zinc-sensitive variety as is present in the original mutant; it appears to be the inhibitor concentration rather than the nature of the enzyme which is altered by the suppressor.

Unfortunately for those who would like a simple theory, not all cases of suppression of the effects of td alleles can be easily explained in this sort of way. An earlier study by Yanofsky [596] on the mutant td^2 provides a case in point. This mutant produces no tryptophan synthetase activity, but does form a variety of the enzyme (CRM) which is now known to be able to catalyse hydrolysis of indoleglycerolphosphate. A suppressor mutation su-2, at a locus different from that of su-24, restored a significant amount of apparently normal enzyme which was produced along with the CRM. More recent work by Yanofsky and his colleagues on tryptophan synthetase of *Escherichia coli* [597, 58] shows conclusively that suppressor mutation can sometimes correct an alteration in amino acid sequence in at least a proportion of the enzyme molecules. This situation should be distinguished from the phenomenon of 'second site reversion' [223] in which a second mutation within the same gene results in the formation of tryptophan synthetase molecules *all* of which have a second mutational alteration which somehow partially compensates for a deleterious effect of the first. We are concerned here with a second mutation in a different gene which restores the *wild type* protein structure to a *significant fraction* of the molecules. A similar kind of suppressor action has been inferred by Garen and Siddiqi [188], working on the genetic control of alkaline phosphatase of *E. coli*, and by Benzer and Champe [30] in bacteriophage.

This seems a clear disproof of the one gene–one polypeptide chain principle. However, although a suppressor mutation may sometimes affect the amino acid sequence of a polypeptide chain primarily controlled by another gene, it is possible that it does so in what is, in a sense, an unspecific and accidental way. The theoretical background to the following explanation is given briefly in Appendix I. The suggestion is [598, 30, 188] that a suppressor mutation of this type enables a DNA coding triplet which, in the original mutant, could not be translated into any amino acid (i.e. was 'nonsense'), or which translated as the wrong amino acid (i.e. 'mis-sense'), to be translated at least some of the time as if it were the corresponding wild type triplet. In other words, the suppressor mutation is envisaged as reducing the specificity of the decoding mechanism so as to allow a proportion of nonsense or mis-sense symbols to read as sense. This reduction of specificity could be a result of a change either in one of the types of transfer RNA or in the corresponding amino acid activating enzyme (cf. Appendix I). Such a mechanism of suppression means either that the mutant coding triplet, the translation of which is being altered, must be a rare one not occurring at all widely in the wild type genome, or that the suppressor mutation must cause a proportion of 'mistakes' in the synthesis of many different proteins. Consistent with the latter alternative is the fact that several suppressors of *tryp-3* mutants in Neurospora do bring about a general depression of growth rate, but even so it seems unlikely that the 'suppressible' triplet could be one in very common use.

An explanation of suppressor action invoking an alteration in the reading of a mutant coding triplet would imply that suppressors acting in this way should be allele-specific but not gene-specific. All mutant alleles sharing the same mutant triplet should be suppressed by the same suppressor no matter to what gene they happen to belong. Suppressors acting on a proportion of the mutant alleles of a variety of different genes have actually been found in yeast by Hawthorne and Mortimer [221]. The surprising thing about these 'super-suppressors', as they were called, was the high proportion of auxotrophic mutants which they would suppress, amounting to almost a half of the mutants tested at each of several different loci. It seems, at first sight, rather unlikely that all these mutants should share the same mutant coding triplet. However, if most totally auxotrophic mutants represent nonsense mutations rather than mis-sense (which could be the case inasmuch as a great many amino acid substitutions do not affect protein function in a crucial way), and if there are very few possible kinds of nonsense (i.e. if nearly all triplets stand for some amino acid), then the suggested explanation is still quite a possible one. Interestingly enough, Hawthorne and Mortimer found that two different suppressors at two different loci affected the same spectrum of mutants.

Some support for the 'nonsense' hypothesis has been obtained by Manney [340] who studied the complementation relationships of 36 *tr-5* mutants of yeast each of which lacked normal tryptophan synthetase. Of 17 non-complementing mutants, ten were suppressed by one of the super-suppressor mutations. Of the remaining 19, which complemented each other in certain combinations, only two were suppressible.

Possible mechanisms of inter-allelic complementation are discussed below; it seems likely that the ability of a mutant allele to complement is usually an indication that that allele forms a mutant protein because of a 'mis-sense' mutation, rather than no protein because of a 'nonsense' mutation. If this is true in the present case then there is a strong correlation between genetic nonsense and responsiveness to super-suppressors.

Complementary action of alleles in enzyme formation

One of the most challenging problems in connexion with the biochemistry of gene action is the phenomenon of inter-allele complementation, already considered in Chapter 6. This kind of complementation occurs at many gene loci known to be concerned with the formation of single enzymes; the basic observation to be explained is that sometimes a pair of alleles in a heterocaryon is able to promote the formation of an enzyme when neither allele could do so alone.

There are three pieces of evidence which suggest an explanation for this seemingly anomalous type of result. Firstly, the amount of enzyme activity formed by inter-allele complementation is never as much as would be found in wild type, or even in a heterocaryon containing equal numbers of wild and mutant nuclei. Different pairs of alleles give different levels of activity ranging from the barely detectable up to around 25 per cent of the typical wild type value. This is particularly well shown in Woodward's work on *N. crassa* adenylosuccinase [589, 590]. Secondly, in those cases where the *kind* of enzyme formed has been looked at carefully, it is clear that most if not all, combinations of alleles give a qualitatively abnormal product. At the *am* locus, for example, two pairs of 'complementary' alleles have been investigated, and in each case the enzyme formed by the complementation has reduced thermostability and certain other peculiar properties [153], the two complementary pairs differing somewhat from each other in the type and degree of the enzymic abnormality. Partridge [393] has demonstrated essentially the same thing at the *ad-4* locus controlling adenylosuccinase, as have Suyama and Bonner [532] and Gross and Webster [211] in their studies of complementation at *tryp-3* and *leu-3* respectively. The stability and other properties of the complementation product in each case depend on the particular pair of alleles showing complementation. The third important piece of evidence is that in three cases in Neurospora, namely those of tryptophan synthetase controlled by *tryp-3* [531], adenylosuccinase (*ad-4*) [589] and glutamate dehydrogenase (*am*) [158], active enzyme can be formed as a result of mixing crude extracts or purified mutant proteins from complementing mutants under conditions in which protein synthesis can certainly not occur. A further relevant fact is that many purified enzymes from animal and bacterial sources have been shown to be compound or aggregate molecules, consisting of two, four or even more identical polypeptide chains. As an extreme example, Yielding and Tompkins [600] have shown that mammalian glutamic dehydrogenase is a tetramer which can be dissociated into four identical sub-units each of which can in turn be dissociated by more drastic treatments into four smaller sub-units. Neurospora glutamate dehydrogenase is also an aggregate protein, probably containing eight (or possibly ten) identical sub-units [158].

The general hypothesis which has been based on the above considerations is as follows. If the enzyme in question is an aggregate of identical sub-units, we may suppose that the function of the corresponding structural gene is to make, or to control the making of, the sub-unit. If in a heterocaryon two mutant alleles are present, each making a sub-unit with a particular defect, one may expect a proportion of hybrid aggregates to be formed. It is certainly conceivable that in such hybrids the deficiency in the structure of each kind of polypeptide sub-unit may be corrected or compensated for by the presence of the homologous normal segment of the other. Crick and Orgel [93] have suggested a number of mechanisms. Most simply, each sub-unit may contribute some essential active grouping lacking in the other. A more plausible general explanation is that at least one member of each pair of complementary mutants produces a variant enzyme which is potentially capable of showing activity but which generally does not do so because its most stable conformation is one in which the active centre is masked. One may suppose that when sub-units of such a defective enzyme are associated with the sub-units of another kind of variant, their abnormal conformation tends to be corrected and their potential activity realized. The situation at the *am* locus tends to support this type of model. In most of the complementary combinations one allele (am^2, am^3, am^{19}, am^{21}, am^{3b}) produces an enzyme variety which is nearly inactive as extracted from the mycelium, but which can be activated by appropriate treatments. In most cases the second component is am^1. The effect of the am^1 protein, which is totally inactive itself, may be to stabilize the potentially active second component in its active conformation. This explanation presupposes some kind of section-for-section association of sub-units so that a normally folded section of one may, in some way, hold in place an abnormal homologous section of the other. Sections which could associate in this way would correspond to the segments of the complementation map. The overlaps in the complementation map would be due to mutational changes which affected the structure of two or more adjacent sections, while the frequent class of non-complementing mutants would produce sub-units so abnormal in structure as to be incapable of polymerising at all. If the enzyme sub-unit is visualized as a more or less linear structure with the relative positions of the various mutational lesions corresponding roughly to the linear order of the mutant sites in the corresponding gene, the general co-linearity of recombinational and complementation maps is to be expected. Since, however, the folding of the polypeptide chain will tend to distort the original linear arrangement of the changes in the primary structure, one would not expect this co-linearity to be without exceptions. Occasional exceptions do, indeed, occur. Figure 41 shows the comparison of the recombinational and complementation maps in the best analysed case at present available, that of the *pan-2* locus of Neurospora investigated by Case and Giles [77]. One should also not be surprised to find occasional instances of the complementation map itself not being linear. Several examples of non-linear complementation maps are indeed known; those of *leu-2* [209] and of *ad-8* [249] are both circular. Kapuler and Bernstein [277] showed that a point-for-point alignment of the complementation map of *ad-8* with the linear recombinational map required that the latter be wound around the circle about one and a quarter times. They suggested that this represented the conformation of the polypeptide chain in the presumably polymeric protein. While this interpretation might seem to be too naive to be true in the simplest sense, the idea behind it may well be valid.

Strong evidence that the complementation product is indeed a hybrid polymeric protein has been obtained in the case of Neurospora *am* mutants (Coddington & Fincham, *J. mol. Biol*, in the press), but the most completely understood example, and one in which a hybrid dimer is the active product, has been provided by the work of Schlesinger and Levinthal on the alkaline phosphatase of *Escherichia coli* [482]. In every case where the necessary experiments have been done [482, 52, 158] the complementation product appears to contain the same number of sub-units as the normal enzyme.

Apart from inter-allelic complementation there is also the possibility of inter-genic complementation in the formation of a single enzyme. This may be expected where the enzyme consists of two or more

14

different kinds of polypeptide chain as is suspected to be the case for the isomerase involved in the pathway of leucine synthesis in *Neurospora*. This enzyme appears to be a compound of the products of *leu-2* and *leu-3* [211]. Two thoroughly documented examples are available from other organisms. The first case is haemoglobin of man [252], where

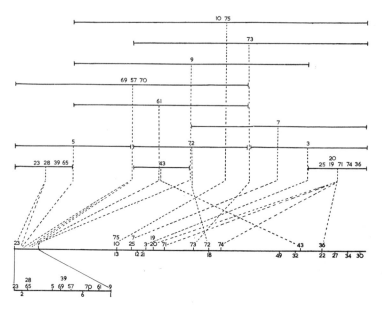

FIG. 41. Comparison of the recombinational and complementation maps of *pan-2* in *Neurospora crassa*. Below is shown the map of the mutant sites based on frequencies of prototrophs from inter-mutant crosses; the left-hand end of the map is expanded for easier representation. Complementing mutants are represented above, and non-complementing mutants below the line. Above are shown the segments of the complementation map. Dashed lines connect the mid-point of each segment to the corresponding sites on the recombination map to show the tendency towards colinearity. Note that mutant 43 is definitely an exception to colinearity, as are mutants 72 and 3. Redrawn after Case & Giles [77].

the protein consists of four polypeptide chains, two of one kind and two of another. The second case is tryptophan synthetase of the bacterium *Escherichia coli*, which is a protein rather easily dissociable into two individually inactive and non-identical sub-units [597]. In each of these examples the structures of the two kinds of sub-unit are determined by different genes, unlinked in the haemoglobin example, and closely linked in the case of the tryptophan synthetase. This kind of situation means that a mutant defective in one sub-unit can complement a mutant defective in the other with the formation of a *qualitatively normal* product. In contrast to the examples in Neurospora in which complementation maps show continuous series of overlaps, the haemoglobin or *E. coli* tryptophan synthetase situation is one in which the protein is controlled by two clearly distinct, functionally non-overlapping genes.

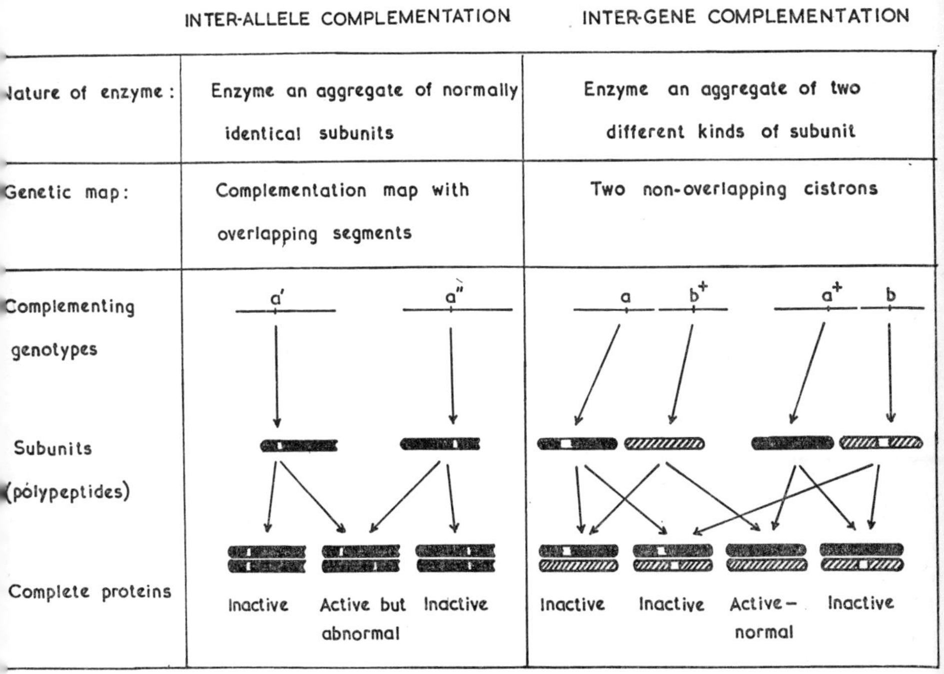

	INTER-ALLELE COMPLEMENTATION	INTER-GENE COMPLEMENTATION
Nature of enzyme:	Enzyme an aggregate of normally identical subunits	Enzyme an aggregate of two different kinds of subunit
Genetic map:	Complementation map with overlapping segments	Two non-overlapping cistrons
Complementing genotypes	a' $\qquad$ a''	a b^+ $\qquad$ a^+ b
Subunits (polypeptides)		
Complete proteins	Inactive $\quad$ Active but $\quad$ Inactive abnormal	Inactive $\quad$ Inactive $\quad$ Active— $\quad$ Inactive normal

FIG. 42. Two mechanisms for genetic complementation in the formation of a single enzyme. The regions of the polypeptide enzyme sub-units which are altered by mutation are shown as white spaces. Different kinds of polypeptide sub-unit are indicated by solid black versus cross-hatching.

It seems possible that such cases as that of the twin *ad-3* genes of *N. crassa*, between which no phenotypic distinction has yet been made, may be of this kind. The distinction between complementation based on interaction between different mutant derivatives of the same sub-unit, and complementation involving different *kinds* of sub-unit, is shown diagrammatically in Fig. 42. If this representation of complementation mechanisms is correct, it is the qualitative normality or otherwise of the gene product (enzyme) which must be the ultimate criterion for deciding whether two complementing mutants should be referred to the same or different genes.

KINDS OF GENE ACTION

During the last few years a comprehensive theory of gene action has been developed, chiefly by Jacob and Monod [253] of the Pasteur Institute, Paris. This theory is based entirely on information from bacteria, and there is little evidence so far for its applicability to other organisms. However, it seems likely to be so influential that we shall outline its main points and consider briefly how it may apply in fungi.

According to Jacob and Monod there are two broad categories of genes, respectively structure-determining and regulatory in function. As we have seen, there is ample evidence for gene action of the first type in Neurospora. Regulatory genes are not supposed to affect protein structure but rather to determine whether and to what extent a given protein shall be formed in a given set of environmental conditions.

There are two ways in which bacterial cells can adapt their enzyme content in response to environmental changes. The first is enzyme induction, in which the synthesis of an enzyme is evoked by the presence in the growth medium of its substrate, which is usually an energy source or nutrient which cannot be utilised without the enzyme in question. The second is enzyme repression, in which the formation of an enzyme or a series of enzymes involved in the synthesis of a certain metabolite is repressed by the presence of the metabolite in the medium. These two phenomena have the effect of giving the bacterial cell those enzymes which it needs while preventing the synthesis of unnecessary proteins. Recent work, for which the Pasteur Institute group have been largely responsible, suggests strongly that induction and repression are two aspects of what is basically the same mechanism. It seems that, in the *Escherichia coli* β-galactosidase system at least, enzyme induction is due to the release of a repression; inducers of enzyme synthesis evidently act by antagonizing the action of a repressor. On the other hand, substances which repress the formation of an enzyme are assumed to in some way activate the repressor. It is an essential part of the theory that repressors are specific, either for single enzymes or for groups of enzymes involved in the same process. In the latter case, where several enzymes are controlled by a common repressor, the structural genes for these enzymes form a closely linked group, at least in some well investigated cases. In the most impressive of these cases, that of the group of enzymes catalysing the entire pathway of histidine synthesis in *Salmonella typhimurium*, the different enzyme activities remain in constant proportion to each other at all levels of repression [6]. The corresponding structural genes are closely linked in an order which broadly reflects the order of action of the enzymes [215, 7]. A group of structural genes subject to co-ordinate repression by a common repressor has been called an *operon* by Jacob and Monod.

The evidence from *E. coli* is that the specific repressors are themselves the products of specific genes. Such genes are recognized by their mutation; in the case of an inducible enzyme mutational loss of the repressor gives a constitutive strain which does not require induction, while in the case of a repressible enzyme a non-repressible strain results.

The nature of the specific repressor substances is not known with certainty, but there is some evidence of a rather indirect kind that they are proteins whose inhibitory effects on the transcription or translation of the genetic message are drastically influenced by combination with the low molecular weight substances which initiate induction or repression. The question of whether the control is exercised at the level of DNA-to-RNA transcription or of RNA-to-protein translation is still an open one.

In the *E. coli* β-galactosidase system mutations are known which render the structural genes constituting the *operon* insensitive to repression even though repressor is being produced in the cell. These mutations fall into a short genetic segment at one end of the operon which has been called the *operator*. Other mutations, which fall near the same end of the operon, have the effect of causing *all* the genes of the operon to be permanently repressed. The implication is that the switch which controls the operon is located at one end and can, so to speak, be jammed either on or off by different kinds of mutation.

According to the theory, then, there are three elements in the genetic control of groups of functionally related enzymes, the tightly linked cluster of structural genes constituting the operon, the operator which generally seems to be at one end of the cluster, and the gene producing the repressor specific for the operon but not necessarily closely linked to it.

If this scheme is true for bacteria, it is difficult to think that it will not be of some relevance for other organisms too. Yet it must be admitted that there is very little evidence as yet for its applicability to fungi. The evidence for the organization of functionally related genes into operons in fungi is so far rather slight. The closest known parallel to the bacterial operon-type situation in a fungus was demonstrated by Douglas and Hawthorne [126] in their studies on mutants of Saccharomyces unable to utilize galactose. They found that the first three enzymes of galactose catabolism (galactokinase, galactose-1-phosphate–UDP-glucose trans-ferase and UDP-galactose epimerase) were controlled respectively by the three closely linked genes *ga-1*, *ga-7* and *ga-10*. The action of these genes was shown to be co-ordinated in two ways. A recessive mutation (*i⁻*) at another unlinked locus caused all three enzymes to be produced constitutively instead of only in response to galactose. This mutation seems quite analogous to the supposed repressor-negative mutations known in the *E. coli* β-galactosidase system. Secondly, another recessive mutation, *ga-4*, at yet another unlinked locus, prevented formation of all three enzymes when it was homozygous. This type of mutation does not correspond to any of the types of operator mutants described by Jacob and Monod but seems rather to cause the loss of something necessary for the induction of the group of enzymes.

In Neurospora it seems likely that some of the genetic regions, such as *his-3*[1] [80, 198] and *pyr-3* (cf. p. 193), which have been regarded as single genes controlling enzymes with complex functions, will turn out to be groups of genes co-ordinated in operons and controlling groups of enzymes acting in sequence. Apart from this possibility, which is only now being explored, the evidence from Neurospora amounts to little more than a number of fairly close linkages between genes of related function. Some relevant cases in *Neurospora crassa* will be seen in the linkage maps shown in Fig. 21. In two of these cases something is known of the enzymes corresponding to the genes concerned. In the fifth linkage group there appear to be two functionally distinct genes some four units apart governing two enzymes catalysing sequential steps in the synthesis of both isoleucine and valine (Wagner *et al.* [557]; Bernstein and Miller [33]). The situation here is still somewhat confused as regards complementation relationships, and there is some indication that the two genes may not be entirely independent in their functioning. In the second linkage group Gross and Fein [210] have shown three genes concerned in the synthesis of aromatic substances in general. One of these, *arom-3*, is only distantly linked to the others, and is concerned with a late step in the synthetic sequence. Of the other two, which are closely linked about 0·3

1 See also Ahmed, Case & Giles in *Brookhaven Symposium in Biology* (1964).

unit apart, *arom-1* is concerned with the enzyme dehydroshikimate reductase (cf. Fig. 37), while *arom-4* is thought to control the enzyme immediately following in the sequence. A further mutant, designated *arom-2*, is non-complementing with both *arom-1* and *arom-4* (though it complements with *arom-3*) and, in addition, lacks 5-dehydroquinate synthetase and 5-dehydroquinase, two enzymes catalysing the steps immediately preceding the reaction catalysed by dehydroshikimate reductase. *arom-2*, in fact, is lacking the enzymes for the whole pathway excepting the earliest known step and the late step lacking in *arom-3*. The most obvious interpretation is that *arom-2* is a deletion covering a cluster of perhaps four genes, all tightly linked and controlling sequential steps in aromatic biosynthesis. Another possibility is that it represents a mutation in an operator gene, and that the entire cluster (including *arom-1* and *arom-4* but excluding *arom-3*) is an operon in Jacob and Monod's sense.

Induction of enzyme formation is well known in Neurospora [240, 208] and mutants are known in which the timing or the extent of the induction of a given enzyme are altered, rather its structure. A good illustration of this is provided by the work of Horowitz *et al.* [240] on *N. crassa* tyrosinase. As we have seen tyrosinase structure in this species depends on a locus T; the alleles T^S and T^L determine thermostable and thermolabile varieties of the enzyme, respectively. But at least two other loci, *ty-1* and *ty-2*, can mutate so as to cause a drastic reduction in the *amount* of enzyme produced during growth on minimal medium. The mutants *ty-1* and *ty-2* each behave as recessive to the corresponding wild type allele, *ty-1$^+$* or *ty-2$^+$*, in a heterocaryon. A heterocaryon $T^S ty-1^+$ $T^L ty-1$ produced a mixture of T^S and T^L type tyrosinases, the latter being indistinguishable from the enzyme formed $T^L ty-1^+$ homocaryons. Thus the type of tyrosinase determined by a nucleus depends on its allele at the T locus, and is not at all affected by the presence of *ty-1*, even though the latter allele, when in a homocaryon, suppresses almost all tyrosinase production. Analogous experiments involving *ty-2* gave similar results.

Strains homocaryotic for *ty-1* or for *ty-2* can be induced to form tyrosinase by the addition of an aromatic amino acid to the culture medium, and in these conditions the tyrosinase formed is of a kind determined by the state of the T locus. Wild type *N. crassa* produces little or no tyrosinase in minimal medium containing relatively high levels of sulphate, but the repressing effect of sulphate can be overcome by the addition of an aromatic amino acid. Thus a type of enzyme induction necessary in wild type only in high sulphate medium, is necessary in *ty-1* and *ty-2* strains at all sulphate levels. It is quite likely that the *ty-1* and *ty-2* loci have their effects through increasing the intracellular concentration of sulphate or of some repressor of tyrosinase derived from sulphate.

Enzyme repression is also known in Neurospora [125], but the effects reported are very unspectacular compared with those common in bacteria. No fungal mutations with clear effects on enzyme repressibility are known to us.

CHAPTER 9

COMPARATIVE GENETICS AND PHYSIOLOGY OF MATING TYPE AND SEXUAL DEVELOPMENT

A great many fungi, comprising the very diverse group called the Fungi Imperfecti, have no known sexual reproduction although some of these may have the possibility of parasexual recombination as discussed in Chapter 5. In many groups, such as the Penicilliaceae, it seems clear that the imperfect species are related to sexually reproducing ones and must have originated from them. However, since so many species have retained the rather elaborate processes involved in the sexual cycle this must mean that they derive some benefit from it. For some fungi the important feature may be that the zygotes or the products of meiosis are highly resistant structures; however the same end can often be achieved more simply by the production of sclerotia or chlamydospores. Undoubtedly the decisive advantage conferred by sexual reproduction is the variability which it generates through genetic reassortment. The reassortment of genetic differences will by definition only occur if the haploid nuclei which fuse are different. Self-fertilization of a haploid organism is genetically equivalent to clonal reproduction. It is therefore not surprising to find in the fungi mechanisms which reduce self-fertilization and encourage outbreeding. What is perhaps surprising is the great variety of methods employed in the different groups of fungi.

Broadly, sexual reproduction in the fungi is subject to two kinds of control. The first is responsible for the differentiation of sex or mating type while the second is responsible for the steps of development leading from the production of gametes, or their equivalents, to meiosis and the liberation of its products.

SEX AND MATING TYPE CONTROLS

There are two means of minimizing or eliminating self-fertilization; one is sexual dimorphism in which the fusing gametes are of two kinds distinguished on the basis of size, motility or the direction of nuclear transfer at the time of fusion, and are produced on different mycelia; the

other is the differentiation of haploid mycelia into two or more morphologically similar cross-compatible, but self-incompatible, groups.

Sexual dimorphism

Although common in the algae, sexual dimorphism is almost entirely
limited to the aquatic Phycomycetes among the fungi. In this group information on the nature of the genetical control is very limited. In some
forms sex is not genotypically determined. As an example we may take
Emerson's [140] studies of sex determination in *Blastocladiella variabilis*.
Male plants each bear a single orange coloured gametangium in which are
formed a number of uniflagellate male gametes. Female plants bear a
similar, but colourless, gametangium in which slightly larger uniflagellate female gametes are formed. Female gametes from a single gametangium may develop parthenogenetically and form new gametophytes some
of which however bear orange male gametangia the others bearing colourless female gametangia. In each of nine generations of transfer of par-
thenogenetic female gametes a proportion of male gametophytes
appeared. Each haploid gametophyte thus carries determinants for maleness and femaleness.

In other forms sex may be genetically determined. Couch [88] observed
that isolated strains of several species of the filamentous aquatic Phyco-
mycete Dictyuchus remained either male or female in reaction. Each
strain formed only sporangia in culture alone but when placed in contact
with a strain of the other sex consistently formed antheridia or oogonia
according to whether it was male or female. Couch was able to germinate
zygotes (oospores) from such crosses and found that they gave rise to
either a branched mycelium or a short hypha bearing a sporangium.
When tested, the mycelia were either wholly male or wholly female or
reacted with tester stocks of both sexes. Different parts of a single mycelium were sometimes found to be of different sexes or of mixed reaction. The same was also true of mycelia arising from the sporangiospores.

It has been suggested [437] that polyploidy or multiple sex factors
might explain these results. An alternative explanation would be that the
mycelia of mixed reaction were heterocaryotic having both male and
female nuclei. Where the mycelium was homocaryotic only one sexual
reaction would be found. The first description of heterocaryosis by
Burgeff [62] was an explanation along similar lines for germinating zygospores of *Phycomyces nitens* which were sometimes either self-sterile but
cross-fertile, or self-fertile. Achlya, another genus of aquatic Phycomycetes showing sexual dimorphism, has been the subject of extensive

studies by Raper (see pp. 227–229). Unfortunately the difficulty of germinating the oospores of this group has completely blocked its genetical investigation.

Among the Ascomycetes, Olive [388] has reported that ascospores of *Ascosphaera apis* give rise to male or female mycelia in a 1:1 ratio. Only the female mycelium produces ascogonia and then only when grown together with the male mycelium. The trichogynes from the ascogonia on the mated female mycelium fuse with male vegetative hyphae.

When male and female gametes are produced on the same plant, *e.g.* *Allomyces arbuscula* and *A. macrogynus* or on genetically identical plants, *e.g. Blastocladiella variabilis*, and there are no other barriers to their copulation, self-fertilization is likely to be very common. While in many fungi there is no barrier to selfing, in others only nuclei of different genotype are able to fuse and form diploid zygotes. These latter fungi are called heterothallic. Dictyuchus and Ascosphaera are probably of this type; the two kinds of nuclei being male and female. In most other heterothallic fungi, however, the differentiation of cross-compatible but self-incompatible groups is independent of the differentiation of sex organs. In many of the heterothallic Phycomycetes and Ascomycetes and in the rusts male and female organs are normally formed by each of the two mating types. In the higher Basidiomycetes no sex organs at all are formed and many more than two mating types may be found.

Heterothallism: two allele systems

The simplest form of heterothallism is that in which two compatible mycelia possess different alleles at a single locus. In *Neurospora crassa* two alleles *A* and *a* control mating. Fertilization of protoperithecia borne on an *A* mycelium will only succeed with conidia or hyphae from an *a* mycelium and *vice versa*. There is no morphological difference between the two mating types. This simple two allele system is very widespread, being found in all groups of fungi except the higher Basidiomycetes.

Several examples of the breakdown of this control are known which are instructive because they reveal some of the functions of the controlling genes and that they may mutate or be overridden.

In the Ascomycete *Chromocrea spinulosa* each ascus produces sixteen spores, eight of which are small and eight large [348]. Cultures produced by the small ascospores are self-sterile but cultures from the large ascospores form scattered perithecia whose asci, though the result of selfing, each have eight large and eight small spores. When a colony from a small spore is paired with a colony from a large spore abundant production of hybrid perithecia occurs along the line of confluent growth. It seems that the gene controlling spore size also determines mating type, and that mutation from 'large' to 'small' in a small proportion of the nuclei in a large colony enables it to produce scattered fertile perithecia. The reverse mutation has not been observed.

Ahmad [1] reported the frequent origin of one mating type allele by mutation of the other in Saccharomyces. More recently Takahashi [537] has described complementary genes which when present in certain pair-wise combinations render single ascospore cultures of Saccharomyces diploid, thus overriding the control of the mating type locus. The diploid stocks from single ascospores were able to sporulate, but retained their mating type reactions as judged by tests carried out with the aid of complementary auxotrophic markers. Only compatible, complementary haploids or diploids gave prototrophic growth on minimal medium [539].

Winge and Roberts [587] described a cross between the heterothallic species *Saccharomyces cerevisiae* and the homothallic *S. chevalieri*. Hybrid asci carried two spores which gave rise to diploid cultures and two spores which gave haploid cultures of either *a* or *α* mating type. Evidently a gene *D*, responsible for diploidization, had segregated in the asci. Hawthorne [219] recently reported that *D* in fact causes the mutation of the mating type allele in a haploid cell to the opposite allele. Under its influence *a* mutates to *α*, or *α* to *a*, at rates which vary from one in two to one in 30 cell divisions. Once the mutant allele is expressed the haploid cells fuse and stable diploids heterozygous for mating type are formed. The heterozygous condition of the mating type alleles apparently blocks any further action of *D*.

In the yeast *Schizosaccharomyces pombe* Leupold [303] described three alleles at the mating type locus *h*: strains carrying h^+ and h^- were heterothallic but those carrying h^{90} were homothallic with some 87 per cent of their cells producing ascospores. Mutation from one allele to another was common. Later work (307) revealed that the mating type locus was complex. Crosses were made between h^+ and h^- stocks carrying the markers *his-7* and *his-2* on either side of the mating type locus. Approximately 0·4 per cent of the progeny were homothallic and 86 out of 103 of these were recombinant for *his-7* and *his-2*, suggesting that recombination within the mating type locus had also recombined the two markers. Some 701 complete tetrads were analysed from the cross $\dfrac{\textit{his-7}\ h^+\ \textit{his-2}}{+\ \ h^-\ \ +}$

Seventeen tetrads produced homothallic spores and these are shown in Table 25.

The tetrads were of two kinds. The first kind had two or four homothallic spores and showed no recombination of the markers. Each homothallic spore had the same marker constitution as the h^+ parent. The second kind, of which there were fifteen, had only one homothallic spore, showing recombination of the markers. Eleven of these resulted from a single exchange and four from double exchanges.

The first group of tetrads could have arisen from homothallic mutants present in the h^+ parent strain. The second group of tetrads can be accounted for by recombination within the mating type region. The h^+ stocks (his-$7\ h^+\ +$) which were reciprocal in marker constitution to the

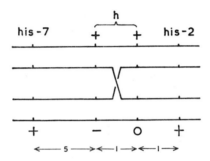

FIG. 43. Interpretation of the origin of homothallic *Schizosaccharomyces pombe* by crossing-over (cf. Table 25).

h^{90} recombinants ($+\ h^{90}\ his$-2) when crossed with h^- failed to yield any homothallic recombinants among 26,000 ascospores tested. There were thus two types of h^+ stocks, parental and recombinant, distinguishable only by a genetic test. This was confirmed by crossing the h^+ recombinant with h^{90}, when in six out of 169 tetrads h^+ parental and h^- ascospores were recovered. There are at least two ways of interpreting this evidence. The first is that the mating type locus has duplicated. The duplicate locus, approximately one map unit away, has two alleles. One independently confers a heterothallic reaction of the h^+ type; the other does not contribute to the mating type reaction. The homothallic condition would result from the combination of a '—' allele at the first locus with a '+' allele at the second. The two kinds of h^+ mating type strains could be represented as 'h^+h^+' and 'h^+h^0'. This interpretation is illustrated in Fig. 43.

The second theory is that a modifier, at a locus approximately one map unit from the mating type locus, converts what would otherwise be an $h-$ strain to a homothallic one. The modifier is allele specific having no effect on $h+$. Both interpretations leave undecided the problem of whether the second locus is to be considered as a part of the same gene determining $h+$ versus $h-$. This problem will arise again in connexion with the Hymenomycetes.

TABLE 25. Tetrads of *Schizosaccharomyces pombe* with homothallic spores, after Leupold [307]

Cross	Markers not recombined		Markers recombined	
$\dfrac{his-7\ h+\ his-2}{+\ \ \ h-\ \ \ +}$	1: $\left\{\begin{array}{l} his-7\ h^{90}\ his-2 \\ his-7\ h^{90}\ his-2 \\ +\ \ \ h-\ \ \ + \\ +\ \ \ h-\ \ \ + \end{array}\right.$	11: $\left\{\begin{array}{l} his-7\ h+\ his-2 \\ +\ \ \ h-\ \ \ + \\ his-7\ h+\ \ \ + \\ +\ \ \ h^{90}\ his-2 \end{array}\right.$	1: $\left\{\begin{array}{l} his-7\ h+\ his-2 \\ +\ \ \ h-\ \ \ + \\ +\ \ \ h+\ \ \ + \\ his-7\ h^{90}\ his-2 \end{array}\right.$	
701 tetrads	1: $\left\{\begin{array}{l} his-7\ h^{90}\ his-2 \\ his-7\ h^{90}\ his-2 \\ his-7\ h^{90}\ his-2 \\ his-7\ h^{90}\ his-2 \end{array}\right.$	2: $\left\{\begin{array}{l} his-7\ h+\ his-2 \\ his-7\ h-\ \ \ + \\ +\ \ \ h+\ \ \ + \\ +\ \ \ h^{90}\ his-2 \end{array}\right.$	1: $\left\{\begin{array}{l} +\ \ \ h+\ his-2 \\ his-7\ h-\ \ \ + \\ his-7\ h+\ \ \ + \\ +\ \ \ h^{90}\ his-2 \end{array}\right.$	

Note: h^{90} gives self-fertility.

Multiple allelomorph heterothallism, bipolar and tetrapolar systems

The two allele system of heterothallism limits compatibility to 50 per cent of random matings not only between the progeny of a single fruit body but also in the population as a whole. If there are more than two alleles at the mating type locus, then, although the frequency of compatible matings among sibs (i.e. among products of the same fruit body) is never greater than 50 per cent, among individuals in the population at large it can approach 100 per cent depending on the number and distribution of alleles. In the Hymenomycetes and Gasteromycetes compatibility is controlled by multiple alleles at one locus, A in the *bipolar* species, or two independent loci, A and B in the *tetrapolar* species. These names derive from the fact that the progeny of a single fruit body in a bipolar form are of two mating types while in a tetrapolar form they are of four mating types. In the tetrapolar species only 25 per cent of random matings between sibs are compatible. Mycelia which carry different A alleles, in the bipolar species, or which differ with respect to both A and B, in the tetrapolar species, are compatible (see Table 26). If they carry alleles in common they are incompatible but can generally form hetero-

SEX AND MATING TYPE

caryons. The product of a compatible mating is a dicaryotic mycelium which is capable of indefinite growth and which forms fruiting structures in a suitable environment (see Fig. 44).

In many tetrapolar Basidiomycetes the two mating type loci A and B control different parts of the process of dicaryon formation. This was established by studying the properties of heterocaryons formed by pairing stocks of various mating types. The four types of pairing possible are summarized in Table 27 for *Coprinus lagopus*. *Schizophyllum commune* [391] and the Gasteromycete *Cyathus stercorareus* [183] behave in essentially the same way. In all three species the formation of clamp connexions is controlled by the A locus, since only when this locus is heterozygous are clamps or false clamps formed. Nuclear migration is controlled by the B locus since it only occurs when the mated mycelia possess different B alleles. Takemaru [541] has claimed a similar functional differentiation between the A and B mating type loci in eight

TABLE 26

a) Matings between progenies of bipolar dicaryons (A_1+A_2) and (A_3+A_4)

	A_1	A_2	A_3	A_4
A_1	−	+	+	+
A_2	+	−	+	+
A_3	+	+	−	+
A_4	+	+	+	−

(b) Tetrad types formed by the tetrapolar dicaryon $(A_1 B_1+A_2 B_2)$ and results of sib matings and matings with progeny of $(A_1 B_1+A_3 B_3)$

		$A_1 B_1$	$A_1 B_2$	$A_2 B_1$	$A_2 B_2$	$A_1 B_3$	$A_3 B_1$	$A_3 B_3$
PD	$A_1 B_1$	−	−	−	+	−	−	+
	$A_1 B_1$	−	−	−	+	−	−	+
	$A_2 B_2$	+	−	−	−	+	+	+
	$A_2 B_2$	+	−	−	−	+	+	+
NPD	$A_1 B_2$	−	−	+	−	−	+	+
	$A_1 B_2$	−	−	+	−	−	+	+
	$A_2 B_1$	−	+	−	−	+	−	+
	$A_2 B_1$	−	+	−	−	+	−	+
T	$A_1 B_1$	−	−	−	+	−	−	+
	$A_1 B_2$	−	−	+	−	−	+	+
	$A_2 B_1$	−	+	−	−	+	−	+
	$A_2 B_2$	+	−	−	−	+	+	+

Note: + indicates dicaryon formation.

other tetrapolar Hymenomycetes. Thus in these species the designation of the mating type loci as A and B is not arbitrary. In the majority of bipolar species the dicaryon bears clamp connexions suggesting that in these forms the mating type locus is homologous with the A mating type locus of the tetrapolar species.

In *C. lagopus* evidence has been recently found that nuclear migration may be facilitated by the conversion of the complex septal pores, described on p. 37 to simple pores [193]. The conversion to simple pores only occurs when mycelia possessing different B alleles are mated, and thus could well be a key function of the B locus.

TABLE 27. Summary of the reactions given by the four different types of mating among monocaryons of *Coprinus lagopus*[535, 536]

$A_1 B_1 \times A_2 B_2$	Nuclear migration		Fruit bodies
	Dicaryon with clamp connexions		
,,	$\times A_1 B_2$	Nuclear migration	Sterile
		Common A heterocaryon	
,,	$\times A_2 B_1$	No, or very limited, nuclear migration	Sterile
		Common B heterocaryon with false clamp connexions	
,,	$\times A_1 B_1$	No nuclear migration	Sterile
		Unstable common AB heterocaryons	

Accounts by Whitehouse [574] and Quintanilha and Pinto-Lopes [432] show that of some 230 species of Hymenomycetes and Gasteromycetes studied before 1950 between 10 and 15 per cent were homothallic. Of the remaining heterothallic species 35 per cent were bipolar and from 49 to 55 per cent tetrapolar. Whitehouse estimated the number of alleles at the mating type loci in natural populations of Hymenomycetes to be in the order of magnitude of a hundred per locus. In the tetrapolar Gasteromycetes, *Crucibulum vulgare* and *Cyathus striatus*, however, there are fewer alleles, probably of the order of ten for each locus. Whitehouse suggested that this is probably because the basidiospores are dispersed together in a coherent mass or peridiolum. The consequent inbreeding would reduce the number of alleles.

In the most extensive survey of allelomorphs in a tetrapolar Hymenomycete yet carried out Raper, Krongelb and Baxter [441] found ninety-six A alleles and fifty-six B alleles in a world wide sample of 114 monocaryons of *Schizophyllum commune*. Estimates of 339 (with 5 per cent probability limits of 216 and 562) A alleles and sixty-four (with 5 per cent limits of 53 and 79) B alleles in the natural population of Schizophyllum were made on the basis of these findings.

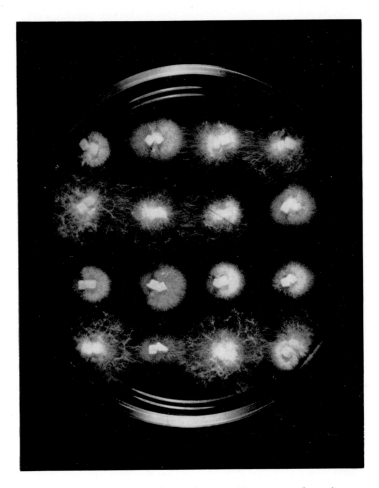

FIG. 44. Matings in *Coprinus lagopus* between 16 segregants from the cross $A_1B_1 \times A_2B_2$ with a tester stock (A_1B_3) carrying a common *A* factor but a dissimilar *B* factor. The tester stock inoculum is placed to the right in each pairing. 8 of the 16 pairings are incompatible due to shared *A* factors: 8 are compatible and show the formation of the vigorous dicaryotic mycelium.

Structure of the A *locus in tetrapolar species*

The structure and function of multiple allelemorphic loci have aroused a good deal of interest. In Schizophyllum Kniep [288] reported the appearance of non-parental mating type factors among the progeny of controlled matings. These were attributed to mutation until Papazian [389] found similar non-parental A specificities in tetrads and random spores which had evidently arisen by recombination between two genetic loci. These findings were confirmed and expanded by Raper and his associates at Harvard [438, 439]. A random sample of thirty-one different A factors was crossed with A_{41} from Massachusetts. Reciprocal non-parental A factors appeared in the progeny of all but two of the crosses in frequencies ranging from 3·2 to 17·6 per cent. The two sub-units of each A factor can be identified by test matings; the procedure is as follows.

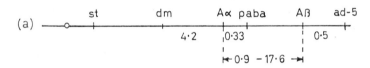

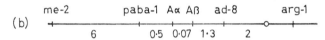

FIG. 45. Linkage maps of the chromosomes carrying the A factors of (*a*) *Schizophyllum commune* [135, 438] and (*b*) *Coprinus lagopus* [111].

The two sub-units of the common parent A_{41} were designated α_1 and β_1 where α and β represent two loci. When A_{41}, which we may now write as $A_{41}\alpha_1—\beta_1$, was crossed with A_{51} from Australia, carrying the morphological marker dm, two non-parental A's were formed showing that A_{51} and A_{41} were different at both α and β. One of the recombinant types was always associated with dm and the other with dm^+. Hence A_{51} can be written $A_{51}\alpha_2—\beta_2$ and the recombinants as dm^+ $A\alpha_1—\beta_2$ and dm $A\alpha_2—\beta_1$; dm is evidently closely linked to the α locus, as shown in Fig. 45. A third allele A_{49} from Canada was crossed with A_{41} carrying the marker dm and also yielded two non-parental A factors one of which was dm^+ and incompatible with $A\alpha_2—\beta_1$. From this we may deduce that A_{49} is $A_{49}\alpha_2—\beta_3$ having the same α specificity as A_{51}. From similar test crosses Raper and his colleagues concluded that in their random sample of thirty-two A factors there were nine $A\alpha$'s and twenty-five $A\beta$'s. Only A factors which have identical α's and β's are incompatible:

15

$$Aa_1 - \beta_1 \quad \times \quad Aa_1 - \beta_1 \text{ incompatible;}$$
$$\text{,,} \quad \times \quad Aa_1 - \beta_2 \text{ compatible;}$$
$$\text{,,} \quad \times \quad Aa_2 - \beta_1 \text{ compatible;}$$
$$\text{,,} \quad \times \quad Aa_2 - \beta_2 \text{ compatible.}$$

It is of interest to note that an apparently unrelated locus, *paba*, is situated between the α and β loci of Schizophyllum.

Preliminary reports, indicate that the *B* locus of Schizophyllum is also composed of at least two sub-units.

In *Coprinus lagopus* Day [108] has also reported a two sub-unit structure for the *A* locus. The sub-units recombine with a frequency of 0·07 per cent and are thus closer together than those of Schizophyllum. Because of the low frequency of recombination Day used closely linked markers on either side of the *A* locus (see Fig. 45b) to improve the yield of *A* recombinants. In crosses of *paba*-1 *ad*-8+ × *paba*-1+ *ad*-8 0·2 to 0·9 per cent of the progeny are prototrophs which may be selected by plating basidiospores on minimal medium. These were tested for non-parental *A* reactions and approximately 4 per cent carried a non-parental *A* factor. The reciprocal class, which is presumably doubly auxotrophic, is lost. By this means eight different *A* factors were found to include four *Aa*'s and at least five different *Aβ*'s [110].

The sub-units of the *A* locus in Coprinus may either be regarded as sites in the same functional region of the chromosome or as distinct genes. The latter view seems more acceptable in Schizophyllum since between the sub-units there is the apparently unrelated *paba* locus, and the sub-units recombine with a comparatively high frequency. In *Collybia velutipes* Takemaru [541] has reported two sub-units at the *A* locus approximately 1·3 map units apart, and two sub-units at the *B* locus approximately 19·4 map units apart. Evidence that one of the mating type factors consists of at least two sub-units has also been found in *Pleurotus ostreatus* [439, 548]. In extensive tests of the bipolar species *Polyporus betulinus*, J. H. Burnett (personal communication) was unable to find non-parental *A* factors which could be interpreted as inter-sub-unit recombinants. In tests of more than 10^5 spores of both Coprinus and Schizophyllum no more than two non-parental *A* specificities have ever been recovered from a single cross apart from forms in Coprinus, discussed below, which seem to be mutants at the *A* locus. Nevertheless it would be surprising if multiple sites did not exist within the *A* locus sub-units of Schizophyllum and Coprinus in view of the well-established complexity of many genes in fungi. It is most improbable that the relatively large numbers of alleles present at each sub-unit in the population at large are all determined at one site. It seems more likely that there are many sites which the relatively crude analyses have not yet resolved.

Mutation of the mating type loci in tetrapolar species

The appearance of non-parental alleles among the progeny of a single fruit body was first attributed to mutation of the mating type locus. As Whitehouse [574] has pointed out, in some of the early literature no tests for contamination with other laboratory stocks are reported. The examples recorded by Kniep [288] in Schizophyllum can, as we have seen, be explained by recombination.

Some other examples which appear to be the result of mutation were found by one of us in studies of the fruiting behaviour of common A heterocaryons of Coprinus [109]. These heterocaryons are formed from matings between mycelia which possess an A factor in common. They are moderately stable and when made up from complementary auxotrophic mutants show prototrophic growth on minimal medium [535].

Common A heterocaryons fruit rarely, but when they do the fruit bodies generally yield mutant alleles at the A locus. Thus a fruit body formed on the heterocaryon $(A_5B_5 + A_5B_6)$ would yield the following spore types:

$$A_5B_5, A_5B_6, A_{5m}B_5 \text{ and } A_{5m}B_6$$

The cultures designated A_{5m} produce false clamps which sometimes contain a nucleus trapped by the failure of the clamp to fuse with its neighbouring hyphal cell. These cultures fruit when inoculated to dung and produce only A_{5m} spores. Mycelia carrying A_{5m} are compatible with A_5 provided the stocks tested carry different B alleles. Some ten A mutants of independent origin were obtained in this way and were designated A_{5m-1}, A_{5m-2} etc. or A_{6m-1}, A_{6m-2} etc. according to whether the allele carried by the original heterocaryon was A_5 or A_6. These mutants closely resemble a mutant of similar origin described by Quintanilha in 1935 [430]. Matings between stocks carrying the same A_m mutant allele but different B alleles resulted in the formation of dicaryons bearing normal clamp connexions. A_m mutants were also crossed with normal alleles and the progenies screened for α—β recombinants and for the original parental A allele. The original A factor was never recovered from a cross of this kind. Six of the ten mutants still had the original α sub-unit but from three of the remaining four mutants neither α nor β were recovered.

This evidence confirms that the mutations either involve or are very close to one or both of the sub-units.

A second type of mutant was recovered from a single fruit body formed on the common A heterocaryon ($A_5B_5 + A_5B_6$). All the progeny of this one fruit body bore false clamps and fruited, but were incompatible with A_5 and B_5 testers. However, when the mutant bearing false clamps was crossed with normal A_6, normal A_5, normal A_6 and A_5 and A_6 types with false clamps appeared in the progeny in equal numbers. Evidently the mutant phenotype was due to a recessive gene, called su–A, which was not linked with A but which suppressed its normal function allowing the development of false clamps.

In Schizophyllum Parag [391] has described three B factor mutants which arose in common B heterocaryons. The mutants were fully compatible with 55 other B alleles. Homocaryons carrying the mutants were similar in morphology to the distinctive common A heterocaryon of Schizophyllum. All three mutants possessed functional α sub-units which suggests that the mutations had occurred in or near the β sub-unit of the B factor.

The Rapers [443, 435] have also found suppressor, or modifier, mutants which disrupt the normal function of the A factor in Schizophyllum. Nine such mutants so far examined are dissimilar to su–A in Coprinus since they are all dominant to their wild type alleles. Most of them are only expressed in the presence of two different B factors in a heterocaryon or disomic, or in the presence of a B factor mutant. However, as in Coprinus, the mutants suppressed all the A factors with which they were tested.

We have been considering examples of changes due to mutation or recombination within the loci controlling heterothallism, or to mutation of suppressor loci, which have either resulted in self-fertility or have increased the proportion of fertile sib matings.   On p. 219 we will consider another way in which self-fertility may be superimposed on heterothallism.

The Buller phenomenon and vegetative recombination

If a dicaryon and homocaryon of a tetrapolar Basidiomycete are placed together the homocaryon generally becomes dicaryotized by the migration of nuclei from the dicaryon into the homocaryon. Discovered by Buller [61], the phenomenon was named after him by Quintanilha.

When a dicaryon of the type ($A_1B_1 \times A_2B_2$) is mated with a homocaryon of the type A_1B_2 or A_2B_1, the homocaryon frequently becomes dicaryotized in spite of the incompatibility between the homocaryon nucleus and both dicaryon nuclei. Two explanations have been found; either the homocaryon nuclei are replaced by the two nuclei of the dicaryon, or the homocaryon is dicaryotized by a recombinant nucleus A_2B_1 or A_1B_2 [94]. Recent investigations [136, 392] point to at least two different means whereby the recombinant nuclei might be formed: (i) a meiotic type of event, involving recombination of mating type and other markers, possibly located on several different chromosome arms rather than on just one as would be expected in ordinary mitotic recombination; and (ii) a recombination event which involves all three nuclei but often results only in the transfer of mating type factors, and no other markers, from the dicaryon nuclei. Evidence for the class (i) type of recombinant comes from the finding that macerates of unmated dicaryons may give rise to low frequencies of recombinant homocaryons recoverable by total isolation [392].

In *Schizophyllum commune*, Ellingboe [135] made the cross ($A_{41}B_{42}$ ad-5 nic-2 + $A_{42}B_{41}$ paba) × $A_{41}B_{41}$ x-11 x-15, where x-11 and x-15 represent unlinked mutations to auxotrophy with unknown requirements. Out of the 86 dicaryons produced on the homocaryon which were analysed, 43 were class (i) and 20 were class (ii). In the latter all the unselected markers came from the homocaryon (i.e. their constitution was $A_{42}B_{42}$ x-11 x-15 + $A_{41}B_{41}$ x-11 x-15). The remarkable feature was that, though all of these recombinant nuclei carried both the $A\alpha$ and the $A\beta$ sub-units of the A_{42} factor of the parent dicaryon, none of them carried the *paba* marker located between $A\alpha$ and $A\beta$, 0·3 map units from the former. The remaining dicaryons included two like class (ii) except that the markers *paba* and *nic* were present in the recombinant nucleus, and 20 which possessed recombinant A factors in one or both nuclei and could have been either class (i) or class (ii).

Because any normal meiotic or mitotic crossing-over mechanism would be expected to give only a very low frequency of class (ii) recombinant nuclei, the authors suggest some other, unspecified, means of genetic transfer. The problem clearly demands further study.

Secondary homothallism

When it is possible to include nuclei of both mating types of a heterothallic fungus in a common cytoplasm the heterocaryon so formed is usually self-fertile. In some fungi two nuclei of opposite mating type are regularly included in the same spore. Asci of *Neurospora tetrasperma* and *Podospora anserina* normally contain only four spores. When they germinate they form a self-fertile mycelium. Occasionally five spores are formed in an ascus, two of which are smaller than the other three. The smaller spores give rise to self-sterile mycelia which fall into two cross fertile groups corresponding to *A* and *a* in *N. crassa*. When inter-crossed the self-sterile mycelia of *N. tetrasperma* or Podospora form perithecia containing four-spored asci.

Neurospora tetrasperma

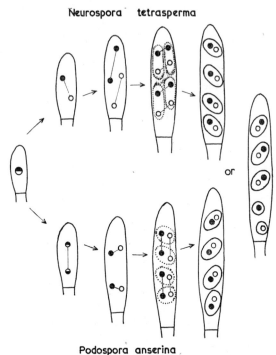

or

Podospora anserina

FIG. 46. Two mechanisms for the maintenance of secondary homothallism: *Neurospora tetrasperma* (top) and *Podospora anserina* (bottom).

Different mechanisms are responsible in the two fungi for ensuring that nuclei of opposite mating type are included in the same spore. In *N. tetrasperma A* and *a* nearly always segregate at the first division [473]; there is rarely an exchange between the mating type locus and the centro-mere. In the second and third divisions in the ascus the spindles overlap each other so that daughter nuclei from different figures pass each other (Fig. 46). The ascospores are delimited around adjacent pairs of nuclei of different mating type.

By contrast, segregation for mating type in Podospora occurs at the second division in 98 per cent of the asci [451]. The second division spindles lie transversely across the ascus. The four third-division spindles are more or less parallel to the long axis of the ascus and are arranged in two pairs. After the third division sister nuclei move apart and the asco-spores are formed around pairs of non-sister nuclei as in *N. tetrasperma* [171] (Fig. 46).

In *Podospora anserina* Esser [147] has examined the mating relation-ships of isolates derived from uninucleate ascospores. As we would expect, strains of this kind, from one geographical locality, fall into two mating type groups + and −. However, when strains from different geographical areas were compared it was clear that another incompati-bility control was overriding the basic mating type system. In its extreme form complete incompatibility between strains which were known to be + and − resulted. A less extreme form, called non-reciprocal or semi-incompatibility, resulted in the formation of fertile perithecia as a result of fertilization of protoperithecia of one parent by spermatia from the other. The reciprocal cross did not occur. Esser gave these pheno-mena the name of heterogenic incompatibility. Four unlinked loci, each with two alleles, control the system. Semi-incompatibility results from the matings $ab \times a_1b_1$ and $cv \times c_1v_1$ and complete incompatibility from a combination of the two, namely $abc_1v_1 \times a_1b_1cv$. All other combinations of the four pairs of alleles lead to no disturbance in reciprocal com-patibility.

When the progeny of the two semi-incompatible matings were examined further effects of these genes were revealed. Thus one of the two recombinant types of haploid homocaryon from each (a_1b and c_1v respectively) showed reduced vigour which could be restored by muta-tion to ab or cv. Also, heterocaryons arising from binucleate ascospores carrying the genes a_1 and b (or c_1 and v) were unstable, apparently because nuclei carrying a_1 (or c_1) were eliminated. Everything happens as if a cytoplasmic product of b is inimical to the survival of nuclei carrying the non-allelic gene a_1, and a similar relationship holds between v and c_1. If the effects of b and v are exerted through the cytoplasm the differences in fertility between reciprocal crosses are understandable. This system seems more akin to heterocaryon incompatibility (see p. 233) than to any of the other sexual incompatibility mechanisms discussed in this chapter.

Secondary heterothallism, of various degrees of efficiency, occurs also in the Hymenomycetes. *Coprinus sassii* (syn. *C. ephemerus f. bisporus*) produces two spores on each basidium. In a study of two collections from

Denmark Lange [295] found that 80 per cent of the spores produced mycelia with clamps. Mycelia from the remaining 20 per cent were without clamps and fell into two self-sterile, cross-fertile, groups. A cytological study [481] had shown earlier that each basidiospore receives two nuclei. *C. sassii* appears to be bipolar and depending on whether the two nuclei included in each spore are of the same or different mating type the spores are homothallic or heterothallic.

Lange studied two four-spored species of Coprinus, *C. plagioporus* and *C. subpurpureus* and found, even in these, as many as 50 per cent homothallic spores, an earlier account [431] of *C. plagioporus* reported approximately 88 per cent homothallic spores. Both species were basically tetrapolar, the mycelia without clamps falling into four self-sterile groups. *C. subpurpureus* was especially interesting for while four collections were of this kind, two others produced only heterothallic spores.

The production of homo- and heterothallic mycelia from spores formed by the same fruit body was called by Lange *amphithallism*. Two explanations are possible for amphithallism in the four-spored species. The first was demonstrated in another agaric *Omphalia flavida* by Sequiera [484] who found that a mitotic division may take place after meiosis either in the basidium or in the basidiospores. When division occurs in the basidium two compatible nuclei may be included in the same spore. The mechanism whereby this can occur in as many as 88 per cent of the spores of a basically tetrapolar form is not clear. It seems likely that there may be some attraction between compatible nuclei so that they pass together from the basidium through the sterigma to the spore. In two-spored forms like *C. sassii* an additional mitotic division following meiosis is unnecessary. It is possible that mechanisms similar to those in *N. tetrasperma* or Podospora ensure that unlike adjacent nuclei are included in the same basidiospore. The second explanation is hypothetical and might apply to examples in which 50 per cent of the spores are homothallic. If self-fertile mutants of the type discovered in laboratory stocks of Coprinus and Schizophyllum also occur in the wild then dicaryons of bipolar and tetrapolar forms, heterozygous for a mutant at the *A* locus would give 50 per cent apparently homothallic spores. These might be distinguished from true dicaryons by possessing false clamps. The remaining 50 per cent would all be self-sterile and cross-sterile with each other by virtue of the shared normal *A* allele. The presence or absence of a *B* factor would only be revealed in crosses with stocks from fruit bodies carrying a different normal *A* allele.

On the other hand, if the original fruit body were heterozygous for an unlinked suppressor gene, like *su–A* in Coprinus, the 50 per cent of the progeny with no clamps, carrying *su–A+*, would fall into two or four groups according to whether the basic control was bipolar or tetrapolar. Lange's collections of *C. subpurpureus* could be explained in this way. Four fruit bodies could have carried *su–A* in the heterozygous condition while two did not, spores with *su–A+* were of the four mating types expected in a tetrapolar Hymenomycete.

The second theory could be best tested by analysing tetrads and observing the mating reactions of the clamp-bearing mycelia.

DEVELOPMENTAL CONTROLS AND POSSIBLE ORIGINS OF HETEROTHALLISM

Those controls of sexual reproduction which are concerned with the steps of development of sex organs and the events which follow nuclear fusion are less amenable to genetic analysis. We shall limit our discussion to some laboratory mutants of three homothallic Pyrenomycetes. These mutants show sexual restrictions which in some respects resemble those imposed by the mating type genes in heterothallic fungi.

Glomerella cingulata

Single ascospore cultures of the plant pathogen Glomerella isolated from nature ordinarily produce fertile perithecia in large glomerate masses or clumps. Many variants including sterile forms have been described from cultured material. So-called plus and minus strains were recorded as early as 1914 [131]. In 1952 Wheeler and McGahen [573] summarized much of the intervening work and reported on the effects of mutants at four loci on sexual reproduction in a strain isolated from Ipomoea (sweet potato). Two mutant alleles were found at each of two loci A and B. Mutants carrying the allele A_1 were sexually sterile, producing only conidia. The mutants A^+B_1 and A^+B_2 produced scattered perithecia but those on A^+B_2 were almost completely sterile because of ascus abortion. The allele A_2 was responsible for perithecial clumps which were smaller than those of wild type when associated with B^+, and enhanced black pigment production but reduced the fertility of perithecia when associated with the alleles B_1 and B_2. Two other mutant genes F_1 and st_1 suppressed the development of perithecia beyond initials, and a mutant dw_1, responsible for dwarf ascospores, caused aborted meiosis in 70–80 per cent of asci.

Wheeler [571] has shown that the effects of the mutants which bring about sterility can be represented as genetic blocks in the sequence of normal development as shown in Fig. 47.

Crosses between any of the self-sterile mutants shown in Fig. 47 result in the production of dense lines of fertile perithecia through complementary action of wild type alleles. Not all of the perithecia are the result of crossing. For example in some matings between self-sterile A_1 stocks and self-fertile stocks 'selfed' perithecia were formed in addition to the 'crossed' perithecia. Ascospores from the selfed perithecia yielded only non-perithecial A_1 cultures. Perithecia yielding only F_1 cultures were also found in a cross between F_1 and A_1. In the crossed perithecia recombination between the mutants produce self-fertile homothallic forms.

It was also shown that culture filtrate from the wild type stock A^+B^+ when applied to young cultures of A^+B_2 increased the percentage of fertile perithecia from ca. 2–3 to from 43 to 94 in different experiments [128].

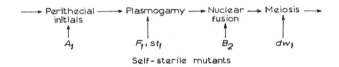

FIG. 47. Genetic blocks in the development of fertile perithecia of Glomerella after Wheeler [571].

Sordaria fimicola

Carr and Olive [74] studied two non-allelic self-sterile mutants of *S. fimicola* designated *st-1* and *st-2*. Mutant *st-1* was spontaneous in origin. It produced a white mycelium bearing protoperithecia and occasional fertile perithecia. Mutant *st-2* was isolated after X-ray treatment of the same stock as that from which *st-1* was derived. Mutant *st-2* was brown and completely sterile showing only slow growth compared with wild type. When mycelia of the two mutants were paired heterocaryotic sectors were formed at the point of contact between them. The heterocaryon was fertile giving rise to perithecia about two-thirds of which were of hybrid origin as judged by the presence or absence of segregation for an ascospore colour marker carried by one of the parents. The remaining third were homozygous for *st-1 st-2+*. The progeny of the crossed perithecia included the class *st-1+ st-2+*, which was homothallic, and three sterile classes including the recombinant *st-1 st-2*.

At the same time nuclear migration occurred from the *st-1 st-2+* parent into the *st-1⁺* st-2 mycelium which, however, remained sterile. It is of interest to note that while the migration heterocaryon was sterile the sector heterocaryon was fertile although both heterocaryons were of the same genetic constitution. The acquisition of the complementary type

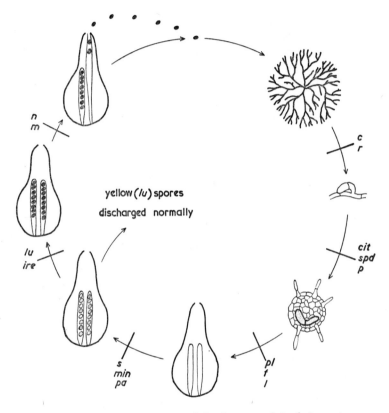

yellow (*lu*) spores
discharged normally

FIG. 48. Genetic blocks in the sexual development of *Sordaria macrospora*.
(From Esser & Straub [148]).

of nucleus by a sterile mycelium failed to produce a change in the already fixed phenotype, although mycelium, formed from the outset under the joint control of the two nuclear types, was fertile. It would be of interest to know if sub-cultures of the migration heterocaryon are sterile or fertile; one would expect them to be fertile.

Stocks carrying the mutant *st-2* were found to be unable to anastomose with each other but matings between *st-2* and *st-2*+ resulted in fertile sector heterocaryons. No selfed perithecia homozygous for *st-2* were ever recovered from such a fertile heterocaryon, so that *st-2* must also block perithecia formation perhaps at the point of nuclear fusion.

Sordaria macrospora

The most complete analysis of mutant blocks in sexual development is probably that of Esser and Straub in *S. macrospora* [148]. They isolated eighteen sterile mutants and four mutants with abnormal perithecial development after X-ray treatment of self-fertile wild type material. The developmental sequence from vegetative mycelium to the liberation of ascospores is shown in Fig. 48. Each of the mutants was found to block development at a certain point. The position of these blocks are shown in Fig. 48 together with symbols for the genes concerned.

Mutants c_1, c_2 and r produced only vegetative mycelia; *cit*, *spd* and *p* only ascogonia and no protoperithecia; *pl*, f_1, f_2, f_3, l_1, l_2, few to many protoperithecia and no perithecia; s_1, s_2, s_3, s_4, *min* and *pa* sterile perithecia; *n*, *ire* and *m* fertile perithecia but with asci which failed to discharge their ascospores. The mutant *lu* produced yellow ascospores which were discharged normally.

Matings between sterile mutants blocked at different points resulted in the production of fertile perithecia where the two mycelia met. Matings between different mutants with the same letter symbol were sterile, or non-complementary, presumably because of allelism, and developed only as far as the blocked step. In many of the complementary matings between self-sterile mutants, normal, selfed perithecia of one or both parents developed in addition to crossed perithecia. No diffusible agents which might account for this were found in the medium and it was concluded that induction of normal development occurred through the conduction of substances through the mycelium.

The mutants of Glomerella and the two species of Sordaria seem to provide the kind of basic controls needed for a two-allelomorph system of heterothallism. There are, however, some important differences between these mutants and mating type genes.

The phenomenon of induced self-fertility found in many of the matings between self-sterile mutants of all three species has no counterpart in heterothallism. Presumably a satisfactory system of self-incompatibility would have to be based on a block in the synthesis of a non-diffusible, cell-limited substance, as could be the case with the *st-2* mutant of *S*.

fimicola. A further difference is that complementary self-sterile mutants, unless they are closely linked, recombine to produce self-fertile forms which again have no counterpart in heterothallism apart perhaps from the homothallic recombinants of Schizosaccharomyces (see p. 210). The chances of producing two complementary, self-sterile mutants blocked at the same step and closely linked appear to be remote. If they were blocked at different steps this might introduce a dimorphism not found in heterothallic fungi apart from rare sexual dimorphism. Olive [388] has discussed models which to some extent by-pass the difficulty of recombination. He suggested that two allelic, but complementary, mutants might arise in a locus controlling a step in sexual development. That mutants may be complementary though allelic is now well established (see Chapter 6). Recently El-Ani and Olive [132] recovered two closely linked mutants in *S. fimicola* which fulfil the requirements for mating type controls. The first mutant, *a-3*, produces asci in which ascospores fail to form while the second, *st-59*, produces asci with hyaline, inviable ascospores. When the two mutants are crossed fertile perithecia containing viable ascospores are formed. Some 504 asci from the cross were all parental ditype, there being no recombination between the mutant factors. Both mutants require arginine and show complementation in a heterocaryon. The mutants show two features not found in typical heterothallic Ascomycetes, namely, the nutritional requirement and the occurrence of the block to development *after* nuclear fusion. However, these results lend encouragement to the search for mutants blocked at an earlier stage in development which more closely counterfeit natural controls.

PHYSIOLOGY OF MATING TYPE DETERMINATION

At the present time very little is known about the physiological basis of mating type determination in heterothallic fungi. We know little or nothing about the biochemistry of compatibility and incompatibility or what kind of processes are controlled by mating type genes. In this section we will attempt to summarize the little that is known together with some current speculations.

Hormonal control

The regulation of sex organ development by substances which diffuse from mycelia and can be isolated and partly purified has been known in the fungi since 1924 when Burgeff [63] demonstrated a sex hormone in *Mucor mucedo*. Examples found since then cover most groups of fungi except the Basidiomycetes. The most complete study was made by Raper on Achlya and was reviewed in 1952 [436]. Raper worked mainly on two sexually dimorphic species of this genus of aquatic Phycomycetes. Male and female strains produce no sex organs when grown in isolation. When grown together in the same culture, in water or on semi-solid agar, the male mycelium sprouts antheridial hyphae, the primordia of the male sex organs. After profuse development of antheridial hyphae, spherical swellings, the oogonial initials, appear on the female mycelium. The antheridial hyphae are attracted to the oogonial initials and grow towards them delimiting antheridia when they touch them. The oogonial initial is then cut off from its hypha becoming an oogonium in which oospheres are differentiated. Fertilization follows shortly after when single male nuclei pass to each oosphere through small tubes which grow from the antheridium into the oogonium.

Each of these stages is initiated by one or more specific diffusible hormones. Several lines of evidence show this to be so. The sequence of development is invariable. Certain strains show a complete development of sex organs, which only stops short of fertilization, when the two thalli are separated by a permeable cellophane membrane. Delimitation of antheridia takes place when the antheridial hyphae touch the membrane but only if oogonial initials are in the immediate vicinity. Culture filtrate experiments revealed that while male plant filtrate has no apparent effect on a female plant, female plant filtrate stimulates a male plant to produce antheridial hyphae. The filtrate from a stimulated male will in turn bring about the formation of oogonial initials in a female plant. An elegant version of this experiment was as follows. Four micro-aquaria were linked together by siphons through which passed a slow stream of water in one direction. Water passing out of the fourth tank was replaced by fresh water entering the first. Each tank contained a single plant. Adjacent tanks contained plants of opposite sex. When the first plant was a female, the adjacent male and the male in the fourth tank produced antheridial hyphae. The third plant, a female, produced oogonial initials and the antheridial hyphae of the fourth plant grew towards the end of the siphon connecting its tank with the third one.

The homothallic species *Achlya americana* and *Thraustotheca clavata* were found to respond to sex hormones from the dimorphic, or heterothallic, species *A. ambisexualis* and *A. bisexualis*. They also secreted substances which stimulated development of antheridial hyphae on male plants. This shows that the homothallic species possess chemical regulators of sex organ formation which are similar to those in the heterothallic species.

In the heterothallic species of Achlya a minimum number of seven hormones were postulated to account for the co-ordination mechanism. None of the hormones have been identified chemically. However, a partial purification of the female hormone A, which stimulates development of antheridial hyphae, produced a final fraction which was active in a dilution of 10^{-13}.

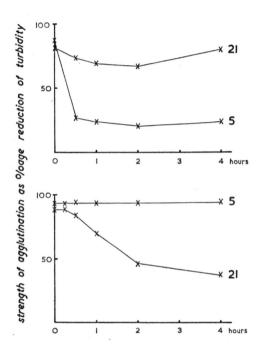

Fig. 49. The effects of (top) periodate oxidation and (bottom) trypsin digestion on agglutinability of strains 5 and 21 of *Hansenula wingei*. Agglutinability was tested against untreated cells of opposite type, and is expressed as the percentage reduction in turbidity of agglutinated mixtures over unagglutinated controls. From Brock [54].

Nothing is known of the genetical control of the hormone mechanism. All attempts to germinate oospores have failed. An additional complication is the existence of many sexual intergrades between the extremes of strong male and strong female. In the heterothallic species these intersex forms are self-sterile but can act as male or female according to the relative maleness or femaleness of their partners.

Agglutination

In the yeast *Hansenula wingei* mass agglutination takes place when cells of two different mating types (strains 5 and 21) are brought together When the cells are in intimate contact fusion occurs with the formation of diploids. The diploid hybrid shows no agglutination reaction with either haploid strain. Brock [54] has shown that the components of the agglutination reaction occur on isolated cell walls which suggests that mating type depends on a reaction between cell surfaces. The agglutinability of strain 21 was lowered after treatment with trypsin while that of strain 5 remained unchanged. Strain 5 however was sensitive to mild oxidation with ·001M sodium periodate while strain 21 was not (Fig. 49). These results are consistent with the hypothesis that the mating component of strain 21 is a protein while the complementary component in strain 5 is a polysaccharide. The mating agglutination appears to be due to a combination between these complementary molecules similar to that between antigen and antibody.

From further studies Brock [55] suggests that the development of conjugation tubes between paired cells follows from the synthesis of wall softening enzymes which formed in both cells as a result of the diffusion of inducers across the region of contact. When either cell type had been treated with ultraviolet light or the antibiotic cycloheximide fusion did not occur, neither did it occur if protein synthesis was inhibited by amino acid analogues. Presumably under these circumstances the wall softening enzymes were not formed in both partners. Evidence was, however, obtained from mating ultraviolet irradiated cells with untreated cells, which were then deagglutinated by autoclaving, that the normal untreated cells were stimulated to bud possibly as a result of the induction of wall softening enzymes. In *Saccharomyces cerevisiae* conjugation tubes were formed by minus mating type cells when they were placed near, but not in contact with, plus cells, or on an agar surface from which interacting plus and minus cells had been removed [308]. In Hansenula fusions only occur between cells brought in contact by the agglutination reaction. The enzyme inducers of Hansenula may be less stable than comparable inducers of Saccharomyces and only transmitted by contact.

Multiple allelomorph control

In a two allelomorph system of heterothallism it is possible to suppose that each mating type supplies some essential function in the sexual process which the other lacks. Such an interpretation is hardly possible for multiple allelomorph systems. Where there are of the order of a hundred alleles, as for example at the β sub-unit of the A locus of Schizophyllum, and a combination of any two different ones will give fertility, it seems fantastic to suppose that there are a hundred separate functions which can be lost individually and yet be controlled by the same apparently indivisible chromosome segment. Yet if we regard multiple mating type alleles as belonging to the same functional gene, how do we explain their complementary action in all heteroallelic combinations?

One line of thought depends on the analogy between the complementary action of mating type alleles and inter-allelic complementation as discussed in Chapters 6 and 8. As we have seen, it is possible to obtain, by mutation of distinct but closely linked sites, a series of alleles deficient with respect to formation of a single enzyme, yet complementary in all combinations. The analogy fails, however, in as much as typical cases of inter-allelic complementation show numerous alleles which are non-complementary with blocks of other alleles, or non-complementary with all the others of the series. It is, of course, this feature which permits the construction of complementation maps. If we wished to represent a Hymenomycete mating type locus in the form of a complementation map, we would have to show a series of non-overlapping segments with no indication as to order. If multiple mating type alleles are regarded as a series of defective but mutually complementing mutant derivatives of a hypothetical fertility gene, we must suppose that all the mutations which showed even partial non-complementation with any of the rest of the series were very efficiently eliminated from the population by selection, since mating type factors with dual or multiple specificity have never been found. Even if we can believe that a complementation map with upwards of a hundred different segments is reasonable on any hypothesis of complementation (see Chapter 8), it still seems improbable that minor overlaps in the map would not persist. However, this kind of hypothesis cannot be altogether ruled out. If the mechanism really is that discussed in Chapter 8, namely the formation of an active enzyme through co-polymerisation of differently defective sub-units, it should be possible in principle to show the formation of the active hybrid in the fertile dicaryon, and, perhaps, the two defective sub-units in the respective monocaryons. The serological approach of Raper and Esser [440], who have been able to demonstrate the presence of proteins in a dicaryon different from those present in either of the parent monocaryons, could lead to a real test of the hypothesis.

An alternative is to suppose that mating type alleles are not comple-

16

mentary at all in any direct way, but merely determine a series of different variants of the same protein which are individually 'recognisable' by some cellular mechanism analogous to antibody production in animals. We can imagine that each fungal genotype is capable of forming antibodies or agglutinins against the whole range of possible variants of the protein *except* that which it produces itself. Such a paralysis of the capacity to react with 'self' components has an obvious analogy with the phenomenon of acquired immunological tolerance in mammals. Whether such a comparison can be valid is not at all clear. In animals the cells which produce antibodies are generally reacting against something outside themselves, and it is possible to make hypotheses in terms of shifts in the composition of cell populations as a result of selection or differential stimulation of the appropriate cell types [298]. No such explanation could apply in fungi, where the whole course of events would have to occur as a result of interaction of components within a single type of cell. Nevertheless, while the mechanism of 'self-recognition' remains so mysterious even in mammals, it is perhaps not too far-fetched to suggest that a somewhat similar mechanism might operate in fungi also. It seems possible to pose some relevant questions which could be investigated experimentally. What, for instance, is the nature of the dicaryon-specific proteins detected by Raper and Esser? If the complementation type of hypothesis is correct they should be molecules formed by aggregation of slightly different but homologous sub-units contributed by the respective homocaryons. If, on the other hand, they are the result of a sort of immune reaction, they should each consist of two quite different components, the specific mating type protein contributed by one homocaryon, and the agglutinin contributed by the other. If each of these components is already present in the unmated homocaryons, it might be possible to demonstrate a precipitation reaction *in vitro*.

The two, necessarily very speculative, hypotheses just considered are representatives of two basically different kinds of theory. The first kind supposes that the mating type substance *promotes* some essential process leading to fruiting, but that complementary action of alleles is necessary before the effective mating type substance is formed. The second supposes that the mating type locus, in any of its allelic forms, *inhibits* some process essential to fruiting, and that the interaction in the dicaryon results in a neutralization of the inhibitor. The first hypothesis would suggest that self-fertile mutants, such as are found in Coprinus, contain mutant or 'suppressed' alleles, analogous to prototrophic revertants in auxotrophs, which can act autonomously without complementation.

The second hypothesis would imply that self-fertile mutants had suffered a *loss* in function of the mating type locus.

GENETIC CONTROL OF HETEROCARYON FORMATION

Where fusion between hyphae of different strains is an essential part of sexual reproduction, as in the Hymenomycetes, there is the possibility of incompatibility between different mycelia leading to cross sterility. An example in *Podospora anserina* was discussed on p. 221.

Genetically controlled heterocaryon incompatibility has been described in *Neurospora crassa* and in *Aspergillus nidulans*. In Neurospora heterocaryosis is irrelevant to sexual reproduction, except in the secondarily homothallic species, but in *Aspergillus nidulans* it is a necessary preliminary to outbreeding. *A. nidulans* is homothallic and the methods for recovering either crossed asci or heterozygous diploids make use of forced heterocaryons (see p. 30).

In *N. crassa* and *N. sitophila* hyphal fusion with the formation of stable heterocaryons only occurs readily between mycelia of the same mating type. In addition, two other loci in *N. crassa*, C and D, have been identified with respect to which two strains must be similar if they are to be heterocaryon compatible [189, 190]. Two alleles are known at each of these loci: C and c, D and d. If paired mycelia differ at either locus or at both, a lethal cytoplasmic incompatibility reaction follows hyphal anastomosis, effectively preventing the formation of a heterocaryon [191]. Wilson *et al* [586] were able to show that extracts of a CD strain caused death when injected into hyphae of a cd strain. The active principle was sensitive to heat and to proteinase digestion. Subsequent studies by Wilson [585] have shown that *N. crassa* heterocaryons between heterocaryon compatible strains may be produced by transplantation of nuclei using the technique of micro-injection. Two other genetic systems have been described which affect the initiation and growth rate of Neurospora heterocaryons [237, 411]. It is of interest that in the secondarily homothallic species *N. tetrasperma* heterocaryosis is unrestricted; heterocaryons between strains of opposite mating type are easily formed, and no other incompatibility factors have been discovered.

Heterocaryon incompatibility in *Aspergillus nidulans* was recognized by the failure of two parental homocaryons, differing with respect to a conidial colour marker, to form a heterocaryon bearing chains of conidia of different colours on a common conidiophore vesicle [206, 207]. The genetic basis for the difference between two completely heterocaryon incompatible strains cannot readily be determined because genetic analysis is precluded. However, Jinks and Grindle [267] used the heterocaryon test (see p. 236) to establish that the difference between two partially heterocaryon incompatible strains was controlled by chromosomal genes. The genetic mechanism in Aspergillus could well be similar to that in Neurospora.

CHAPTER 10

EXTRA-NUCLEAR INHERITANCE

Most of the genetic determinants we have discussed so far have been genes located on the chromosome inside the fungal nucleus. These genes express themselves through the surrounding cytoplasm, modifying its structure and function by their action. There is, however, a considerable body of evidence for the existence of determinants which are located outside the nucleus in the cytoplasm. Since these are not carried by the chromosomes the pattern of their inheritance differs from that of nuclear genes. In this chapter we will first consider some of the general features of extra-nuclear inheritance, then the main categories into which the examples can be grouped together with accounts of some of the clearest examples, and will finish with a discussion of the nature of extra-nuclear inheritance.

The most striking feature of extra-nuclear inheritance, shown by many examples, is the failure of segregation among the progeny from a cross between two stocks which differ by virtue of cytoplasmic determinants. The progeny are generally all alike with respect to the character difference, all like one parent or all like the other. When the same progeny shows normal segregation of other, chromosomal, markers carried by both parents, the absence of segregation of the primary difference cannot be explained by the failure of fertilization. In many heterothallic fungi the volume of cytoplasm contributed by one of the parents at sexual fusion may be much greater than that contributed by the other. This is so in those Ascomycetes and Basidiomycetes in which the fertilising element is a microconidium, a spermatium, or a nucleus which migrates through a compatible mycelium. A cytoplasmic difference in these forms is often revealed as a difference between reciprocal crosses, the progeny resembling the parent which has contributed the bulk of the cytoplasm. The failure of one parent to transmit a trait to the progeny of a cross could be attributed to the trait being dependent on a combination of several genes, not present in the other parent. In such a case, genetic reassortment during meiosis could lead to the necessary gene combination being present in very few of the meiotic products. This explanation will not hold where there is a difference between reciprocal crosses. Even where reciprocal crosses cannot be made, the polygenic explanation can be tested and often ruled out by what Jinks [258, 259, 261] has called the *heterocaryon test*. This involves combining the two parents in a hetero-caryon so that the two types of nuclei are present in a common cyto-plasm. The heterocaryon is then resolved into its component homocar-yons, which are identifiable by other nuclear markers. If these homo-caryons continue to show the original differences it may be concluded that these were controlled by determinants which were bound to if not inside the nuclei. If the homocaryons are alike except for their nuclear markers, or if the phenotypes under investigation show segregation independent of that of the nuclei, it may be concluded that the original difference lay in the cytoplasms of the two parents. This test is one of the most convenient for cytoplasmic determinants and is, of course, particularly useful for analysing imperfect fungi and homothallic forms.

Some other features of extra-nuclear inheritance which are not com-mon to all examples are discussed below.

CONTINUOUS CYTOPLASMIC VARIATION

Extra-nuclear or cytoplasmic variation is of two kinds; continuous and discontinuous. In the former all grades occur between the phenotypic extremes of the range of variation, while in the latter the character in question can be scored as either present or absent in one of two relatively clearly defined classes. The rôle of the cytoplasm in continuous variation has been demonstrated by the work of Jinks in Birmingham. In *Asper-gillus glaucus* Jinks [260, 262] has shown by means of the heterocaryon test that variation in such characters as spore germination, growth rate, pigmentation and perithecial density is under cytoplasmic control during differentiation and ageing. Selection for any of these characters produced marked changes within the limits imposed by the inviability of the most extreme forms. The selections were carried out by plating conidia of the line to be studied and selecting at random twenty colonies which were then scored for the four characters. Colonies showing the high and low extreme expressions of each character were kept and put through the same process again. Response to selection was so rapid that marked changes were obtained within three cycles. As an example, the response to selection for high and low growth rate is shown in Fig. 50. The original phenotypes were recoverable by back selection. Evidently none of the cytoplasmic elements were lost; their balance in the cytoplasm had merely been changed. A striking feature of this morphological variation were the correlated responses to selection. All the characters changed when any one was selected. The change in the germination rate of conidia following selection for either high or low growth rates is shown in Fig. 50. The correlated responses could mean either that all the determinants are bound to one cytoplasmic element or that a single pleiotropic determinant is involved.

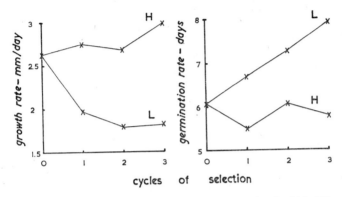

cycles   of   selection

FIG. 50. *a*. Changes in growth rate resulting from selection for high (H) and low (L) growth rate in a homocaryon of *Aspergillus glaucus*. *b*. The changes in the rate of germination of conidia of the selections for high (H) and low (L) growth rate in the same homocaryon. This is an example of correlated response to selection. From Jinks [260].

Differences in adaptability within clones of *A. glaucus* to mercuric chloride and unfamiliar sugars (galactose, arabinose, xylose and lactose) are also controlled by cytoplasmic systems similar to those involving morphological variation [262]. In these examples, however, the differences between parental clones in adaptability were shown by means of the heterocaryon test to be in large part nuclear in origin. The variation under nuclear control may transcend that determined by the cytoplasm.

A. *glaucus* and *A. nidulans* may be propagated in the laboratory by means of hyphal tips, conidia or ascospores. Propagation by cutting off single hyphal tips from a cytoplasmically continuous mycelium can hardly be considered a natural form of asexual reproduction and it is not surprising to find that the greatest spontaneous variation of cytoplasmic origin was shown among different hyphal tips isolated from the same colony. Single conidia showed less variability while single ascospores gave uniform colonies. Jinks [259] has concluded that these differences reflect the cytoplasmic heterogeneity of the three kinds of samples. This heterogeneity could well result from the different cytoplasmic requirements for the formation of the three kinds of organ. The ability to form perithecia is readily lost during vegetative propagation suggesting that only a 'complete' cytoplasm will produce the sexual apparatus. Unbalanced hyphae would be unable to differentiate it and would thus make no contribution to ascospore formation. Similar but less stringent requirements could well apply to the formation of conidia.

Continuous cytoplasmic variation is undoubtedly of considerable importance in many fungi. As Jinks [263] has pointed out many of the features of the *dual phenomenon* in imperfect fungi, which have been explained as the segregation of a heterocaryon into its homocaryotic components, can also be explained in terms of cytoplasmic variation.

DISCONTINUOUS CYTOPLASMIC VARIATION

Sharpe [486], one of the first workers to appreciate the potentialities of Aspergillus for studies of extra-nuclear inheritance, described a spontaneous cytoplasmically controlled variant of *A. glaucus* which, unlike those studied by Jinks in the same organism, arose in one step from the wild type as a sharply distinct sector. This variant, called *A*, differs from wild type in growth rate, conidiation, perithecium formation and pigmentation. When propagated by large inocula *A* remains stable, but when propagated by single conidia or ascospores, varying proportions give rise to normal colonies, the remainder being *A*. All perithecia formed in marked crosses between *A* and the normal type give a proportion of *A* ascospores, whether they show segregation for the nuclear markers originally associated with *A* or not. When *A* mycelia are placed in contact with normal mycelia a progressive change to *A* is observed which appears to be due to cytoplasmic infection, since the altered mycelium is homocaryotic for the original nuclear markers carried by the normal mycelium. In this example it is still possible to recover normal from mutant mycelia and yet the difference between the two is as clear cut as in the examples of discontinuous variation discussed below. Again the infective behaviour of the mutant type resembles that of the suppressive mutants, discussed below, but in this case does not suppress the capacity of the mycelium to produce normal segregants.

Petite colony yeast

Unlike those we have just considered, many cytoplasmic variants are stable. Normal or wild type forms cannot be recovered from them by selection. One of the best known examples is that of petite yeast studied extensively by Ephrussi and his colleagues. The following account of various petite colony variants in yeast introduces a number of important features of extra-nuclear inheritance. In haploid or diploid cultures of *Saccharomyces cerevisiae* about one per five hundred single cells give rise to colonies which are smaller than normal. These smaller colonies are called petites. Their cells are unable to utilize oxygen and have an anaerobic metabolism even under aerobic conditions; this deficiency is explained by the absence of a number of important respiratory enzymes including succinic dehydrogenase and cytochrome oxidase. The absence of the absorption bands characteristic of cytochromes *a* and *b* in the spectra of petite forms provides a means for their identification in addition to colony size [498]. The frequency of petite cells may be increased to 100 per cent by growing vegetative cells on a medium containing one part in 300,000 of acriflavine [145]. Lindegren and Hino [317] claimed that the frequency of respiratory deficient mutants is increased after a period of anaerobic growth although this finding has been disputed by other workers [480, 322].

Genetic analysis revealed that the petite character was inherited as though cytoplasmically determined. Diploid petites do not form asci and ascospores, but the hybrid diploid cells, formed by fusion between normal haploid cells and petite haploids of the kind we are considering at the moment, are normal with respect to respiration and are able to sporulate. When ascospores are cultured all four from each ascus are normal and there is no segregation of the petite character except for occasional petites which arise with the same frequency as in the normal parent. Chromosomal markers distinguishing the two parents show a normal 2:2 segregation in each ascus. Petites which behave in this way are called *neutral petites*. It is assumed that the cytoplasm of a normal yeast cell contains a small number of autocatalytic particles required for the synthesis of respiratory enzymes. In the neutral petite these particles are either not present or have mutated to a functionally inactive form. When a neutral petite is mated with a normal haploid, the diploid cell which is formed receives normal particles which are shared out among the daughter cells all of which are thus normal.

It has been shown that the expression of the cytoplasmic element is dependent on a gene in the nucleus. This was revealed by the discovery of a *segregational petite* variant [85]. The segregational petite has the same phenotype as the neutral petite and when mated with a normal haploid produces normal diploid cells. However, when placed on sporulation medium the asci formed by these diploids gave a 2:2 segregation of normal and petite clones. The petite phenotype was determined by a recessive gene. When segregational and neutral petites were intercrossed the diploid cells were normal and formed asci showing a 2:2 segregation for normal and segregational petite. This result can be explained if we assume that the segregational petite possesses normal cytoplasmic particles but these, although able to reproduce, are inactive in the presence of the recessive gene. In the diploid hybrid cells the normal particles derived from the segregational petite are physiologically active in the presence of the dominant gene derived from the neutral petite parent. The segregation of the recessive gene occurs at ascus formation.

James and Spencer [255] have described spontaneous petite mutants which resemble neutral petites. In their yeast strain putative respiratory deficient mutants died very rapidly but could become viable as a result of gene mutation. Such viable forms behaved like neutral petites in crosses with the normal strain, but the segregation of the mutant gene could be observed because viable petite forms arose directly from the clones carrying it.

A third type of petite, called *suppressive*, is cytoplasmically determined and phenotypically indistinguishable from the neutral petite but behaves quite differently in crosses [146]. The diploid cells produced by crossing a highly suppressive petite with normal all give rise to petite colonies. If these diploid cells are placed on sporulation medium soon after mating, they are able to form asci all the spores of which give rise to petite colonies. Evidently the newly formed diploid cells contain respiratory enzymes which enable them to sporulate, but which are lost when they undergo further division. The suppressive petites apparently contribute something to the zygotes which suppresses the perpetuation of the normal cytoplasmic factor derived from the normal parent. If the diploid cells are placed on sporulation medium after incubation overnight, during which time they have undergone not more than six budding cycles, only a small minority form asci, and the resulting spores are all normal. Ephrussi [146] explained this by assuming that a small proportion of neutral petite cells occur in suppressive petite cultures and that these when mated with normal cells would form normal diploids. After a period of incubation only these cells would be able to sporulate and would thus be selected. The explanation is strengthened by the finding that neutral petites may be isolated from populations of suppressive petites. A further complication is that suppressive petites vary in the extent of their suppressiveness. The degree of suppressiveness characteristic of a given strain may be measured as the percentage of zygotes which give rise to diploid petite clones following a mating with a normal strain. Two explanations of the different degrees of suppressiveness have been put forward. The first is that the different strains each have a certain probability, or efficiency, of suppression which has different values in different strains. There would then be as many cell types as there are degrees of suppressiveness. The second explanation is that each strain consists of a mixture of completely suppressive petites and neutral petites occurring in different proportions in the different strains. The more neutral petites there were in a strain the lower its suppressiveness would be. At the present time there is no information as to which of these two explanations is correct. Some other suppressive systems are discussed at the end of this chapter.

Neurospora—poky

In *Neurospora crassa* one of the best known cytoplasmically inherited mutants is *poky*, first described by Mitchell and Mitchell [361]. This is a spontaneous mutant characterized by a reduced growth rate compared with wild type. The cytochrome absorption spectrum of the mutant shows the absence, or extreme reduction, of the cytochrome *a* and *b* bands and a greater amount of cytochrome *c* per unit of dry matter compared with wild type. The growth rate of *poky* cannot be restored to normal by the addition of supplements. When *poky* was crossed with wild type it was found that the character was only transmitted if *poky* was the protoperithecial parent. All the progeny were then *poky* except a small percentage which were accounted for by assuming that the conidial parent had produced a few protoperithecia during the course of the mating. Here then the progeny resemble the parent which has contributed the bulk of the cytoplasm. The mutant *poky* is maternally inherited.

Several other mutants with more or less similar phenotypes were recovered and their biochemical properties compared with those of *poky* [363]. One of these, *mi-3* (*mi* stands for maternally inherited) showed the same pattern of inheritance as *poky* but produced less cytochrome *c* than *poky* and more *b*. Two others, C115 and C117, were nuclear gene mutants showing normal segregations in crosses with wild type. C115 was like *mi-3* while C117 produced no detectable cytochromes *c* or *a* but more *b* than wild type. A third cytoplasmic mutant, *mi-4*, was discovered by Pittenger [409] as a spontaneous mutant in a lysine requiring stock. It was similar to *poky* but sterile when used as a protoperithecial parent. When *mi-4* was used as conidial parent in crosses the mutant phenotype was not transmitted.

A good deal of interest was centred on the behaviour of these mutants in mycelia containing either mixtures of normal and mutant, or two different mutant cytoplasms. A mixture of this kind is the cytoplasmic equivalent of a heterocaryon. The mixed cytoplasms were prepared by first obtaining auxotrophic markers in the *poky* and *mi-3* cytoplasms. This was done by crossing them, as protoperithecial parents, with stocks carrying the required nuclear markers. The markers were then selected from among the progenies which, of course, possessed uniformly mutant cytoplasms. The female sterile *mi-4* strain already required lysine. Heterocaryons were prepared by co-inoculating complementary mutants on minimal medium. It was assumed that the hyphal anastomosis and exchange of nuclei that gave rise to the heterocaryon also allowed cytoplasmic mixing. The growth rates, appearance and absorption spectra of the heterocaryons were compared with wild type. The results of tests carried out by Gowdridge [201] and Pittenger [409] for all combinations of mutant and wild type, except *mi-3* + *mi-4* are shown in Table 28.

TABLE 28. Properties of heterocaryons of *Neurospora crassa* with mixed cytoplasms

Heterocaryon	Growth	Spectrum	Reference
poky + wild type	normal	normal	[201, 409]
mi-3 + wild type	{ sometimes *mi-3* sometimes normal	mutant normal	} [201]
mi-4 + wild type	normal	normal	[409]
mi-3 + *poky*	{ usually *mi-3* sometimes *poky*	} mutant	[201]
mi-4 + *poky*	normal	mutant	[409]

Although Pittenger found that growth of the heterocaryons *poky* + wild type and *mi-4* + wild type was initially normal, and that their conidia produced normal colonies, he reported that the cultures eventually showed mutant phenotypes. This suggests that there was a competitive replacement of wild type elements by *poky* or *mi-4*. This was not observed by Gowdridge for *poky* or *mi-3*. The reason for this apparent lack of agreement is not known.

While there was no evidence that *poky* and *mi-3* interacted to give a growth rate faster than either parent, *poky* and *mi-4* did interact to give normal growth rate. The mixed cytoplasm, however, always gave a mutant absorption spectrum and, although relatively stable in growth tubes over distances up to 1,000 mm or more, eventually gave growth which was characteristic of either *poky* or *mi-4*. That the interaction was not the result of complementation between the nuclei of the heterocaryon was shown by recovering homocaryotic colonies which continued for some time to show a normal growth rate on suitably supplemented minimal medium.

The synergistic action of *poky* and *mi-4* can most easily be explained by cytoplasmic complementation rather than co-operation to form a normal cytoplasmic element since no stable normal strains were ever recovered from the mixture. The mutant spectrum of the mixture is also indicative of the absence of normal elements. The heterocaryon experiments also showed that all three cytoplasmic mutants could be recovered from heterocaryotic mixtures as monocaryons with either parental nucleus. Clearly each mutant must be determined by an altered element in the cytoplasm and cannot be explained simply by the absence of such elements.

A partial suppressor of *poky*, called *f*, was found which restores the growth rate of *poky* nearly to normal but leaves the defective cytochrome system unchanged [362]. When *f* is replaced by its wild type allele the full *poky* phenotype is expressed. The suppressor has no effect on *mi-3* nor on the gene mutants C115 and C117. The effect of *f* may be to enhance the activity of a compensating enzyme, which is only active in *poky*, so as to increase the growth rate and leave unaffected the cytochrome content.

Cytoplasmic differences between Neurospora species

We have seen that *poky* and other similar cytoplasmic mutants of *N. crassa* can be combined with desired nuclear markers either in crosses or by resolving heterocaryons. The mutants are not dependent on a particular nuclear genotype except that *poky* is only fully expressed in the absence of *f*. A corollary of this is the demonstration that the original nuclear genotypes of *poky* and *mi-4* can support normal cytoplasm [409]. This is probably also true of *mi-3*. Here then is a method for examining the relationships and interactions between nucleus and cytoplasm by synthesizing combinations of genetically marked cytoplasm and nuclei. It has been used by Srb [507] to investigate nuclear-cytoplasmic relationships among three species of Neurospora and an unidentified strain.

The cytoplasmic marker which Srb used was an acriflavine-induced mutant called *slow germination* or *SG*. *SG*, originally isolated in *N. crassa* is maternally inherited. Ascospores carrying it germinate more slowly and their germ tubes and hyphae show initially a slower growth rate, than wild type. The same is also true, but to a lesser extent, of the germination and subsequent growth of *SG* conidia. The cytochromes of *SG* mycelia appear to be normal from spectroscopic examination. *SG* was combined with the nuclear genotype of *N. sitophila* by a series of ten backcrosses to this species following the initial interspecific cross. *N. sitophila* was always used as the conidial parent. The cytoplasmic marker showed entirely normal maternal inheritance throughout the series of backcrosses. The effectiveness of the nuclear substitution was evident from increases in the fertility of matings and the number of viable ascospores produced as back-crossing proceeded. The final generation was indistinguishable in these respects from intercrosses among strains of *N. sitophila*, and showed a second division segregation frequency for mating type (57 per cent) similar to that of *N. sitophila* (58 per cent) and quite different from that of *N. crassa* (about 15 per cent). *SG* was also successfully combined with the genomes of a Philippine Island strain of Neurospora and *N. intermedia*.

During the substitution of the *sitophila* genome for the *crassa* genome in *SG* cytoplasm, Srb noted the segregation of a character he called aconidial or *ac*. This took place in a reciprocal cross between the parents in the seventh back-cross generation in which *N. sitophila* was used as protoperithecial parent. Aconidial cultures produced a few small clusters of conidia beneath a mass of sterile, pigmentless hyphae that grew towards the top of the culture tube. The character appeared to be controlled by a nuclear gene but was only expressed in *sitophila* cytoplasm. It was not expressed in *crassa-SG* cytoplasm nor in wild type *crassa* cytoplasm. The cytoplasmic property which interacts with *ac* to give normal conidiation is thus independent of *SG*. In the crosses involving *SG* and the Philippine Island strain a gene controlling colony size was found which was also cytoplasm-dependent. This gene, *S*, determined small colony size when combined with cytoplasm of the Philippine Island strain, but not when combined with *N. crassa* cytoplasm. An allelic gene *s* gave small colonies in combination with either cytoplasm.

It seems from the maintenance of *SG* and the property which interacts with *ac* that the substitution of one genome for another by back-crossing is not seriously complicated by the introduction of contaminating cytoplasm from the conidial parent. The interactions between nuclear genes and cytoplasm in stocks which, except for *SG*, are wild type suggest that the deliberate testing of a variety of nuclear gene markers in foreign cytoplasms may reveal many more such interactions. Some further examples of gene-cytoplasm interaction are considered below.

strain B (10 per cent *alba*) revealed a third allele f_{10}. Tests of the effect of environment on the expression of these alleles revealed that, while reducing the incubation temperatures from 37°C to 25°C had little effect on three of the strains, it lowered the frequency of *alba* in strain D to about 12 per cent.

The precise role of *f* is not clear. Suffice it to say for the moment that it could determine the rate of reproduction of a cytoplasmic particle which was entirely lost in stable *alba* strains or below a threshold density in revertible strains. Alternatively *f* might affect the relative stabilities of two different metabolic states within the cytoplasm, one producing the *alba* phenotype and the other wild type. When both are present a revertible *alba* phenotype might result. Further comment will be deferred until the end of the chapter.

Although it is in some ways a less straightforward example of cytoplasmic inheritance than *poky* or petite, the phenomenon of barrage in *Podospora anserina* is of considerable interest in the present connexion. It was first described by Rizet [450] and subsequently by Ephrussi in his book on nucleo-cytoplasmic relations in microorganisms [145]. Podospora, a Pyrenomycete, is normally secondarily homothallic (see p. 220) producing four binucleate spores in each ascus. Occasional uninucleate ascospores are formed which give rise to self-sterile mycelia of two mating types + and —. When these monocaryotic mycelia are paired, and it does not matter whether they are of the same or different mating types, two kinds of interaction may be observed where the hyphae come into contact. The interactions are controlled by two allelic genes *S* and *s*. In pairings involving the same allele, $S \times S$ or $s \times s$, the mycelia grow into each other forming a thin zone of overlapping and anastomosing hyphae. In $S \times s$ pairings a reaction similar to that shown between heterocaryon incompatible strains of *N. crassa* (see p. 233) takes place. Leading hyphae from either side, which have anastomosed with each other, become disorganized and die back to a cross wall where a branch is formed [454]. This reaction leaves a clear zone between the two mycelia but it does not interfere with perithecial production. If the *S* and *s* strains are of opposite mating type two lines of perithecia are formed, one on each side of the barrage. Crosses in which the relative amounts of cytoplasm contributed by the two parents could be varied were made by spreading microconidia of one mating type over established mycelia of the other. When the two parents differed with respect to *S* and *s* the resultant spores could be tested by pairing the colonies to which they gave rise with *S* and *s* testers. Crosses of this kind revealed that, regardless of

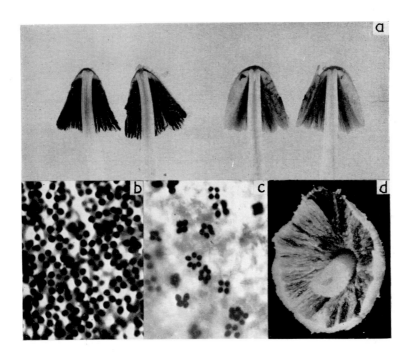

Fig. 51. *a*. Normal and pale-gill fruit bodies of *Coprinus lagopus* in longitudinal section.   The fruit bodies were produced by reciprocally constituted dicaryons having the same nuclei but different cytoplasms. *b*. Normal gill surface. *c*. Pale gill surface with normal and abnormal tetrads. *d*. sectored pale fruit body viewed from beneath.

   a. ca. natural size, *b*., *c*. ca. × 450, *d*. ca. × 3. From Day[106].

whether the maternal parent was S or s, S segregated normally giving a barrage with the s tester. However, sister spore colonies, which were expected to be s, did not give a barrage with the S tester. These colonies were called modified s and denoted by the symbol s^S. However when s was used as maternal parent a small proportion (5–10 per cent) of the ascospores gave rise to s colonies [453].

When s^S was crossed with s the type of spores produced depended entirely on the direction of the cross. Thus when s^S was the conidial parent all the progeny were s, while when s was the conidial parent all the progeny were s^S. This finding clearly showed that the change from s to s^S in the presence of S was not due to directed nuclear mutation but involved the cytoplasm.

s^S stocks sometimes reverted spontaneously to s and they could invariably be induced to do this by contact with an s stock. Following reversion or contact a progressive change from s^S to s could be demonstrated throughout the colony as though it were being invaded by migrating elements. These were shown not to be nuclei by appropriate genetic tests, and it was concluded that a cytoplasmic element was spreading through the mycelium.

From the most recent work [453, 454, 27] it seems that the results can be most easily interpreted in the following way. The genes S and s control the multiplication of cytoplasmic elements which differ from each other and are responsible in some way for the barrage reaction. The gene S is epistatic to s and prevents the multiplication of the s element. Thus in ascogenous hyphae or asci heterozygous for S/s the s cytoplasmic elements would disappear or be diluted to a low level, and the great majority of resulting ascospores, when they contain the gene s, will give rise to modified s mycelia. If s elements arise spontaneously, or are introduced into the s^S mycelium, they rapidly multiply converting it into an s mycelium. The production of S cytoplasmic elements was postulated when it was found that, following anastomosis with an S hypha, an s^S hypha was temporarily changed so that it behaved like S when paired with s. This property was soon lost presumably because the S elements were unable to multiply in the absence of gene S.

Cytoplasmic inheritance in the Basidiomycetes

Among the major groups other than the Ascomycetes the only well established examples of cytoplasmic inheritance are found in the Basidiomycetes and in the Actinomycetes [203], a group with closer affinities with the Bacteria than the Fungi and consequently outside the scope of this book. In the Phycomycetes some indications are known from work on *Blastocladiella emersonii* [71]. It is to be expected that recent demonstrations of heterocaryosis in two of the Mucorales [see p. 24] may lead to further examples being found in the Phycomycetes.

The first indications of the rôle of the cytoplasm in Hymenomycetes were obtained by Harder [214] who recovered the component monocaryons from a dicaryon of *Pholiota mutabilis* by using the micro-surgical technique described in Chapter 2, p. 43. Recovered monocaryons with the same nuclear type showed a range of morphological variation. Moreover the dicaryon used was found to produce a pigment when associated with other mycelia. Neither of the parental monocaryons possessed this property but both types of recovered monocaryons produced the pigment presumably because of their cytoplasmic constitution. Variation comparable to the continuous cytoplasmic variation discussed earlier in this chapter was found between recovered monocaryons of *Collybia velutipes* [9]. Some of these and other experiments have been reviewed by Papazian [390]. In none of these accounts is there any control of the relative amounts of the two parental cytoplasms in the mixture, comparable to that in the demonstration of maternal inheritance in other fungi.

In many Hymenomycetes a dicaryon is produced as a result of nuclear migration throughout both mated mycelia. A strain donating nuclei therefore contributes little if any cytoplasm to that part of the dicaryon farthest from the region of contact between the parents. At the same time the reciprocally constituted dicaryon with the same two nuclei in the other cytoplasm is produced on the opposite side of the mating. Day [106] made use of this feature in a study of a mutant form of *Coprinus lagopus* called *pale gill*. The mutant condition, which appeared only to affect the fruit body, reduced the density of tetrads from about 3,000 per square mm of gill surface to between 0–600. The basidiospores were normally pigmented and the pale colour of the gills was due to their lower density. At the same time about 13 per cent of the basidia of the mutant fruit body bore more or less than the normal number of four spores. These included basidia with from two to as many as eight basidiospores. Some of these features are illustrated in Fig. 51. The condition was first recorded when it was noticed that a particular dicaryon gave rise to normal fruit bodies but the reciprocal dicaryon, fruited at the same time, gave rise to pale fruit bodies. Both dicaryons were resolved into their monocaryon components and these, when used as cytoplasm parents in further crosses with normal stocks, for the most part followed their parent dicaryons in the type of fruit bodies they produced. Some of the exceptional fruit bodies were sectored instead of uniformly pale. These bore mostly pale gills but also bore some gills on which the frequency of tetrads was normal (see Fig. 51*d*). From genetical evidence it appears that the Hymenomycete fruit body starts development from one or at the most a few cells of the same genotype. The sectored fruit bodies must have resulted from the segregation of a cytoplasmic element during the course of development.

Suppressive systems

Until now all the examples we have considered, with the exception of suppressive petite in yeast and *poky* and *mi*-4 in Neurospora, have been cytoplasmic variants which when mixed in heterocaryons with normal cytoplasm have not dominated the wild type elements to the extent of making the recovery of the latter impossible. Indeed the reverse has sometimes been true and the mutants have been 'suppressed' by wild type elements as for example in neutral petite, *alba* and s^S. For *SG* in Neurospora we have no record of heterocaryon tests. In the examples discussed below the mutant cytoplasms appear to have competitive advantages over wild type and are therefore more or less suppressive.

The distinction between suppressive and non-suppressive mutants is interesting since it raises the question how a non-suppressive mutant may survive and be recovered if it appears in a cell containing wild type elements. We can, of course, explain this readily by assuming that, for example, neutral petite, *alba* and s^S are all mutants in which a wild type element was not present at the moment of origin, and that its very absence was the cause of the mutant condition.

An interesting example of a suppressive system in *Aspergillus nidulans* comes from Roper's [461] study of three mycelial strains *M*1, *M*2, *M*3. These were isolated from genetically marked stocks resistant to acriflavine following treatment with acriflavine. All three strains failed to produce conidia except after prolonged incubation and never produced perithecia. Each mycelial mutant was combined as a heterocaryon with a normal strain having complementary markers. If we write the two complementary nuclear genotypes as *A* and *B* and the mycelial and normal characters as [M] and [M+] then the heterocaryon may be represented as *A* [M] + *B* [M+]. Each of the heterocaryons was similar in appearance to the control heterocaryon *A* [M+] + *B* [M+] between two normal strains. Heterocaryons and diploids formed from *M*1 and *M*3 and *M*2 and *M*3 were unambiguously mycelial. The mycelial strains did not complement each other to give normal cytoplasm. When conidia from the heterocaryons involving mycelial and normal strains were plated, the following three types of colony were recovered : A [M], B [M+] and B [M]. The recombinant A [M+] was never recovered.

The non-appearance of A [M+] could be explained by assuming that the M determinant had become attached to the A nuclei, and was transmitted to all daughter A nuclei through nuclear division. To explain the association of M with some B nuclei an infection-like process may be postulated.

Strain $M1$, which was the most fully analysed, was crossed with a normal strain, and the progeny were obtained by dissecting asci from crossed perithecia. While some range in the expression of mycelial was noted, the two classes mycelial and normal were clearly defined and occurred in the proportion 3:1. Roper explained this segregation on the basis of two unlinked genes m_1 and m_2 independently determining the mycelial character so that only a strain with m_1^+ and m_2^+ would be normal. When normal strains A[M1 ?] from the progeny were tested by making heterocaryons with a known normal strain B[M+] they were found to carry the mycelial cytoplasmic determinant. This was shown by recovering the recombinant B[M1] as well as the two parental types A[M1 ?] and B[M1+] from conidia of the heterocaryon. Mitotic analyses of diploids made up from mycelial and normal strains confirmed the role of the two genes in controlling the expression of the mycelial character, and showed that the alleles were recessive. Thus the heterocaryon and derived diploid from $m_1 m_2^+$ [M1] $+ m_1^+ m_2$ [M1] were normal in appearance but the diploid later showed mycelial sectors. These sectors were either homozygous for m_1 and/or m_2 or haploid segregants carrying m_1 and/or m_2.

Here then we have an example of a competitive cytoplasmic element which behaves in some ways as if bound to the nucleus. Although it is not expressed in an unfavourable genotype it is none the less able to multiply so that it does not become diluted out and lost during growth.

The mycelial mutants of *A. nidulans*, recovered after treatment with acriflavine, were probably induced rather than selected by this reagent. Some other suppressive mutants of spontaneous origin have been reported. One of these is the mutant *vegetative death* which Jinks found to occur sooner or later in all vegetatively propagated stocks of *Aspergillus glaucus* [264]. *Vegetative death* first appears as a gap in the edge of a growing colony caused by the death of one or a few adjacent hyphal tips which have swollen and burst. As the colony grows the gap becomes a sector which is not invaded by the nearby healthy hyphae; in fact the sector spreads sideways. The death of many hyphae is accompanied by the formation of an intense brown pigment. Conidia taken from apparently healthy regions of the colony invariably give rise to colonies which develop similar symptoms within a short time.

A heterocaryon test, using mutant and normal healthy mycelia, revealed that vegetative death is a suppressive cytoplasmic condition. Only by repeated and intensive selection was it possible to recover any normal colonies from the heterocaryon. The mutant condition can in fact only be maintained by repeated transfer to healthy strains which become infected and eventually die.

A comparable condition, called *senescent*, occurs in stocks of *Podospora anserina* and was described by Rizet *et al.* [453]. Senescence is a progressive attenuation of the vigour of a vegetatively propagated mycelium which eventually reaches the point where growth is completely arrested. Crosses between young normal stocks and stocks that are senescent, but not yet dead, revealed that the character is maternally inherited. When a normal stock was used as maternal parent all the progeny were normal but when the senescent stock was maternal parent some of the progeny were senescent. All the spores from the same perithecium were similar, either normal or senescent. Further evidence of the nature of senescence was obtained from experiments in which isolated senescent and young normal hyphae were paired and allowed to anastomose. The normal hyphae became infected; pieces cut from these hyphae gave more or less senescent colonies depending on their distance from the point of anastomosis. The degree of senescence was determined by measuring the growth of each colony before death. The possible explanation that senescent nuclei had invaded the normal mycelium was excluded by using chromosomal markers.

If senescent mycelia were placed in conditions limiting growth such as low temperature, absence of nutrients or culture under mineral oil, they were rejuvenated to an extent proportional to the length of treatment. One explanation may be that under these conditions strong selection occurred for non-senescent parts of the mycelium. There is evidence suggesting that regions of this kind occur since a small proportion (less than 1 per cent) of fragments, prepared by macerating a senescent culture, give rise to healthy mycelia [453].

POSSIBLE EXPLANATIONS OF EXTRA-NUCLEAR HEREDITY

Although, as the foregoing examples have shown, there is no doubt that various states of the cytoplasm can have the property of self-perpetuation, the interpretation which should be given to this is still very much open to dispute. Some (for example, Darlington and Mather [96]) have attributed extra-nuclear heredity to *plasmagenes*, which are regarded as having properties analogous to those of chromosomal genes. The parallels between nuclear and extranuclear genetic systems have been well stated by Jinks [265]. That these parallels are due to basic similarities in mechanism is, however, still doubtful.

Evidence for the particulate nature of cytoplasmic-determinants

The main evidence for cytoplasmic genetic determinants being particles of some kind is that, in several cases, segregation with respect to the character in question can occur during vegetative growth. For example, occasional buds in yeast clones lose their normal respiratory system and give petite colonies, while their sister cells remain normal. Again in Wilkie's examples in *Aspergillus nidulans*, different conidia from the same mycelium give rise to normal and *alba* colonies. The more recent investigation, by Arlett, Grindle and Jinks [8], of the 'red' cytoplasmic variant of the same species, isolated following ultraviolet irradiation of conidia, provides an excellent example of the ability of alternative cytoplasmic elements to segregate from a heterocaryon independently of nuclear markers. If the cytoplasmic factors responsible for the character differences in these cases were small molecules of which enormous numbers would be present even if their concentration was by ordinary standards extremely low, they would be very unlikely to be distributed all to one cell and none to another. If, on the other hand, they were microscopically visible structures like mitochondria, and only present in numbers of the order of tens per cell, random distribution at cell division could easily give occasional cells lacking the structures altogether. The well established genetic continuity and segregation of mutant chloroplasts in green plants provides an obvious model here [353]. One should remember, however, that the observation of segregation really tells us nothing about the nature of the factor involved, but only, perhaps, about the number present per cell or the asymmetry of distribution at cell division. One should also bear in mind that segregation is only significant if the different cytoplasms were actually mixed in the first place. In Aspergillus, at any rate, where balanced heterocaryons tend to be less readily formed and less vigorous once formed than in Neurospora, there is some suspicion that mixing is often minimal, with extensive cytoplasmic domains retaining, more or less, their original compositions, though this could hardly explain persistent segregation from single conidia as of *red* in Aspergillus [8]. One should bear in mind that the appearance of a segregant type may not depend on the total absence of a cytoplasmic factor but rather on a drop in its concentration to a certain critical level below which it is no longer able to maintain itself.

The idea that mitochondria might have genetic properties was given serious consideration following the earlier work on respiratory deficiencies in petite colony yeast. The enzymes and pigments lacking in these mutants are all associated with mitochondria, as are those affected by the cytoplasmic abnormalities in *poky* and allied variants in Neurospora. It was attractive to think that cytoplasmic respiratory deficiencies were due to loss or mutation of mitochondria. As Yotsuyanagi has shown [601], neutral petite colony yeast has mitochondria which are normal looking though lacking cytochrome oxidase, so it seems more tenable to postulate mutation rather than loss of mitochondria here. However, it has never been possible to obtain any evidence that these deficiencies have their origin in the mitochondria, although deficiencies in mitochondrial enzymes are certainly the most striking symptom. No convincing evidence for self-replication of mitochondria can be cited.

What other cytoplasmic particles might be involved in extra-nuclear inheritance? The most attractive analogy is with viruses, particularly those containing RNA and not DNA as genetic material. In a number of cases, of which tobacco mosaic virus is the best known, it has been shown that the viral RNA is infective, and when introduced into the host cell can bring about the formation of more of itself and of specific viral protein. Since the work on RNA viruses has shown beyond doubt that RNA *can*, in certain systems, at least, be responsible for genetic continuity and transmission of information on protein structure, there seems to be no reason to doubt the possibility of cytoplasmic RNA, native to the organism, carrying out genetic functions as a sort of supplementary system to the nuclear DNA. Auto-reproduction by RNA which is clearly a normal part of a cell, rather than a viral infection, remains to be demonstrated. RNA in the cytoplasm is mostly in the ribosomes, and it is not out of the question that some ribosomes, or constituents of ribosomes, act as genetic material. Another possibility may be that genetic determinants are sometimes detached from the chromosomes and can be transmitted in the cytoplasm. Elements of this kind, known as episomes, which are almost certainly DNA, have been demonstrated in recent work on the genetics of the bacterium *Escherichia coli* [254].

Models based on self-perpetuating metabolic systems

The distinctive feature of the nuclear genetic material is not only that it reproduces itself with great precision but that it carries information—more particularly, according to our present knowledge, information about the amino acid sequences of proteins. The genetic role of the cytoplasm would be greatly clarified if one could tell whether any information of this sort is carried outside the nucleus. Too little detailed biochemical work on cytoplasmic mutants has been done to permit any decision on this point, but so far it has not been shown that extra-nuclear factors play any part in determining protein structure. It is noteworthy that auxotrophy, which characteristically occurs as a result of mutational alteration of enzyme structure, has never been shown to be due to extra-nuclear mutation, at least in fungi. So far as we know, all cytoplasmic mutant phenotypes could be due to variation in the *quantities* of enzymes or substrates rather than in the *kinds* of enzymes which the cell is capable of producing.

It is in fact possible to make a variety of models of self-perpetuating cytoplasmic states without invoking the presence of information, of the type carried in DNA, in the cytoplasm. What is required is simply that the presence of some substance (it could be a relatively simple metabolite) in the cytoplasm sets up a chain of reactions which result in the concentration of that substance being maintained. This kind of system, which would be an example of positive feed-back, is most easily imagined as involving the phenomena of enzyme induction and repression, already discussed in Chapter 8. One could suppose, for example, that in one cell type the production of a particular enzyme is repressed by a specific repressor normal to the cell, and that enzyme induction (or de-repression) could be brought about by a substance which was itself, directly or indirectly a product of the activity of the enzyme. In this sort of system the inducing substance, if it were subject to metabolic breakdown, could easily be ineffective below a certain concentration, inducing too little enzyme to boost its own concentration to the critical level. Above the critical level, however, enzyme induction and concomitant build-up of the inducer concentration could occur very rapidly, giving the effect of a sudden 'switch-on' or mutation. Thus two-cell types might each have quite a high degree of stability and might coexist in the same environment. This model is capable of taking account of many observed features of extra-nuclear inheritance. For example, any change in nuclear genotype which tended to alter the level of either the inducing substance or any metabolically closely related substance, could well influence the relative stabilities of the two cytoplasmic states. Any frequency of mutation from one state to the other could be explained as due to a greater or lesser degree of fortuitous variation in the level of the inducing substance. Finally, 'infective' or 'suppressive' transmission could well be explained as due to the introduction from one cell type to the other of a small amount of inducer which then sets off a chain reaction of enzyme de-repression and inducer production in the recipient cells. Many detailed variants of this kind of model can be imagined, and one has actually been demonstrated in bacteria. Novick and Wiener [383] showed that when *Escherichia coli* cells, potentially capable of forming β-galactosidase but requiring induction by lactose, are exposed to extremely low lactose concentrations, most remain for a long time unable to form the enzyme. The reason for this is that uptake of lactose from very dilute solutions is extremely slow in the absence of a specific uptake factor, called galactoside-permease, which is itself only induced to form when the cell already contains some lactose. A few cells, however, become enzyme-positive and produce clones in which this character is transmitted in a quasi-genetic fashion. It seems clear from the evidence that these clones originate from cells which have, presumably by passive diffusion, chanced to acquire one or a few molecules of lactose which then induce formation of a little permease, which catalyses the uptake of more lactose, and so on.

If, as many now believe [253], induction-repression mechanisms act directly on the nuclear genetic material itself, a cytoplasmic system based on the kind of positive-feedback just discussed is even less independent of the nucleus than cytoplasmic determinants have been generally supposed to be. Though a permease molecule, in the example just cited, would behave like a self-replicating extra-nuclear particle, it is, in fact, just part of a system, including both nuclear and extra-nuclear components, necessary for the maintenance of the activity of a chromosomal gene. The relevance of this kind of speculation to the general problem of cell differentiation in higher organisms is obvious.

Criteria for distinguishing between different models

The distinction between a piece of genetic material governing its own replication by virtue of its specific macromolecular pattern (base sequence of DNA or RNA), and a self-maintaining cycle of metabolic reactions, may seem clear. The self-replicating properties of the former are contained within a discrete structure, while those of the latter would depend, presumably, on a relatively diffuse ensemble of reactions, not localized in anything which one could possibly call a gene. However, there is no reason why a self-perpetuating metabolic cycle could not be contained within a discrete cytoplasmic structure, which would thus have a degree of genetic continuity without necessarily carrying any genetic information in the sense in which DNA carries information. The only really satisfactory way of distinguishing between different chemical mechanisms of extra-nuclear inheritance would be through actually doing some chemistry. This is perhaps not so hopeless a task as it might appear. Modern work, especially with bacteria and viruses, has emphasized the possibility of isolating genetically active DNA or RNA free of other materials, and demonstrating its genetic properties by 'infecting' living cells with it. The development of micro-injection techniques in fungi by Wilson [584] and others has made possible the introduction of any kind of isolated material into fungal hyphae, even if they will not take up certain substances from the medium for themselves. It seems possible to attack experimentally such questions as, for example, exactly what it is in s *Podospora anserina* which can infect and convert s^s strains so rapidly. If the infection can be carried out by micro-injection it should be relatively easy to find out whether the active principle has the properties of a relatively large and chemically complex cytoplasmic particle, or whether it consists entirely of DNA or RNA or protein, or even a substance of low molecular weight.

CHAPTER 11

GENETICS OF PATHOGENICITY

Plant pathogenic fungi can be expected to share most if not all of the genetical properties of the saprophytic fungi we have been considering in this book. This is certainly so for those which can be cultured, and is probably also true of the obligate parasites. What features do they possess which makes it desirable to reserve a chapter solely for their consideration? Undoubtedly the most interesting feature is their property of pathogenicity or ability to overcome the natural barriers to infection by foreign organisms possessed by their hosts. How they are able to do this we cannot yet say, but from genetical analyses of this property, plant pathologists and plant breeders have been able to reach a number of important conclusions which bear on this problem. In this chapter some of these conclusions are discussed together with some of the special genetical problems encountered in plant pathogenic fungi.

The diseases of crop plants are generally controlled either by the use of toxic materials applied in the form of sprays, dusts or seed dressings, etc. or by breeding for disease resistance. The first method aims at killing or weakening the parasite before it enters a potential host plant and the materials employed must do this without causing harm to the plant they are designed to protect. When carried out rigorously this control method is usually successful and very rarely fails because of the development of resistant variants [496]. The second method makes use of natural defence mechanisms often present in wild related host species which have no or few other economically important features. These defence mechanisms, known collectively as disease resistance, kill or limit the growth of the parasite after it has gained entry to the host. The genes controlling this resistance are introduced into the cultivated varieties by slow, often tedious, breeding work. When first released the new resistant varieties can be extremely successful but all too often, after two or three seasons of widespread use, they succumb to new or previously undetected variants of the pathogen which appear to be unaffected by the disease resistance mechanism. These new forms of the pathogen, distinguished from each

257

other only by the spectrum of differential host varieties on which they can grow, are called *physiologic races*. Their origin and genetic control form the central problem of this chapter.

The two terms pathogenicity and virulence are commonly used to describe ability to incite disease. In this chapter we have adopted the following distinction; pathogenicity is a general attribute of a plant pathogen while virulence defines the pathogenicity of a particular pathogen strain in relation to a particular host genotype. For example an adenine requiring mutant of a pathogen might be non-pathogenic on all host varieties but a given adenine independent physiologic race might be virulent on some varieties but avirulent on others.

THE GENETICAL BASIS OF PHYSIOLOGIC SPECIALIZATION

For a simple example of physiologic specialization we may turn to the tomato leaf mould fungus *Cladosporium fulvum*. Varieties of tomato carrying various combinations of three different genes for resistance to Cladosporium have been widely grown in Ontario [14] and in England and Wales [102]. The eight different races of Cladosporium which can be distinguished with these host varieties are shown in Table 29. The three resistance genes are dominant and independent and although their resistance phenotypes differ, as is shown in Fig. 52, they can each be

TABLE 29. The eight races of *Cladosporium fulvum* that can be differentiated with tomato varieties carrying three resistance genes in various combinations the total number of which is eight. The most efficient differentiation can be made with the three monogenic resistant varieties. Race 1, 3 is hypothetical, never having been found. − = resistant, S = susceptible. From Day [103].

Dominant resistance genes present in the host	Physiologic races							
	0	1	2	3	1, 2	(1, 3)	2, 3	1, 2, 3
−	S	S	S	S	S	S	S	S
Cf_1	−	S	−	−	S	S	−	S
Cf_2	−	−	S	−	S	−	S	S
Cf_3	−	−	−	S	−	S	S	S
$Cf_1\ Cf_2$	−	−	−	−	S	−	−	S
$Cf_1\ Cf_3$	−	−	−	−	−	S	−	S
$Cf_2\ Cf_3$	−	−	−	−	−	−	S	S
$Cf_1\ Cf_2\ Cf_3$	−	−	−	−	−	−	−	S

classified as either effective (—) or ineffective (S). Seven of the races have been identified but the eighth, race 1,3, is hypothetical. The races are named in such a way that they indicate the host resistance genes towards which they are virulent. Thus race 1,2,3 can grow on any of the host varieties listed in the table while race 0 will only grow on varieties which carry no resistance genes. Three host varieties each carrying a single resistance gene are all that are needed for the most efficient classification of the races.

The method of naming races used in the Cladosporium example was first introduced by Black *et al* [39] for races of potato late blight, *Phytophthora infestans*. In the potato some six major host genes for blight resistance are now known [38] and with the same classification of phenotypes into two classes, resistant or susceptible, 2^6 or sixty-four races of Phytophthora may be recognized in theory.

Both of these examples are free from the complication of scoring and taking into account intermediate reactions. These however do occur and are often important particularly in the cereal rusts. To this extent, therefore, the races of *C. fulvum* and *P. infestans* represent categories within which some variation is to be expected. Detailed genetical analyses of *C. fulvum* and *P. infestans* have yet to be made but the relationship between races and host genotypes in both of these fungi strongly suggests that genetic systems, comparable to those controlling resistance in the hosts, control virulence in the two pathogens.

In flax rust, *Melampsora lini*, the genetic control of virulence has been worked out in some detail by Flor in the United States [164]. Flor has described some 179 races of flax rust characterized by their reactions on eighteen differential host varieties [162]. During more than 25 years' work on flax rust he has investigated the inheritance of resistance in the host and the inheritance of virulence in the rust. In one example, he studied the inheritance of virulence on the flax varieties Ottawa and Bombay in rust races 22 and 24, and at the same time studied the inheritance of resistance in these two varieties to the same two races. The results are shown in Table 30.

The Table shows that the F_1 hybrid between the two rust races is avirulent on both host varieties. Avirulence is dominant to virulence. The F_2 shows a segregation which is a fair approximation to the 9:3:3:1 ratio which would be expected if the two races are each homozygous for one of two different recessive virulence genes which segregate independently. We may note here that since, in rusts, the haploid products of meiosis are not scored directly but are paired and give rise to dicaryotic clones, the genetic ratios encountered are those more usually characteristic of diploid organisms (cf Fig. 10). We can also see that the F_1 hybrid between the two flax varieties is immune to both races. Resistance is evidently dominant to susceptibility and the F_2 again shows a fair approximation

TABLE 30

a. The inheritance of virulence on two resistant flax varieties in races 22 and 24 of *Melampsora lini*.

Host	Race 22	Race 24	Rust F_1	Rust F_2			
	$p_1 p_2^+$	$p_1^+ p_2$	$p_1^+ p_2^+$	$p_1^+ p_2^+$	$p_1 p_2^+$	$p_2^+ p_2$	$p_1 p_2$
Ottawa (R_1)	S	I	I	I	S	I	S
Bombay (R_2)	I	S	I	I	I	S	S
				85	: 23	: 21	: 4

b. The inheritance of resistance to two races of flax rust in the flax varieties Ottawa and Bombay.

Race	Ottawa	Bombay	Flax F_1	Flax F_2			
	$R_1 r_2$	$r_1 R_2$	$R_1 R_2$	$R_1 R_2$	$R_1 r_2$	$r_1 R_2$	$r_1 r_2$
22 (p_1)	S	I	I	I	S	I	S
24 (p_2)	I	S	I	I	I	S	S
				110	: 32	: 43	: 9

S = susceptible; I = immune or resistant.
The genetic symbols describing the rust races and host varieties refer to phenotypes. From Flor [164].

to the 9:3:3:1 ratio which would be expected if the two varieties were each homozygous for one of two different dominant resistance genes which segregate independently. From evidence of this kind Flor advanced the concept of a *gene-for-gene relationship* between host and pathogen. In general terms the relationship can be stated as follows: the ability of a pathogen to grow and produce disease symptoms on a host bearing major genes for resistance is determined by alleles governing virulence at corresponding loci in the pathogen. Put another way: for each gene in the host capable of mutating to give resistance there exists a gene in the pathogen capable of mutating to overcome that particular resistance. This generalization offers little encouragement to those trying to breed plants for disease resistance. Evidence of a similar nature to that from flax rust has been found by Moseman [367] for the powdery mildew of barley (*Erysiphe graminis f. sp. hordei*).

If one accepts the gene-for-gene relationship, the reactions of susceptibility or resistance given by a range of host varieties to a range of genetically defined pathogen races can be used to construct hypothetical host genotypes even in the absence of direct genetic evidence. The work of Boone and Keitt [46] on the inheritance of virulence in apple scab (*Venturia inaequalis*) on a number of cultivated apple varieties shows how the gene-for-gene relationship can be used. Genetical analysis of Venturia has shown that genes controlling virulence show simple segregation in the ascus. Seven different genes controlling virulence were identified and seven different apple varieties were selected to differentiate between them. The reactions controlled by alleles for avirulence, of each of the seven Venturia genes, on the differential host varieties are shown in Table 31.

TABLE 31. Resistant, or avirulent, reactions ($\times$) controlled by alleles for avirulence at seven loci in *Venturia inaequalis* on a group of seven apple varieties. The probable apple genotypes for resistance based on a gene-for-gene relationship are shown in the last column. From Boone and Keitt [46].

Apple variety	Reactions of Venturia stocks carrying alleles for avirulence at loci							Putative host genotype
	1	2	3	4	5	6	7	
McIntosh	$\times$							R_1
Yellow Transparent	$\times$		$\times$	$\times$				$R_1\,R_3\,R_4$
Haralson		$\times$		$\times$				$R_2\,R_4$
Red Astrachan			$\times$					R_3
Hyslop				$\times$	$\times$			$R_4\,R_5$
Grimes Golden						$\times$		R_6
Prairie Spy		$\times$					$\times$	$R_2\,R_7$

The stocks carrying an avirulence allele of gene 1 were avirulent on the varieties McIntosh and Yellow Transparent. This can most simply be explained if we assume that these two varieties carry a common resistance gene R_1. Since McIntosh is not resistant to stocks carrying any of the other six avirulence alleles we may conclude that, at least as far as these seven Venturia genes are concerned, it carries no other resistance gene. On the other hand, Yellow Transparent is also resistant to stocks carrying avirulence alleles of genes 3 and 4 and it seems reasonable to conclude that this variety carries three resistance genes R_1, R_3, and R_4. By extending this reasoning we can build up putative resistance genotypes for all of the apple varieties and these are shown in the last column of the Table. The genotypes are incomplete since they do not of course reveal whether the host varieties are homozygous or heterozygous for the genes concerned.

The implications of the gene-for-gene relationship have been most fully explored by Person [406]. Starting from a theoretical model with five genes in both host and pathogen, Person has listed the various groupings and inter-relationships of races and host varieties which such a system would be expected to generate. The most efficient way of identifying the physiologic races of a pathogen, as we have seen from Table 29, is to use a set of host varieties each of which carries only one of the resistance genes. A set of n such varieties enables us to differentiate 2^n different races so that with five such genes, R_1 to R_5, we can differentiate thirty-two races. If, however, one of the monogenic differentials is missing from the set then all the races characterized by this variety will fail to be identified and only sixteen instead of thirty-two races will in fact be recognized. On the other hand, if one variety carries two of the resistance genes, the set of five varieties (for example R_1, R_2, R_3, R_4 and R_1R_5) will distinguish twenty-four races. Another important feature emerges from a consideration of varieties with more than one resistance gene. In the theoretical model there are host varieties with 5, 4, 3, 2, 1 or 0 genes for resistance. The number of races which are virulent on each of these varieties increases in the geometric series 1, 2, 4, 8, 16, 32. Thus only race 1, 2, 3, 4, 5 will attack the first variety, and only this race and one other will attack a variety with four genes, and so on until we reach the variety with no genes for resistance which all the races can attack.

These, and other rules derived from the gene-for-gene relationship, Person applied to Flor's data on rust races. His analysis revealed the hitherto unsuspected probability that there were two or more genes for resistance in many of the differential flax varieties. For example the varieties Leona, Abyssinian, Akmolinsk and Bison formed one geometric series and Leona, Koto and Bison another. Thus thirty-five of the races were virulent on Leona all of which, with an additional eighteen, were virulent on Abyssinian; again these fifty-three races, with an additional forty-eight, were virulent on Akmolinsk and Bison. All thirty-five races virulent on Leona were, together with another fifty-nine races, virulent on Koto and Bison. The races virulent only on Abyssinian, Akmolinsk and Bison were avirulent on Koto and those virulent only on Koto were avirulent on Abyssinian and Akmolinsk. These relationships are expressed in the diagram in Fig. 53. They are most easily explained by assuming that Leona has three resistance genes (ABC), Abyssinian has two of these (AB), and Akmolinsk and Koto have different single genes (A and C respectively), the one carried by Koto (C) being different from the two carried by Abyssinian. A variety with the single gene B was apparently not represented among Flor's differential host varieties since there is no variety other than Akmolinsk which is attacked by the fifty-three races to which Abyssinian is susceptible. The race ratios depart significantly from an ideal geometric series. The reason for this

Cf₁

Cf₂

Cf₃

FIG. 52. Young leaves of the tomato variety Ailsa Craig (above left) and the three monogenic resistant varieties Leafmould Resister No. 1 (*Cf*₁), Vetomold (*Cf*₂) and V-121 (*Cf*₃) inoculated twelve days previously with conidia of race 0 of *Cladosporium fulvum*. The leaf of Ailsa Craig shows the fully susceptible phenotype of a variety with no genes for resistance. The other three show resistant phenotypes, that of *Cf*₂ being the most extreme. The susceptible phenotypes of the resistant leaves, after inoculation with race 1,2,3, would all be similar to that of the first leaf.

[*facing p.* 262

may be that the rust strains tested were not a random selection of all possible genotypes; indeed, in the absence of the host tester type *B*, some of these genotypes could not even have been recognised. The discovery that some of the flax varieties appear to have more than one gene for resistance provides an alternative to Flor's earlier conclusion that a number of the genes for rust resistance are allelic or tightly linked. The undetected possession of a common resistance gene by two different, supposedly monogenic, resistant varieties would result in F_2 progenies which were uniformly resistant to a race avirulent on both parents. The absence of susceptible segregants under just these circumstances was construed by Flor as evidence of tight linkage or allelism [161]. However,

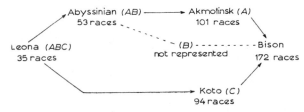

FIG. 53. Two ·'geometric series' of flax varieties showing the number of different rust races which can attack each, and the postulated complements of resistance genes. All races listed on the left of the diagram can attack varieties to the right to which they are connected by lines. From Person [406].

later results [167] meet this criticism and support Flor's claims for two of the series of very closely linked genes for resistance.

While the simplest explanation for the kinds of host-parasite relationship exemplified by Fig. 53, is the existence of a gene-for-gene relationship, the data are equally consistent with the possibility that some of the unit differentials in host or parasite have a more complex basis than a single gene, always provided that they act independently of each other. Thus the occurrence of complementary or duplicate gene action in either host or parasite would not necessarily be incompatible with the patterns and numerical regularities described by Person. To this extent the internal consistency of the analysis does not constitute a critical test of the gene-for-gene hypothesis. Indeed, there can be no conclusive demonstration of the correctness of the hypothesis except through conventional genetic analysis of both host and parasite.

The concept of physiologic specialization which we have used so far may appear to be a rather rigid one tied to the acceptance of well defined races and differential host varieties. For the purposes of the gene-for-gene relationship it has excluded consideration of intermediate de-

TABLE 32. Representative examples of the generation of new physiologic races of plant pathogenic fungi by various means

	Phycomycetes	Ascomycetes	Basidiomycetes		Fungi Imperfecti
			Rusts	Smuts	
Sexual recombination	*Phytophthora infestans* [502]	*Venturia inaequalis* [278] *Erysiphe graminis* [367] *Cochliobolus (Helminthosporium)* spp. [374]	*Melampsora lini* [160] *Puccinia graminis tritici* [379]	*Ustilago avenae,* [213, 239]	
Heterocaryosis			*P. graminis tritici* [375]		*Fusarium oxysporum pisi* [65]
Mitotic recombination		*Helminthosporium sativum* [550]	*P. graminis tritici* [561]		*F. oxysporum pisi* [65] *F. oxysporum cubense* [67] *Verticillium albo-atrum* [216]
Spontaneous mutation	*P. infestans* [185]		*P. graminis tritici* [559] *P. recondita tritici* [472]		
Induced mutation	*P. infestans* [580]		*M. lini* [163, 165, 483] *P. coronata avenae* [204] *P. graminis tritici* [468]		*Cladosporium fulvum* [104]
Cytoplasmic variation			*P. graminis tritici* [269]		

See also general reviews by Buxton [66], Day [107], Holton [238] and Johnson [270]; and for *P. infestans* Gallegly and Niederhauser [187] and for *Venturia inaequalis* Keitt et al. [281].

grees of resistance. Races of some cereal rusts are defined on the basis of three intermediate reactions as well as the two extremes of resistance and susceptibility. However the reactions of crown rust of oats, *Puccinia coronata avenae*, can be grouped into only two classes with a consequent limitation on the number of recognizable physiologic races [494]. A promising start for wheat stem rust was made by Loegering and Powers [327] who assigned putative rust resistance genotypes to 4 wheat varieties based upon the behaviour of 108 F_2 segregants from a cross between two rust races whose F_1 was heterozygous for at least 8 virulence loci.

ORIGIN OF PHYSIOLOGIC RACES

The gene-for-gene relationship implies a more or less strict genetic complementarity between each host and pathogen which has evolved with the development of the association between the two organisms. When the host genotype changes towards increase in resistance by genetic recombination or mutation we can expect the pathogen to respond in a similar way by an increase in pathogenicity so that the balance is restored. We would therefore expect to find that the genetic systems we have discussed in the preceding chapters, which include sexual recombination, heterocaryosis, mitotic recombination, mutation and cytoplasmic variation, may all be employed by plant pathogenic fungi to this end. This is certainly borne out by greenhouse and laboratory studies and, to a lesser extent, by practical experience in the field. To give an indication of the possibilities we have listed in Table 32 representative examples of the generation of new physiologic races in the main groups of plant pathogenic fungi. Some of these are discussed below.

Phytophthora infestans

For a long time the sexual stage of the potato blight organism was considered unimportant and of rare occurrence. It now seems that in North America and Europe this is because only one mating type is present in the populations which are consequently sexually sterile. In Central Mexico both mating types are equally common and oospores are frequently produced [186]. As in many Phycomycetes, the problem of oospore germination has hampered genetical investigation. In the only reported example a culture from a single germinated oospore obtained from a cross between race 1,2 and race 3 proved to be race 0 [502]. Oospore germination under natural conditions has not been recorded but may occur more readily than in the laboratory. Central Mexico is a more prolific source of physiologic races of *P. infestans*, a fact which Niederhauser [380] has attributed to the occurrence of the sexual stage, permitting recombination, and the presence of resistant wild host species responsible for the selection of many genes for virulence.

Such a situation provides the plant breeder with an opportunity to assess resistant potato varieties under the most rigorous conditions. Indeed the tests carried out in central Mexico have caused a trend among potato breeders toward the abandonment of major blight resistance genes with all or none effects, and a shift towards breeding for *field resistance* [217]. Field resistance is generally controlled by many genes of individually small effects. Although allowing infection and growth of the pathogen, it prevents extensive host damage and restricts spread of the pathogen through the crop. The growth rate and rate of sporulation of the parasite are thus important factors in field resistance. Some recent studies have examined the variation of *P. infestans* in this respect. In Birmingham Jinks and his colleagues collected isolates from named potato varieties and recorded their growth rates in tubers and rates of sporulation on leaves over a period of several years [256, 266]. They observed that fresh isolates grew significantly better on the tubers of the varieties from which they had originally been collected than on tubers of other varieties. During transfers on an artificial medium this specific adaptation to one variety was slowly lost but could usually be restored by training the isolates on leaves of tubers of that variety. However, the period of training induced no other change than the restoration of the original phenotype. Training on tubers of another variety had no effect. These results were confirmed by Paxman [398] who found that 81 passages through tubers of three potato varieties over a two-year period induced no adaptive changes in three different isolates of *P. infestans*.

While some of the changes that Jinks *et al.* recorded are most likely due to spontaneous gene mutation, for example the appearance on artificial media of some variants whose original phenotypes could not be restored by training, the authors suggest that the changes which were brought about by training were specifically induced by the host. The response of the strain to the host was evidently determined by its genotype. The results of these experiments would seem to encourage the use of field resistance, since they show that the fungus is much less plastic in its adaptive responses than the potato breeder may have feared. Whether field resistance will prove to be broken down by further evolution of the pathogen remains to be seen. Experience with existing field resistant varieties is conflicting; some varieties appear to have declined in resistance while others, in central Mexico, have not [187]. In laboratory studies *P. infestans* shows an appreciable rate of spontaneous mutation from one race type to another [185, 38]. Changes involving the host resistance gene R_4 appear to be especially frequent. These changes may occur in either direction, to virulence or to avirulence, and have been reported in cultures started from single zoospores. Induced changes from race 1 to race 1,2 and race 1,4 following treatment with ultraviolet light have also been reported by Wilde [580]. None of these spontaneous or induced mutants have so far been subject to genetical analysis. Little is known in *P. infestans* of the possible roles of heterocaryosis and mitotic recombination. Wilde claimed to have produced race 1,2,4 after inoculating non-resistant plants with a mixture of races 1,2 and 1,4. Varying proportions of the strains recovered from foliage of plants carrying the resistance genes R_2 and R_4, which had been inoculated with spores produced by the earlier mixed inoculations, were able to infect R_1, R_2 and R_4 plants. Unfortunately these supposedly recombinant strains were not tested on plants carrying either all three genes, or the two genes R_2 and R_4. The possibility that they were mixtures of races 1,2 and 1,4, and not a relatively stable heterocaryon or mitotic recombinant, cannot be excluded.

The smut fungi

Physiologic specialization in the smuts is complicated by the fact that the pathogenic dicaryon is not propagated clonally. When diploid brandspores from the host germinate they immediately undergo meiosis to form the usually saprophytic haploid stage, which in different smuts may be sporidial or mycelial. Pathogenic dicaryons are then reformed as the result of fusion between compatible haploids. Because of the obligatory intervention of meiosis between successive infections there must be a constant reassortment of genes controlling virulence. Occasionally meiosis fails and diploid sporidia are formed which are *solopathogenic*, that is they will produce infections when inoculated singly.

Heterothallism in the majority of smut fungi is controlled by a simple bipolar mating system with two alleles. Whitehouse [575] has suggested that more complex interpretations have arisen from a confusion between the mating controls and genes controlling pathogenicity. In *Ustilago maydis* Rowell and DeVay [467] and Rowell [466] subsequently showed that brandspore production in the host is controlled by two genes; *a*, with two alleles, and *b*, with multiple alleles. The fusion between haploid sporidia on maize extract agar was shown to be controlled by the *a* gene and was independent of *b*. More recently the distinction between *a* and *b* was further clarified by Holliday [233] who synthesized pathogenic diploids and recovered from them diploids homozygous for various markers using the methods described on page 106. He found that diploids homozygous for a_1, but heterozygous for *b*, were pathogenic but would not fuse in the hosts with a_2 haploids even if the haploids carried a third *b* allele. Presumably the presence of two different *b* alleles in the diploid nucleus prevented fusion. Diploids homozygous for *b* and heterozygous for *a* were not pathogenic but would readily fuse in the host with haploids with a different *b* allele to form a pathogenic heterocaryon. Evidently the *a* gene controls fusion between haploid sporidia while the *b* gene controls pathogenicity so that only sporidia, or heterocaryons, heterozygous for *b* are pathogenic.

The genetical control of pathogenicity in the smuts as revealed by sexual analysis has recently been reviewed by Holton [238] and by Johnson [270]. These analyses have largely been concerned with interspecific crosses to study the inheritance of aspects of pathogenicity such as the form of the groups of brandspores or sori and the range of host species attacked. Halisky [213] and Holton and Halisky [239] have shown, from crosses of different physiologic races of *U. avenae*, that virulence on two oat varieties is controlled by two independent genes and that virulence is recessive to avirulence.

The parasexual analysis of *U. maydis* discussed in Chapter 5 suggests that this method may prove fruitful in other smuts. An approach of this kind would first involve the maintenance of several race types as pairs of compatible, haploid lines on artificial media. The induction, isolation and characterization of a variety of suitable marker mutants would be a further step. The use of nutritional and other markers that can be studied in haploids, in addition to characters which can only be studied in dicaryons on the host, would greatly speed up the genetical analysis of other smuts.

The rusts

Genetic studies in the rusts date from Craigie's [90] discovery in 1927 of the function of the pycnia. Since that time the sexual analysis of the inheritance of virulence has been undertaken in several species. The subject has been reviewed by Johnson [268, 270] and Flor [166]. More recently attention has been directed to other means of genetic recombination and variation in these fungi.

A rust dicaryon is composed of two haploid nuclei of opposite mating type so that we might expect that if two different dicaryons were able to reassort their nuclei by hyphal anastomosis only two recombinant dicaryons would be produced. Working with *Puccinia graminis tritici* Watson [560] and Watson and Luig [561] mixed uredospores of a red spored race 111 with uredospores of an orange spored race NR-2. The orange rust was virulent on four host varieties for which the red rust was avirulent. Cultures derived from red pustules on these four varieties after inoculation with the mixture were analysed, and among one hundred heterocaryons tested eleven different races were found. More recombinant races than the two expected on the basis of nuclear reassortment had been produced. Similar results were obtained in further experiments with races of *P. graminis tritici* and a culture of *P. graminis secalis* [562]. The release of variation can only be explained satisfactorily by mitotic recombination. Beyond the reported occurrence of hyphal anastomosis between uredospore germ tubes in *P. graminis tritici* [579], and of tri- and quadrinucleate uredospores in heterocaryotic mixtures [373], we have no indications as to how such recombination may take place. There can be little doubt that the present interest in mitotic recombination in the Hymenomycetes (see pp. 218–9) will throw light on the phenomena here. As Ellingboe [134] has commented, it is striking that no reports exist of mitotic recombination in a single heterozygous rust race propagated clonally by uredospores despite the fact that variation is readily detected when two different strains are mixed. It seems likely that following hyphal anastomosis the introduction of foreign nuclei and cytoplasm may result in a localized breakdown in the stability of either dicaryon. The association of more than two complementary nuclei in a cell could lead to an unbalanced condition in which new genetic combinations would have some chance of being established.

Spontaneous mutations from avirulence to virulence on the wheat variety Lee have been observed in *P. graminis tritici* by Watson [559], who used four different races carrying spore colour and virulence markers More attention has been paid to induced changes of this kind. The first experiments were carried out by Flor [163, 165] using uredospores of *Melampsora lini*. Flor argued that if the rust dicaryon were heterozygous for avirulence on a monogenic resistant host variety, then the mutation of the dominant allele for avirulence to the recessive allele for virulence should allow recovery of the virulent mutants as susceptible lesions when treated spores are inoculated to the resistant host. Flor chose for his experiments an F_1 hybrid dicaryon between race 1 and race 22 heterozygous for avirulence on sixteen supposedly monogenic resistant flax varieties, but homozygous for avirulence on three other varieties. Uredospores of the hybrid were treated with ultraviolet light to kill approximately 90 per cent. Groups of one hundred plants of each of twenty-two flax varieties were inoculated with the treated spores. Each set of one hundred plants screened at least 40,000 viable spores. Seven mutants were recovered. In another experiment on a larger scale, uredospores of the same F_1 hybrid were treated with X-rays at two dosage levels and a further 154 mutants were recovered. In this experiment each set of plants screened about 200,000 viable spores. Of the 154 mutants 94 were tested further. The results of these experiments are summarized in Table 33. The numbers opposite each variety refer to the number of mutant lesions isolated from that variety. The lines bracketing the varieties into groups indicate that mutants are virulent on all varieties within the group.

While the results might be explained by assuming that the relevant alleles for avirulence had mutated to the recessive virulent state there are other possibilities. The simplest of these is that the treatments resulted in chromosome breakage which gave rise to deletions. Where lost chromosome fragments carried the dominant allele for avirulence the recessive allele for virulence would be uncovered in the resultant hemizygote.Quite large deletions, providing they are in the heterozygous condition, are likely to be viable and would not affect pathogenicity since the haploid monocaryon is pathogenic on flax.

A curious feature of the data of Table 33 is the fact that the mutants selected on Leona and Koto were virulent on both these varieties and on Abyssinian. Although, from the information of Fig. 53, mutants selected for virulence on Leona might well be virulent also on Abyssinian *or* on Koto, they would not be expected to attack all three varieties, nor should mutants selected on Koto be expected to attack Abyssinian, except as a result of double mutation or deletion of two linked genes. Other possible examples of the same kind are also apparent in the table. While chromosomal deletions, which are known to be readily induced by X-rays, are perhaps the simplest explanation of these data, another possibility exists. As Day [107] has suggested, some of the mutants could have arisen as a result of mitotic recombination induced by the radiation. It has been reported that the rate of occurrence of diploids in heterocaryons of *Aspergillus oryzeae* and *A. sojae* is enhanced by a factor of 10^4 when conidia are treated with ultraviolet light [250, 251]. It is conceivable that the frequency of diploid nuclei and hence of recombinant diploid nuclei or heterocaryons in irradiated uredospores or even in incipient avirulent

TABLE 33. Results of experiments on mutation to virulence in *Melampsora lini*.

The Table shows the flax varieties used to select mutants, and the numbers of mutants recovered. Varieties bracketed by the vertical line show the same susceptibility to the indicated mutants. Further explanation in the text.

Reference	Flor (163)		Flor (165)	
Treatment	Ultraviolet light		X-ray	
Numbers screened	40,000 spores per 100 plant unit		200,000 spores per 100 plant unit	
	Dakota	3	Dakota	15
	Cass	1	Wilden	29 | 1
	Polk	1	Barnes	
	Wilden |	1	Birio	14
	Barnes |	1	Cass	20
	Birio |		Polk	8 | 1
			Marshall	
			Koto	2
			Leona	2
			Abyssinian	
			Towner	2
			Kenya	
			P. Blue Cr.	
			B. Golden Sel.	

infections, might be high enough for them to be recovered as virulent lesions on a selective host.

Griffiths and Carr [204] recovered a mutant of *Puccinia coronata avenae* from uredospores irradiated with ultraviolet light. The mutant was virulent on six oat varieties, all possessing different factors for resistance, to which the original race was avirulent. The mutant was also avirulent on another variety to which the original race was virulent. As the authors point out, it is difficult to suppose that all of these differences were the result of seven separate but simultaneous mutations.

Schwinghamer [483] has carried out mutation studies using race 1 of *M. lini*, known to be heterozygous for a gene controlling virulence on the flax variety Dakota. Comparisons were made between the effects of fast neutrons, X-rays and ultraviolet light, the average maximum frequency of mutants being 2·0, 1·5 and 0·3 per cent of infections respectively for these treatments. Since a mutation frequency of the order of several per cent for one specific gene would be extremely high, it seems probable, as Schwinghamer suggests, that most of the mutants induced by the ionizing radiations, if not by the ultraviolet light, were the result of chromosomal deletions.

THE BASIS OF PHYSIOLOGIC SPECIALIZATION

Perhaps the simplest situation to consider is that of two physiologic races and two host varieties. One race is virulent on only one of the varieties, the other is virulent on both, and the difference between the races is determined by one gene. Similarly one variety is susceptible to both races, the other to only one race, and the difference between the varieties is determined by one gene. Such a situation might be represented by races 0 and 1 with hosts rr and R_1R_1. It would permit comparisons between races and between varieties which should ultimately give information about the functions of the differentiating genes. Surprisingly few plant diseases offer opportunities for such simple comparisons.

Attempts to discover the nature of the difference between our two hypothetical races have followed several lines. Pathogenic fungi which can be cultured on defined media enable comparisons to be made of the nutritional requirements of different races. So far no consistent differences have been found in *Phytophthora infestans*, *Cladosporium fulvum*, *Venturia inaequalis*, *Colletotrichum lagenarium* and several other pathogens. Lewis [312] has pointed out that while *in vitro* differences in nutrition may not be detected there may be *in vivo* differences. This is almost certainly the case if we give nutrition its widest meaning, but *in vivo* nutrition is an unhelpfully complex subject.

A different approach to the nutritional aspect of the problem has been to produce a variety of auxotrophic mutants in a single strain of pathogen and to test these mutants on a wide spectrum of host varieties. The object of such tests is to find some alteration in the pattern of host reactions shown by the parent strain which could be correlated with the mutant requirement. Such alterations have been found in *Venturia inaequalis* [47], *Cladosporium fulvum* [105], *Cochliobolus sativus* [551] and *Colletotrichum lagenarium* [129]. Certain mutants of all three fungi were non-pathogenic but were so on *all* the host varieties tested. Some mutants of *V. inaequalis* had their pathogenicity restored if the host leaves were supplemented with the mutant requirement but others remained non-pathogenic [287]. The mutations affected pathogenicity rather than virulence.

A more promising approach with facultative saprophytes would be to isolate a number of race 1 mutants from race 0 using R_1R_1 plants to select them. Physiological comparisons between the virulent mutants and the original avirulent parent might then reveal significant properties peculiar to the mutants. One mutant of this kind was recovered from *Cladosporium fulvum* [104] following X-ray treatment.

Attempts have been made to differentiate physiologic races by serological means. Probably because of the heterogeneity of the races which have been compared the results from these studies have not been very meaningful. They had not, for example, until recently given any indication that certain antigens are associated with one allele for virulence and not with another. One of the most interesting among recent studies is the comparison made by Doubly *et al.* [124] of the globulin antigens of four races of *M. lini* with those of the rust susceptible variety Bison and three other resistant varieties which had been backcrossed to Bison through seven generations. While preliminary in nature, the results suggest that susceptibility of a host variety to a particular race of rust depends on the host plant containing, as a minor antigenic component, a protein serologically related to one of the proteins of the parasite. The implication is that the resistant reaction in the host plant depends on some sort of recognition of a specific fungal protein, which may fail if the latter resembles too closely one of the host's own proteins.

ON DEOXYRIBONUCLEIC ACID (DNA)

If this book were to confine itself strictly to fungal genetics there could be no place in it for a discussion of DNA, since the fungi are remarkable for having contributed almost nothing to our knowledge of this material. However, the genetic significance of DNA is so firmly established in bacteria and viruses, and some knowledge of its structure is of such importance for the understanding of current genetic theory, particularly in relation to genetic replication, mutation and recombination, that a brief account of it seems desirable.

Structure

Nucleic acids in general are long linear polymers of nucleotide units, each nucleotide consisting of the combination nitrogenous base-pentose sugar-phosphate. The base-sugar compound without the phosphate is known as a nucleoside. Ribonucleic acids (RNA) contain the sugar ribose, two purine bases, adenine and guanine, and two pyrimidine bases,

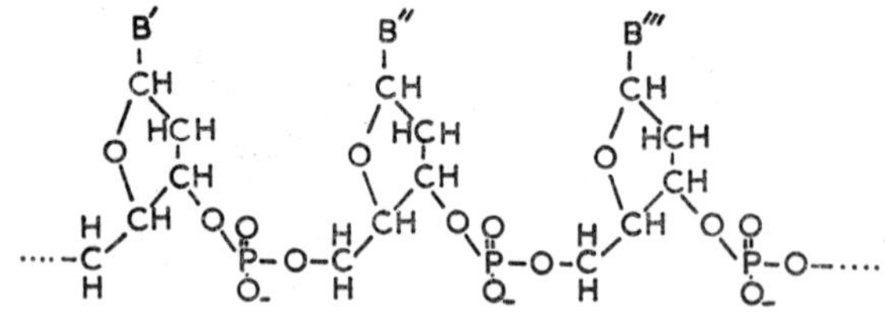

FIG. 54. Structure of a single polynucleotide chain of DNA

uracil and cytosine. DNA differs in containing deoxyribose instead of ribose, and in containing the pyrimidine base thymine to the exclusion of uracil. DNA consists of polynucleotide chains with the structure shown in Fig. 54. B′ etc. are bases, with any sequence of the four bases possible.

RNA differs from this formula only in having a free hydroxyl group on the second carbon atom of the pentose residue instead of one of the two hydrogen atoms.

In DNA it is found that though the adenine/guanine ratio varies considerably from one organism to another, the adenine/thymine and guanine/cytosine ratios are always very close to unity. The explanation of this regularity in composition is that each DNA molecule consists of two polynucleotide chains intertwined in a double helix, with the bases directed inwards [563, 149]. The two chains of each molecule are of opposite polarity; i.e. the 5' (side-chain) carbon atoms of the pentose residues are orientated towards opposite ends of the molecule in the two chains (see

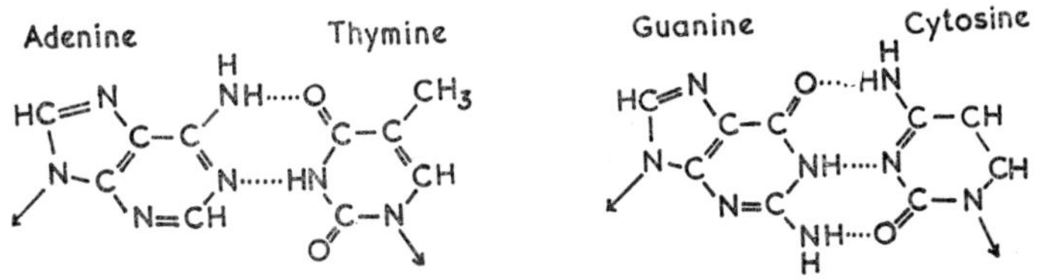

Fig. 55. Specific pairing of DNA bases. Arrows indicate the atoms attached to the deoxyribose of the polynucleotide backbone, while dotted lines show hydrogen bonds.

Fig. 54 for the polarity of one chain). Each adenine on one chain is paired, through the formation of hydrogen bonds, with a thymine on the other chain, and each guanine is similarly paired with a cytosine as shown in Fig. 55. It will be seen that adenine will not pair with cytosine, nor guanine with thymine, because of lack of correspondence between the H-donor and H-acceptor groups. Purines will not pair with purines because such a pairing would occupy too much space, while a pyrimidine-pyrimidine pairing would not occupy enough space to bridge the distance between the two chains.

DNA as ordinarily isolated has a molecular weight of the order of 10^7 (i.e. of the order of 10^4 nucleotide pairs per molecule), but there is evidence that very careful isolation can result in the recovery of DNA of a size some ten times larger than this [64]. DNA molecules can be seen in electron micrographs as fibres about 20 Å in diameter but of relatively enormous length [26, 68].

Replication

Fig. 56 shows a schematic representation of a DNA molecule. Watson and Crick, who proposed the double helical structure [563], were also the first to point out that this structure suggests a mechanism by which the specific base sequence of a DNA molecule could be transmitted as a hereditary character [564]. The suggestion is that DNA synthesis involves the unwinding of the two strands of a previously existing DNA molecule, and the synthesis of a new polynucleotide strand alongside each one. Since the base sequence of each new strand would have to be comple-

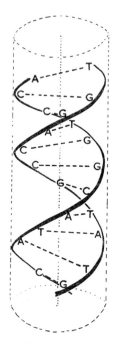

Fig. 56. Schematic representation of a section of a duplex DNA molecule. The solid helices represent the polydeoxyribonucleotide backbones, and the broken lines attachment by hydrogen bonds of complementary base-pairs. The structure is drawn as if viewed obliquely from slightly above the upper end.

mentary to that of its partner, the two new duplex molecules will each be identical with the parental molecule.

Since this mechanism was first proposed powerful support for it has come from two sources. Firstly, Kornberg and his associates have isolated an enzyme from bacteria capable of synthesizing DNA from a mixture of the triphosphates of all four deoxyribonucleosides, *provided* that DNA is present in the system to begin with. The DNA which is synthesized mimics in its base composition [300] and base sequence [271] the

DNA added as primer. This strongly supports the idea that new DNA molecules are copied from pre-existing ones.

Secondly, Meselson and Stahl [351] have shown that in intact cells of *Escherichia coli* DNA molecules are synthesized in the 'semi-conservative' fashion predicted by Watson and Crick. Each newly formed molecule was shown to consist of one half of material derived from pre-existing DNA, and one half of material newly synthesized from the constituents of the culture medium. This demonstration was accomplished through the use of heavy nitrogen (^{15}N) to label pre-existing DNA. The presumption is that the two halves of the molecule thus demonstrated are the single strands of a Watson-Crick duplex, but this has not been absolutely proved.

The evidence for DNA as genetic material

Studies on bacteria (particularly *Diplococcus pneumoniae* [170] and *Bacillus subtilis* [225]) show that practically any genetic trait can be transferred from one strain to another of the same species by treating the recipient cells with purified DNA derived from the donor. In bacterial viruses the evidence is strong that the injection of viral DNA into the host cell is both necessary and sufficient for infection [226]. Although such clear-cut evidence is not available in fungi or higher organisms, DNA is invariably a major component of the chromosomes, which are certainly the main, if not the only, carriers of genetic information; DNA is not usually found outside the cell nucleus.

Information transfer

Though there is much evidence to suggest that protein structure is determined by genetic instructions stored in the DNA of the cell, every present indication is that DNA is not directly involved in protein synthesis. Recent work on the mechanism of protein synthesis has shown that the free amino acids, the precursors of proteins, are combined each with a specific kind of soluble RNA ('transfer RNA') molecule before being assembled into polypeptide chains [230]. The assembly process occurs in association with the ribosomes, which are granules of RNA combined with protein in the cytoplasm. It seems, at least in bacteria, that the bulk of the RNA of the ribosomes is unspecific, and that it is a comparatively minor fraction of the ribosomal RNA, which has the property of being continually broken down and renewed very rapidly, which confers on the ribosomes the capacity for assembling specific kinds of proteins [53].

This rapidly turning-over fraction of RNA is considered to be the 'messenger' which carries information from the genetic DNA.

Although no direct demonstration has been made of the determination of the structure of messenger RNA by DNA in a living system, it has been shown that both animal and bacterial cells contain an enzyme which, given the four ribonucleoside triphosphates, will synthesize RNA provided some DNA is present [567, 184, 83]. The base ratio, and probably the base sequence, of the RNA synthesized in this type of cell-free system is apparently always nearly the same as that of the primer DNA (with uracil instead of thymine) when double stranded DNA is used, or complementary to it when single-stranded DNA is used [83, 184, 567, 568]. This suggests that RNA can be copied from DNA in essentially the same way as DNA can replicate itself. The dominant current theory is that messenger RNA is synthesized with DNA as a template, and then finds its way from the nucleus to the ribosomes where it determines the kinds of proteins which will be synthesized there.

Coding

If the DNA base sequence is to determine amino acid sequence in proteins, the one sequence has to be translated to the other by some kind of code. It is easy to imagine that the translation depends mainly on the specific hydrogen-bonded pairing of nucleotide bases, with the messenger RNA base sequence derived directly from that of the genetic DNA. Then short sequences of bases along the length of the messenger molecule could pair specifically with corresponding short key sections of the various transfer RNA molecules, each of the latter being distinguished by a specific base sequence in the critical region. Thus amino acid residues, carried on transfer RNA, could become aligned in a linear sequence depending on the base sequence in the messenger. Sequences of at least three bases each are required to provide a unique code symbol for each of twenty amino acids, and some recent experiments by Crick, Brenner et al [92] have led to the proposal of a 'triplet code' on purely genetic grounds. An essential requirement for the correct translation of the code must be a series of twenty different enzymes catalysing the transfer of the twenty amino acids each to its own specific transfer RNA molecule. The existence of amino-acid-specific RNA molecules [384], and of specific enzymes transferring the amino acids to these molecules [31], are both now clearly established. A most exciting recent development has been the report of Nirenberg and Matthaei [381] concerning a cell-free preparation from *Escherichia coli* which is stimulated to incorporate isotopically labelled amino acids into protein by the addition of polyribo-nucleotides. Natural RNA stimulates incorporation of all amino acids, but polyuridylic acid *specifically* promotes the incorporation of phenyl-alanine into what appears to be polyphenylalanine. Thus, assuming a triplet code for the sake of argument, U-U-U appears to be the code for phenylalanine. With the increasing availability of artificial mixed polynucleotides of controlled composition the complete elucidation of the RNA-protein code seems a real possibility for the near future, and some preliminary conclusions regarding the triplet coding for most of the amino acids have already been drawn at the time of writing [505].

LIST OF GENE SYMBOLS USED IN NEUROSPORA GENETICS

Except where otherwise indicated, the descriptions refer to **nutritional** requirements of the corresponding mutants.

a/A	mating type	*nit*	non-nitrate utilising
ac	acetate	*nt*	nicotinic acid or trypto-
ad	adenine		phan
al	albino (white conidia)	*os*	sensitive to high osmotic
am	α-amino nitrogen (deficient		pressure
	in glutamic dehydrogenase)	*ox-D*	deficient in D-amino acid
amyc	amycelial (an extreme mor-		oxidase
	phological variant)	*pab*	p-aminobenzoic acid
arg	arginine	*pan*	pantothenic acid
arom	aromatic (tyrosine + phen-	*pdx*	pyridoxine
	ylalanine + tryptophan +	*prol*	proline
	p-aminobenzoic acid)	*pyr*	pyrimidine
asp	asparagine	*rib*	riboflavin
aur	aurescent (probably an	*ser*	serine
	allele of *al*-1)	*sfo*	sulphonamide-requiring
can	canavanine resistance	*su*	suppressor
chol	choline	*suc*	succinate or other Krebs
col	colonial morphology		cycle intermediate
cys	cysteine	*thi*	thiamin
cyt	altered cytochrome system	*thr*	threonine
for	formate	*tryp*	tryptophan
hist	histidine	*tyr*	tyrosine
hs	homoserine	*T*	modified tyrosinase struc-
inos	inositol		ture
iv	isoleucine + valine	*val*	valine
leu	leucine	*vis*	visible (used for various
lys	lysine		morphological mutations)
me	methionine	*ylo*	yellow conidia
nic	nicotinic acid		

GLOSSARY

allele: one particular form of a gene, differing from other alleles of the gene at one or more mutational sites.

allelic: used to describe the relationships between alleles of the same gene, or mutations affecting the same gene.

aneuploid: with a chromosome number other than a multiple of the haploid number normal for the species.

bivalent: an associated pair of homologous chromosomes at the first division of meiosis.

centromere: the point in a chromosome which is the last to divide during mitosis, which does not divide during the first division of meiosis, and which appears to pull the rest of the chromosome during anaphase movement. Genetically defined as a point in the map of a linkage group which always segregates at the first division of meiosis.

chiasma: a point of attachment between homologous chromosomes at diplotene to first metaphase of meiosis. In organisms with large chromosomes two chromatids are seen to cross reciprocally from one chromosome to the other at each chiasma. Chiasmata seem, in general, to be the visible manifestation of genetic crossing-over.

chromatid: one of the two visibly distinct chromosome strands present in a chromosome arm during mitosis or meiosis before the centromere has divided.

chromomere: a short chromosome region distinguished by its relatively heavy stainability, especially at the pachytene stage of meiosis.

chromosome: a self-propagating thread-like structure visible in stained preparations of the nucleus during and immediately before nuclear division.

chromosome arm: one of the two parts into which a chromosome is divided by its centromere.

cistron: a segment of a linkage map containing a number of sites of mutations which show no complementation among themselves[23]. Subsequently[24] redefined to allow mutants in the same cistron to show *incomplete* complementation to give a still partially mutant phenotype. In the revised definition is synonymous with *gene* as used in the book.

clone: a genetically uniform population derived from a single nucleus by mitotic divisions.

complementation: complementary action of genes. Most commonly used in connexion with *inter-allelic complementation:* an interaction in a diploid or heterocaryon between two alleles of the same gene to give a phenotype closer to the wild type than either allele could give by itself.

conversion: replacement of a genetic site or short genetic segment of a chromatid with material of a type corresponding to that of the homologous chromosome in a diploid nucleus, *without* any reciprocal genetic transfer; especially at meiosis.

crossing-over: the process whereby homologous segments of genetic material are exchanged at the first division of meiosis or, occasionally, at mitosis; the effect is as if homologous chromatids broke symmetrically and rejoined cross-wise (cf. *chiasma*).

dicaryon: a mycelium consisting of cells each containing two nuclei usually of unlike mating type.

diploid (=2*n*): containing twice the haploid chromosome complement in each nucleus.

disomic (=*n*+1): containing the haploid set of chromosomes plus a second representative of one member of the set.

dominant: an allele which, in a diploid or heterocaryon, masks the phenotypic effect of another allele of the same gene (see *recessive*).

epistasis: the process whereby the presence of one gene renders the phenotype insensitive to substitution or mutation in another gene.

euploid: with each chromosome of the haploid set represented the same number of times (cf. *aneuploid*).

exchange: the combination in a structurally normal chromosome of genetic material derived, or descended, from two different homologous chromosomes. In most contexts *exchange* is synonymous with *cross-over*; fine structure analysis, however, may reveal non-reciprocal exchanges (see *conversion*).

gamete: a cell (or, in some contexts, nucleus) which fuses with another to give a zygote.

gene: a short section of a linkage group, acting as a functional unit. See also under *cistron*, and fuller discussion in Chapter 6.

genome: a haploid chromosome complement.

genotype: genetic constitution (cf. *phenotype*).

haploid: with a single set of chromosomes; i.e. each chromosome distinct in homology and function.

heteroallelic: heterozygous with respect to some specified gene; usually refers to the presence of two different but allelic mutations.

heterocaryon: a cell or (usually) mycelium containing more than one genetically distinct type of nucleus in a common cytoplasm.

heterozygous: with a pair of dissimilar though homologous chromosomal loci or segments, specified by context.

homoallelic: homozygous with respect to some specified gene.

homocaryon: a cell or (usually) mycelium containing only one genetic type of nucleus.

homologous (applied to chromosomes or parts of chromosomes): specifically pairing at zygotene of meiosis, and functionally similar.

homozygous: with a pair of identical chromosomal loci or segments, specified by context.

interchange: an exchange of position, apparently by breaking and rejoining, of two non-homologous chromosome segments.

interference: an effect of one cross-over in decreasing the probability either that a second cross-over will be formed at all (*chiasma interference*), or that the second cross-over, if formed, will involve the same strands as the first (*chromatid interference*). In *negative interference* the probability in question is *increased*.

inversion: of a section of a linkage group or chromosome relative to the material on either side of it.

linkage: the tendency for certain combinations of genes to segregate together during meiosis.

linkage group: a group of genetic loci which can be placed in a linear order representing the different degrees of linkage which they show among themselves. Specific linkage groups can be shown to correspond to specific chromosomes by special cytogenetic methods (see Chapter 4).

locus: a point in a linkage group recognizable through being the site of an allelic difference. Has come to be used as more or less synonymous with *gene*, a closely linked array of sites concerned in a single function. In this sense *locus* means the *position* in the linkage group or chromosome, rather than any particular allele which can occur in that position.

meiosis: two successive divisions of a nucleus accompanied by only one division of the chromosomes, the chromosome arms generally dividing at the first division but not at the second, and the centromeres generally dividing at the second but not at the first. As a result the chromosome number of a diploid nucleus is halved to give the haploid complement in each of the four products.

mitosis: division of the nucleus with concomitant division of the chromosomes to give two daughter nuclei each with the original chromosome complement.

monocaryon: a cell containing a single nucleus or (usually) a mycelium composed of uninucleate cells. A monocaryon is always a homocaryon, but the converse is often untrue.

mutation: an abrupt inheritable change, generally in a chromosome.

phenotype: all the characteristics of an organism observable without genetic analysis.

plasmogamy: fusion of gametes.

pleiotropy: multiple effects of a single gene.

recessive: an allele whose effect is masked by the presence, in a diploid or heterocaryon, of a dominant allele of the same gene.

recombination: the combination, in a single haploid nucleus, of genes descended from different nuclei; usually as a result of nuclear fusion and meiosis, but occasionally through the parasexual cycle (see Chapter 5).

recombinant: a cell or culture showing recombination; usually a product of meiosis.

segregation: the separation of homologous pairs of unlike alleles into different cells; usually as a result of meiosis, but sometimes through mitotic crossing-over in a heterozygous diploid strain.

sister strands: the two strands arising by division of a chromosome arm, before division of the centromere.

site (=mutational site): the smallest element in a gene capable of independent mutation, *or* the smallest element in a gene which can be separated from neighbouring elements by genetic exchange. So far as is known these alternative definitions describe the same entity.

strand: equivalent to *chromatid,* but usually refers to a structure inferred from genetic evidence, rather than microscopically observed. Not to be confused with one strand in a DNA molecule; a single genetic strand consists of DNA which is 2-stranded in the chemical sense.

suppressor: an allele masking the effect of a mutation in another gene (cf. *epistasis*).

tetrad: the four products of meiosis.

tetraploid: (=4n): with four times the haploid complement of chromosomes in each nucleus.

translocation: the transfer, by breaking and rejoining, of a chromosome segment to another position in the chromosome complement.

triploid (=3n): with three times the haploid complement of chromosomes in each nucleus.

trisomic (=2n+1): with a diploid chromosome complement plus a third representative of one chromosome type.

wild type: a strain approximating to the more or less standard form of the species as found in the wild; without any known abnormalities of mutational origin.

zygote: the stage in the sexual cycle immediately following nuclear fusion.

Other examples of gene-cytoplasm interaction

In every case of gene-cytoplasm interaction it is possible to say either that the cytoplasm is influencing the expression of a gene or *vice versa*. In any particular situation it may be convenient to use one formulation rather than the other but it should be borne in mind that the distinction is somewhat arbitrary.

The first example, from *Aspergillus nidulans*, illustrates the dependence of the cytoplasmic equilibrium upon a nuclear gene. Mahoney and Wilkie [337] examined four strains of *A. nidulans* and found that ascospores and conidia of any one strain both gave rise to roughly similar frequencies of mutant colonies which failed to produce perithecia and the reddish brown pigment characteristic of sexually active colonies. These mutant colonies were called *alba* and they arose with different but characteristic frequencies in each of the four strains. For example in strain A about 4 per cent of the spores gave rise to *alba* colonies nearly all of which were stable. About one in twenty of the *alba* colonies reverted to a sexual condition producing a few perithecia after twenty-eight days' incubation. In strain D about 83 per cent of the spores gave rise to *alba* colonies.

Three spontaneous nuclear markers were obtained two of which, yellow (*y*) and biotin (*bi*), were linked. Using these markers to identify crossed and selfed perithecia, they found that *alba* behaved in crosses as though cytoplasmically determined. The crosses were set up by streaking conidia of the parental strains in lines a few millimetres apart. When *alba* was crossed with wild type, selfed perithecia, derived from nuclei of the *alba* strain, arose well behind the junction of the two strains on the *alba* side of the cross. Ascospores from these perithecia produced normal and *alba* colonies in the frequency characteristic of the wild type strain from which the *alba* mutant had been derived. This was interpreted to mean that a cytoplasmic element had invaded the *alba* mycelium and initiated perithecial production.

Heterocaryons between *alba* variants of the different strains yielded no perithecia, showing that they were non-complementary. Crosses between marked stocks of strain A (4 per cent *alba*) and strain D (83 per cent *alba*) gave progenies from crossed perithecia made up of 60 per cent normal and 40 per cent *alba* colonies. Ascospores or conidia from individual normal progeny colonies showed either the low frequency (4 per cent) or the high frequency (83 per cent) of *alba*. There were equal numbers of each type. Evidently the frequency of *alba* mutants was determined by a nuclear gene. This gene, designated *f* with alleles f_4 and f_{83}, was shown to be linked with *y* and *bi*. Crosses between D and a third

BIBLIOGRAPHY

[1] AHMAD M. (1953) The mating system in *Saccharomyces. Ann. Bot., N.S.* **17**, 329
[2] ALLEN R.F. (1934) A cytological study of heterothallism in flax rust. *J. agric. Res.* **49**, 765
[3] AMES B.N. (1957) Enzymes of histidine biosynthesis in histidine-requiring mutants of Neurospora. *Fed. Proc.* **16**, 145
[4] AMES B.N. (1957) The biosynthesis of histidine; L-histidinol phosphate phosphatase. *J. biol. Chem.* **226**, 583
[5] AMES B.N. (1957) The biosynthesis of histidine: D-*erythro*-imidazole-glycerol phosphate dehydrase. *J. biol. Chem.* **228**, 131
[6] AMES B.N. & GARRY B. (1959) Coordinate repression of the synthesis of four histidine biosynthetic enzymes by histidine. *Proc. nat. Acad. Sci., Wash.* **45**, 1453
[7] AMES B.N. & HARTMAN P.E. (1963) The histidine operon. *Cold Spring Harb. Symp. Quant. Biol.,* **28**, 349
[8] ARLETT C.F., GRINDLE M. & JINKS J.L. (1962) The 'red' cytoplasmic variant of *Aspergillus nidulans. Heredity,* **17**, 197
[9] ASCHAN K. (1952) Studies on dediploidisation mycelia of the basidiomycete *Collybia velutipes. Svensk. Bot. Tidskr.* **46**, 366
[10] ATWOOD K.C. & MUKAI F. (1953) Indispensable gene functions in Neurospora. *Proc. nat. Acad. Sci., Wash.* **39**, 1027
[11] ATWOOD K.C. & MUKAI F. (1953) High spontaneous incidence of a mutant of *Neurospora crassa. Genetics* **38**, 654. (*Abstr.*)
[12] AUERBACH C. (1959) Spontaneous mutations in dry Neurospora conidia. *Heredity* **13**, 414. (*Abstr.*)
[13] BACKUS M.P. (1939) The mechanics of conidial fertilisation in *Neurospora sitophila. Bull. Torrey bot. Cl.* **66**, 63
[14] BAILEY D.L. (1950) Studies in racial trends and constancy in *Cladosporium fulvum* Cooke. *Canad. J. Res.* (C) **28**, 535
[15] BAKERSPIEGEL A. (1959) The structure and manner of division of the nuclei in the vegetative mycelium of *Neurospora crassa. Amer. J. Bot.* **46**, 180
[16] BARNETT W.E. & DE SERRES F.J. (1963) Fixed genetic instability in *Neurospora crassa. Genetics,* **48**, 717
[17] BARRATT R.W. (1962) Altered proteins produced by mutation at the amination (*am*) locus in Neurospora. *Genetics,* **47**, 941
[18] BARRATT R.W. & GARNJOBST L. (1949) Genetics of a colonial microconidiating mutant strain of *Neurospora crassa. Genetics* **34**, 351
[19] BARRATT R.W., NEWMEYER D., PERKINS D.D. & GARNJOBST L. (1954) Map construction in *Neurospora crassa. Advanc. Genet.* **6**, 1
[20] BARRATT R.W. & STRICKLAND W.N. (1961) Linkage maps of *Neurospora crassa.* In *Handbook on Growth.* Office of Biological Handbooks, Washington
[21] BAUMAN, N. & DAVIS B.D. (1957) Selection of auxotrophic bacterial mutants through diaminopimelic acid or thymine deprival. *Science* **126**, 170

[22] BAUTZ, E. & FREESE E. (1960) On the mutagenic effect of alkylating agents. *Proc. nat. Acad. Sci., Wash.* **46**, 1585

[23] BEADLE G. W. (1948) The genes of men and molds. *Sci. Amer.*, Sept. 1948.

[24] BEADLE G.W. & COONRADT V.L. (1944) Heterocaryosis in *Neurospora crassa*. *Genetics* **29**, 291

[25] BEADLE G.W. & TATUM E.L. (1945) Neurospora II. Methods of producing and detecting mutations concerned with nutritional requirements. *Amer. J. Bot.* **32**, 678

[26] BEER, M. (1961) Electron microscopy of unbroken DNA molecules. *J. mol. Biol.* **3**, 263

[27] BEISSON-SCHECROUN J. (1962) Incompatibilité cellulaire et interactions nucléocytoplasmiques dans les phenomènes de 'barrage' chez le *Podospora anserina*. *Annales de Génétique*, **4**, 1

[28] BENZER, S. (1958) The elementary units of heredity. In *The Chemical Basis of Heredity*, 70. (Ed. by McElroy W. D. & Glass B.), Johns Hopkins Press, Baltimore

[29] BENZER S. (1961) *Genetic Fine Structure* Harvey Lectures, 1960–61. Academic Press, New York

[30] BENZER S. & CHAMPE S.P. (1962) A change from nonsense to sense in the genetic code. *Proc. Nat. Acad. Sci., Wash.*, **48**, 1114

[31] BERGMANN F. H., BERG P. & DIECKMANN M. (1961) The enzymatic synthesis of amino acyl derivatives of ribonucleic acid. II. The preparation of leucyl-, valyl-, isoleucyl- and methionyl ribonucleic acid synthetases from *Escherichia coli*. *J. biol. Chem.* **236**, 1735

[32] BERNSTEIN H. (1964) On the mechanism of intragenic recombination. II *Neurospora crassa*. *J. Theoret. Biol.*, **6**, 347

[33] BERNSTEIN H. & MILLER A. (1961) Complementation studies with isoleucine-valine mutants of *Neurospora crassa*. *Genetics*, **46**, 1039

[34] BEUKERS, R. & BERENDS, W. (1960) Isolation and identification of the irradiation product of thymine. *Biochim. biophys. Acta*, **41**, 550

[35] BEUKERS, R., IJLSTRA, J. & BERENDS W. (1960) The effect of ultraviolet light on some components of the nucleic acids. VI. The origin of the ultraviolet sensitivity of deoxyribonucleic acid. *Rec. Trav. chim. Pays-Bas* **79**, 101

[36] BEVAN E.A. & WOODS R.A. (1962) Complementation between adenine requiring mutants in yeast. *Heredity* **17**, 141. (Abstr.)

[37] BISTIS, G. (1956) Studies on the genetics of *Ascobolus stercorarius*, *Bull. Torrey bot. Cl.*, **83**, 35

[38] BLACK W. (1960) Races of *Phytophthora infestans* and resistance problems in potatoes. *Ann. Rep. Scot. Pl. Breed. Stat.* pp. 29–38

[39] BLACK W., MASTENBROEK C., MILLS W.R. & PETERSON L.C. (1952) A proposal for an international nomenclature of races of *Phytophthora infestans* and of genes controlling immunity in *Solanum demissum* derivatives. *Euphytica* **2**, 173

[40] BLAKESLEE A.F. (1906) Zygospore germinations in the Mucorineae. *Ann. mycol. Berl.* **4**, 1

[41] BOLE-GOWDA B.N., PERKINS D.D. & STRICKLAND W.N. (1962) Crossing-over and interference in the centromere region of linkage group I of Neurospora. *Genetics*, **47**, 1243

[42] BONNER D. (1946) Production of biochemical mutations in Penicillium. *Amer. J. Bot.* **33**, 788

[43] BONNER D.M., SUYAMA Y. & DEMOSS J.A. (1960) Genetic fine structure and enzyme formation. *Fed. Proc.* **19**, 926

[44] BONNER D.M., YANOFSKY C. & PARTRIDGE, C.W.H. (1952) Incomplete genetic blocks in biochemical mutants in Neurospora. *Proc. nat. Acad. Sci., Wash.* 38, 25

[45] BOONE D.M. & KEITT G.W. (1956) *Venturia inaequalis* (Cke.) Wint. VIII. Inheritance of color mutant characters. *Amer. J. Bot.* 43, 226

[46] BOONE D.M. & KEITT G.W. (1957) *Venturia inaequalis* (Cke.) Wint. XII. Genes controlling pathogenicity of wild-type lines. *Phytopathology* 47, 403

[47] BOONE D.M., KLINE D.M. & KEITT G.W. (1957) *Venturia* inaequalis (Cke.) Wint. XIII. Pathogenicity of induced biochemical mutants. *Amer. J. Bot.* 44, 791

[48] BOONE D.M., STAUFFER J.F., STAHMANN M.A. & KEITT G.W. (1956) *Venturia, inaequalis* (Cke.) Wint. VII. Induction of mutants for studies on genetics nutrition and pathogenicity. *Amer. J. Bot.* 43, 198

[49] BOYCE R.P. & HOWARD-FLANDERS P. (1964) Release of ultraviolet light-induced thymine dimers from DNA in *E. coli* K12. *Proc. Nat. Acad. Sci., Wash.*, 51, 293

[50] BRACKER C.E. & BUTLER E.E. (1963) The ultrastructure and development of septa in hyphae of *Rhizoctonia solani. Mycologia,* 55, 35

[51] BRACKER C.E. & BUTLER E.E. (1964) Function of the septal pore apparatus in *Rhizoctonia solani* during protoplasmic streaming. *J. Cell. Biol.,* 21, 152.

[52] BRAYMER H.D. & WOODWARD D.O. (1962) Density gradient studies with adenylosuccinase from inter-allelic heterocaryons in Neurospora. *Genetics,* 47, 944

[53] BRENNER S., JACOB F. & MESELSON M. (1961) An unstable intermediate carrying information from genes to ribosomes for protein synthesis. *Nature, Lond.* 190, 576

[54] BROCK T.D. (1959) Biochemical basis of mating in yeast. *Science* 129, 960

[55] BROCK T.D. (1961) Physiology of the conjugation process in the yeast *Hansenula wingei. J. gen. Microbiol.* 26, 487

[56] BROCKMAN H.E. & DE SERRES F.J. (1963) Induction of *ad-3* mutants of *Neurospora crassa* by 2-aminopurine. *Genetics,* 48, 597

[57] BRODIE H.J. (1936) The occurrence and function of oidia in the Hymenomycetes. *Amer. J. Bot.,* 23, 309

[58] BRODY S. & YANOFSKY C. (1963) Suppressor gene alteration of protein primary structure. *Proc. Nat. Acad. Sci., Wash.,* 50, 9

[59] BROWN J.S. (1951) The effect of photoreactivation on mutation frequency in Neurospora. *J. Bact.* 62, 163

[60] BROWN S.W. & ZOHARY D. (1955) The relationship of chiasmata and crossing over in *Lilium formosanum. Genetics* 40, 850

[61] BULLER A.H.R. (1931) *Researches on Fungi* IV. Longmans, Green & Co., London

[62] BURGEFF H. (1914) Untersuchungen über Variabilität, Sexualität und Erblichkeit bei *Phycomyces nitens* I. *Flora., Jena.* 107, 259

[63] BURGEFF H. (1924) Untersuchungen über Sexualität und Parasitismus bei Mucorineen. I. *Bot. Abh.* 4, 5

[64] BURGI E. & HERSHEY A.D. (1961) A relative molecular weight series derived from the nucleic acid of bacteriophage T2. *J. mol. Biol.* 3, 458

[65] BUXTON E.W. (1956) Heterokaryosis and parasexual recombination in pathogenic strains of *Fusarium oxysporum. J. gen. Microbiol.* 15, 133

[66] BUXTON E.W. (1960) Heterokaryosis, saltation and adaptation. In *Plant Pathology.* Ed. by Horsfall, J.G. & Dimond, A.E. Acad. Press, New York. Vol. 2., ch. 10, pp. 359–405

[67] BUXTON E.W. (1962) Parasexual recombination in the banana-wilt Fusarium. *Trans. Brit. mycol. Soc.* **45**, 274

[68] CAIRNS J. (1961) An estimate of the length of the DNA molecule of T2 bacteriophage by autoradiography. *J. mol. Biol.* **3**, 756

[69] CALEF E. (1957) Effect on linkage maps of selection of cross-overs between closely linked markers. *Heredity* **11**, 265

[70] CANTINO, E.C. (1961) The relationship between biochemical and morphological differentiation in non-filamentous aquatic fungi. *Symp. Soc. gen. Microbiol.* **11**, 243

[71] CANTINO E.C. & HORENSTEIN E.A. (1954) Cytoplasmic exchange without gametic copulation in the water mold *Blastocladiella emersonii. Amer. Nat.* **88**, 143

[72] CANTINO E.C. & TURIAN G.F. (1959) Physiology and development of lower fungi (Phycomycetes). *Annu. Rev. Microbiol.* **13**, 97

[73] CARR A.J.H. (1954) Variation in the homothallic fungus *Sclerotinia trifoliorum. Proc. 8th Int. Bot. Congr., Paris. Sect.* **19**, 72

[74] CARR, A.J.H. & OLIVE L.S. (1959) Genetics of *Sordaria fimicola.* III. Cross-compatibility among self-fertile and self-sterile cultures. *Amer. J. Bot.* **46**, 81

[75] CASE M.E. & GILES N.H. (1958) Recombination mechanisms at the *pan-2* locus in *Neurospora crassa. Cold Spr. Harb. Sym. quant. Biol.*, **23**, 119

[76] CASE M.E. & GILES N.H. (1958) Evidence from tetrad analysis for both normal and aberrant recombination between allelic mutants in *Neurospora crassa. Proc. nat. Acad. Sci., Wash.* **44**, 378

[77] CASE M.E. & GILES N.H. (1960) Comparative complementation and genetic maps of the pan-2 locus in *Neurospora crassa. Proc. nat. Acad. Sci., Wash.* **46**, 659

[78] CASE M.E. & GILES N.H. (1964) Allelic recombination in Neurospora. Tetrad analysis of a three-point cross within the *pan-2* locus. *Genetics,* **49**, 529

[79] CATCHESIDE D.G. (1954) Isolation of nutritional mutants of *Neurospora crassa* by filtration enrichment. *J. gen. Microbiol.* **11**, 34

[80] CATCHESIDE D.G. (1960) Complementation among histidine mutants of *Neurospora crassa. Proc. roy. Soc.,* B, **153**, 179

[81] CATCHESIDE D.G. & OVERTON A. (1959) Complementation between alleles in heterocaryons. *Cold. Spr. Harb. Sym. quant. Biol.* **23**, 137

[82] CAVALIERI L.F. & ROSENBERG B.H. (1963) Nucleic acids and information transfer. *Prog. Nucleic Acid Res.,* **2**, 1

[83] CHAMBERLIN, M. & BERG P. (1962) Deoxyribonucleic acid directed synthesis of ribonucleic acid by an enzyme from *Escherichia coli. Proc. nat. Acad. Sci., Wash.* **48**, 81

[84] CHEN S.Y. (1950) Sur une nouvelle technique de croisement des levures. *C.R. Acad. Sci., Paris,* **230**, 1897

[85] CHEN S.Y., EPHRUSSI B. & HOTTINGUER H. (1950) Nature génétique des mutants à déficience respiratoire de la souche B-II de la levure de boulangerie. *Heredity* **4**, 337

[86] COLSON B. (1934) The cytology and morphology of *Neurospora tetrasperma. Ann. Bot., Lond.* **48**, 211

[87] COTTER R.U. (1960) Fertilization of pycnia with urediospores and aeciospores in *Puccinia graminis. Phytopathology* **50**, 567

[88] COUCH J.N. (1926) Heterothallism in Dictyuchus, a genus of the water molds. *Ann. Bot. Lond.* **40**, 848

[88a] Cove, D. J. & Pateman, J. A. (1963) Independently segregating genetic loci concerned with nitrate reductase activity in *Aspergillus nidulans*. *Nature*, **198**, 262

[89] Coyle M.B. & Pittenger T.H. (1964) Somatic recombinants from pseudowild types of Neurospora recovered from 4:4 asci. *Genetics*, **50**, 242

[90] Craigie J.H. (1927) Discovery of the function of the pycnia of the rust fungi. *Nature. Lond.* **120**, 765

[91] Craigie J. H. & Green G.J. (1962) Nuclear behaviour leading to conjugate association in haploid infections of *Puccinia graminis*. *Canad. J. Bot.* **40**, 163

[92] Crick F.H.C., Barnett, L., Brenner, S. & Watts-Tobin R.J. (1961) General nature of the genetic code for proteins. *Nature, Lond.* **192**, 1227

[93] Crick F.H.C. & Orgel L.E. (1964) The theory of interallelic complementation. *J. Mol. Biol.*, **8**, 161

[94] Crowe L.K. (1960) The exchange of genes between nuclei of a dikaryon. *Heredity* **15**, 397

[95] Cutter V. M. Jr. (1942) Nuclear behaviour in the Mucorales. II. The Rhizopus, Phycomyces and Sporodinia patterns. *Bull. Torrey bot. Cl.* **69**, 592

[96] Darlington C.D. & Mather K. (1949) *The Elements of Genetics* pp. 446. Allen & Unwin, London

[97] Davis B.D. (1948) Isolation of biochemically deficient mutants of bacteria by penicillin. *J. Amer. Chem. Soc.*, **70**, 4267

[98] Davis R.H. (1962) A mutant form of ornithine transcarbamylase found in a strain of Neurospora carrying a pyrimidine-proline suppressor gene. *Arch. Biochem. Biophys.* **97**, 185

[99] Davis R.H. (1963) *Neurospora* mutants lacking an arginine-specific carbamyl phosphokinase. *Science*, **142**, 1652

[100] Davis R.H. & Thwaites W.M. (1963) Structural gene for ornithine transcarbamylase in Neurospora. *Genetics*, **48**, 1551

[101] Davis R.H. & Woodward V.W. (1962) The relationships between gene suppression and aspartate transcarbamylase activity in *pyr-3* mutants of Neurospora. *Genetics*, **47**, 1057

[102] Day, P.R. (1954) Physiologic specialization of *Cladosporium fulvum* in England and Wales. *Plant Pathol.* **3**, 35

[103] Day, P.R. (1956) Race names of *Cladosporium fulvum*. *Tomato Genet. Co-oper. Rep.* **6**, 13

[104] Day, P.R. (1957) Mutation to virulence in *Cladosporium fulvum*. *Nature, Lond.* **179**, 1141

[105] Day, P. R. (1957) Mutants of *Cladosporium fulvum*. *Ann. Rep. John Innes Hort. Instn.* 1956 pp. 15–16

[106] Day P.R. (1959) A cytoplasmically controlled abnormality of the tetrads of *Coprinus lagopus*. *Heredity* **13**, 81

[107] Day, P.R. (1960) Variation in phytopathogenic fungi. *Annu. Rev. Microbiol.* **14**, 1

[108] Day P.R. (1960) The structure of the *A* mating type locus in *Coprinus lagopus*. *Genetics* **45**, 641

[109] Day P.R. (1963) Mutations of the *A* mating type factor in *Coprinus lagopus*. *Genet. Res., Camb.*, **4**, 55

[110] Day P.R. (1963) The structure of the *A* mating type factor in *Coprinus lagopus*: Wild alleles. *Genet. Res., Camb.*, **4**, 323

[111] Day P.R. & Anderson G.E. (1961) Two linkage groups in *Coprinus lagopus*. *Genet. Res., Camb.*, **2**, 414

[112] DAY P.R., BOONE D.M. & KEITT, G.W. (1956) *Venturia inaequalis* (Cke.) Wint. XI. The chromosome number. *Amer. J. Bot.* **43**, 835

[113] DE DEKEN R.H. (1963) Biosynthèse de l'arginine chez la levure. I. Le sort de la N-α-acetyl ornithine. *Biochem. Biophys. Acta* **78**, 606

[114] DELAMATER E. D. (1950) The nuclear cytology of the vegetative diplophase of *Saccharomyces cerevisiae*. *J. Bact.* **60**, 321

[115] DESBOROUGH S. & LINDEGREN G. (1959) Chromosome mapping of linkage data from Saccharomyces by tetrad analysis. *Genetica* **30**, 346

[116] DE SERRES, F.J. (1956) Studies with purple adenine mutants in *Neurospora crassa*. I. Structural and functional complexity in the *ad-3* region. *Genetics* **41**, 668

[117] DE SERRES F.J. (1960) Studies with purple adenine mutants in *Neurospora crassa*. IV. Lack of complementation between different *ad-3A* mutants in heterocaryons and pseudo wild types. *Genetics* **45**, 555

[118] DE SERRES F.J. (1964) Genetic analysis of the structure of the *ad-3* region of *Neurospora crassa* by means of irreparable lethal mutations. *Genetics* **50**, 21

[119] DE SERRES F.J. & KÖLMARK H.G. (1958) A direct method for determination of forward mutation rates in *Neurospora crassa*. *Nature, Lond.*, **182**, 1249

[120] DE SERRES F.J. & OSTERBIND R.S. (1962) Estimation of the relative frequencies of X-ray induced viable and recessive mutations in the *ad-3* region of *Neurospora crassa*. *Genetics* **47**, 793

[121] DIXON P. A. (1959) Life history and cytology of *Ascocybe grovesii* Wells. *Ann. Bot., N.S., Lond.* **23**, 509

[122] DODGE B.O. (1929) The nature of giant spores and the segregation of sex factors in *Neurospora*. *Mycologia* **21**, 222

[123] DODGE B.O. (1930) Breeding albinistic strains of the Monilia bread mould. *Mycologia* **22**, 9

[124] DOUBLY J.A., FLOR H.H. & CLAGETT C.O. (1960). Relation of antigens of *Melampsora lini* and *Linum usitatissimum* in resistance and suceptibility. *Science* **131**, 229

[125] DONACHIE W.D. (1964) The regulation of pyrimidine biosynthesis in *Neurospora crassa* I. End-product inhibition and repression of aspartate carbamoyl transferase. II. Heterokaryons and the role of the 'regulatory mechanisms'. *Biochem., Biophys. Acta.* **82**, 284 and 293

[126] DOUGLAS H.C. & HAWTHORNE D.C. (1964) Enzymatic expression and genetic linkage of genes controlling galactose utilization in Saccharomyces. *Genetics* **49**, 837

[127] DOWDING E.S. & BAKERSPIEGEL A. (1954) The migrating nucleus. *Canad. J. Microbiol,* **1**, 68

[128] DRIVER C.H. & WHEELER H.E. (1955) A sexual hormone in Glomerella. *Mycologia* **47**, 311

[129] DUTTA S.K., HALL C.V. & HEYNE G. (1960) Pathogenicity of biochemical mutants of *Collectotrichum lagenarium*. *Bot. Gaz.* **121**, 166

[130] ESPOSITO, R. E.* & HOLLIDAY R. The effect of 5-fluorodeoxyuridine on genetic replication and somatic recombination in synchronously dividing cultures of *Ustilago maydis*. *Genetics,* **50**, 1009

[131] EDGERTON C.W. (1914) Plus and minus strains in the genus Glomerella. *Amer. J. Bot.* **1**, 244

[132] EL ANI, A.S. & OLIVE L.S. (1962) The induction of balanced heterothallism in *Sordaria fimicola*. *Proc. nat. Acad. Sci., Wash.* **48**, 17

[133] EL ANI, A.S., OLIVE L.S. & KITANI Y. (1961) Genetics of *Sordaria fimicola*. IV. Linkage Group I. *Amer. J. Bot.* **48**, 716

* nèe R. EASTON.

[134] ELLINGBOE A.H. (1961) Somatic recombination in *Puccinia graminis* var *tritici*. *Phytopathol.* **51,** 13

[135] ELLINGBOE A.H. (1963) Illegitimacy and specific factor transfer in *Schizophyllum commune*. *Proc. Nat. Acad. Sci., Wash.* **49,** 286

[136] ELLINGBOE A.H. & RAPER J.R. (1962) Somatic recombination in *Schizophyllum commune*. *Genetics* **47,** 85

[137] ELLIOTT C.G. (1960) The cytology of *Aspergillus nidulans*. *Genet. Res., Camb.* **1,** 462

[138] EMERSON M.R. (1954) Some physiological characteristics of ascospore activation in *Neurospora crassa*. *Plant Physiol.* **29,** 418

[139] EMERSON R. (1941) An experimental study of the life cycles and taxonomy of Allomyces. *Lloydia* **4,** 77

[140] EMERSON R. (1950) Current trends of experimental research on the aquatic Phycomycetes. *Annu. Rev. Microbiol.* **4,** 169

[141] EMERSON, R. & WILSON C.M. (1949) The significance of meiosis in Allomyces. *Science* **110,** 86

[142] EMERSON R. & WILSON C.M. (1954) Interspecific hybrids and the cytogenetics and cytotaxonomy of Euallomyces. *Mycologia* **46,** 393

[143] EMERSON S. (1949) Competitive reactions and antagonisms in the biosynthesis of amino acids by Neurospora. *Cold. Spr. Harb. Symp. quant. Biol.* **14,** 40

[144] EMERSON S. (1954) Methods in the biochemical genetics of micro-organism. In *Handbuch der Physiologisch-Chemischen Analyse*. Springer-Verlag, Heidelberg. Vol. 2

[145] EPHRUSSI B. (1953) *Nucleo-Cytoplasmic Relations in Micro-Organisms*. Oxford Univ. Press, London. Pp. 127

[146] EPHRUSSI B., DE MARGERIE-HOTTINGUER H. & ROMAN H. (1955) Suppressiveness: a new factor in the genetic determinism of the synthesis of respiratory enzymes in yeast. *Proc. nat. Acad. Sci., Wash.* **41,** 1065

[147] ESSER K. (1962) Die Genetik der sexuellen Fortpflanzung bei den Pilzen. *Biol. Zentrallblatt,* **81,** 161

[148] ESSER K. & STRAUB J. (1958) Genetische Untersuchungen an *Sordaria macrospora* Auersw., Kompensation und Induktion bei genbedingten Entwicklungsdefekten. *Z. Vererblehre* **89,** 729

[149] FEUGHELMAN M., LANGRIDGE R., SEEDS W.E., STOKES A.R., WILSON H.R., HOOPER C.W., WILKINS M.F.H., BARCLAY R.K. and HAMILTON L.D. (1955) Molecular structure of deoxyribosenucleic acid and nucleoprotein. *Nature, Lond.* **175,** 740

[150] FINCHAM J.R.S. (1951) A comparative study of the mating type chromosomes of two species of Neurospora. *J. Genet.* **50,** 221

[151] FINCHAM J.R.S. (1953) Ornithine transaminase in Neurospora and its relation to the biosynthesis of proline. *Biochem. J.* **53,** 313

[152] FINCHAM J.R.S. (1957) A modified glutamic acid dehydrogenase as a result of gene mutation in *Neurospora crassa*. *Biochem. J.* **65,** 721

[153] FINCHAM J.R.S. (1959) On the nature of the glutamic dehydrogenase produced by inter-allele complementation at the *am* locus of *Neurospora crassa*. *J. gen. Microbiol.* **21,** 600

[154] FINCHAM J.R.S. (1959) The role of chromosomal loci in enzyme formation. *Proc. X Internat. Congr. Genetics* I, 335 (Montreal, 1958, Univ. Toronto Press)

[155] FINCHAM J.R.S. (1962) Genetically determined multiple forms of glutamic dehydrogenase in *Neurospora crassa*. *J. mol. Biol.* **4,** 257

[156] FINCHAM J.R.S. & BOND P.A. (1960) A further genetic variety of glutamic acid dehydrogenase in *Neurospora crassa*. *Biochem. J.* **77,** 96

[157] FINCHAM J.R.S. & BOYLEN J.B. (1957) *Neurospora crassa* mutants lacking argininosuccinase. *J. gen. Microbiol.* **16**, 438

[158] FINCHAM J.R.S. & CODDINGTON A. (1963) Complementation at the *am* locus of *Neurospora crassa*: a reaction between different mutant forms of glutamate dehydrogenase. *J. Mol. Biol.* **6**, 361

[159] FLAVIN M. & SLAUGHTER C. (1960) Purification and properties of threonine synthetase of Neurospora. *J. biol. Chem.* **235**, 1103

[160] FLOR H.H. (1942) Inheritance of pathogenicity in *Melampsora lini*. *Phytopathology* **32**, 653

[161] FLOR, H.H. (1947) Inheritance of rust reaction in flax. *J. agric. Res.* **74**, 241

[162] FLOR H.H. (1954) Identification of races of flax rust by lines with single rust-conditioning genes. *U.S.D.A. Tech. Bull. No.* 1087, pp. 25

[163] FLOR H.H. (1956) Mutations in flax rust induced by ultraviolet radiation. *Science* **124**, 888

[164] FLOR H.H. (1956) The complementary genic systems in flax and flax rust. *Adv. Genet.* **8**, 29

[165] FLOR H.H. (1958) Mutation to wider virulence in *Melampsora lini*. *Phytopathology* **48**, 297

[166] FLOR H.H. (1959) Genetic controls of host-parasite interaction in rust diseases. Chapter 14 in *Plant Pathology: Problems and Progress, 1908–1958* (Edited by Holton C.S. *et al.*) Univ. Wisconsin Press.

[167] FLOR H.H. (1962) Linkage of genes controlling resistance to rust in flax. *Phytopathol.* **52**, 732

[168] FOSTER J.W. (1949). *Chemical activities of the fungi.* Academic Press, N.Y., pp. 648

[169] FOWELL R.R. (1951) Hybridisation of yeasts by Lindegren's technique. *J. Inst. Brewing* **57**, 180

[170] FOX M.S. & HOTCHKISS R.D. (1960) Fate of transforming deoxyribonucleate following fixation by transformable bacteria. *Nature, Lond.* **187**, 1002

[171] FRANKE G. (1957) Die Cytologie der Ascusentwicklung von *Podospora anserina*. *Z. indukt. Abstamm. Vererblehre* **88**, 159

[172] FREESE E. (1957) The correlation effect for a histidine locus of *Neurospora crassa*. *Genetics* **42**, 671

[173] FREESE E. (1958) The arrangement of DNA in the chromosome. *Cold Spring Harbor Symp. Quant. Biol.* **23**, 13

[174] FREESE E. (1961) The molecular mechanism of mutations. *Vth Internat. Congr. Biochem. Moscow* (Pergamon Press), Symposium No. 1

[175] FREESE E., BAUTZ-FREESE E. & BAUTZ E. (1961) Hydroxylamine as a mutagenic and inactivating agent. *J. mol. Biol.* **3**, 133

[176] FRIES L. (1953) Factors promoting growth of *Coprinus fimetarius* (L) under high temperature conditions. *Physiol. Plant.* **6**, 551

[177] FRIES N. (1945) Two X-ray induced auxo-heterotrophies. *Svensk. Bot. Tidskr.* **39**, 270

[178] FRIES N. (1947) Experiments with different methods of isolating physiological mutations of filamentous fungi. *Nature, Lond.* **159**, 199

[179] FRIES N. (1948) Viability and resistance of spontaneous mutations in Ophiostoma representing different degrees of heterotrophy. *Physiol. Plant.* **1**, 330

[180] FRIES N. (1950) The production of mutations by caffeine. *Hereditas* **36**, 134

[181] FRIES N. (1953) Further studies on mutant strains of Ophiostoma which require guanine. *J. biol. Chem.* **200**, 325

[182] FROST L.C. (1961) Heterogeneity in recombination frequencies in *Neurospora crassa*. *Genet. Res., Camb.* **2**, 43

[183] FULTON I.W. (1950) Unilateral nuclear migration and the interactions of haploid mycelia in the fungus *Cyathus stercoreus. Proc. Nat. Acad. Sci., Wash.* **36,** 306

[184] FURTH J.J., HURWITZ J. & GOLDMAN M. (1961) The directing role of DNA in RNA synthesis. Specificity of the deoxyadenylate-deoxythymidylate copolymer as a primer. *Biochem. biophys. Res. Comm.* **4,** 431

[185] GALLEGLY M.E. & EICHENMULLER J.J. (1959) The spontaneous appearance of the potato race 4 character in cultures of *Phytophthora infestans. Amer. Potato J.* **36,** 45

[186] GALLEGLY M.E. & GALINDO J. (1958) Mating types and oospores of *Phytophthora infestans* in nature in Mexico. *Phytopathology* **48,** 274

[187] GALLEGLY M.E. & NIEDERHAUSER J.S. (1959) Genetic controls of host-parasite interactions in the Phytophthora late blight disease. Chapter 17 in *Plant Pathology: Problems and Progress,* 1908–1958. (Edited by Holton C.S. *et al.*), Univ. Wisconsin Press

[188] GAREN A. & SIDDIQI O. (1962) Suppression of mutations in the alkaline phosphatase cistron of *E. coli. Proc. Nat. Acad. Sci., Wash.* **48,** 1121

[189] GARNJOBST L. (1953) Genetic control of heterocaryosis in *Neurospora crassa. Amer. J. Bot.* **40,** 607

[190] GARNJOBST L. (1955) Further analysis of genetic control of heterocaryosis in *Neurospora crassa. Amer. J. Bot.* **42,** 444

[191] GARNJOBST L. & WILSON J.F. (1956) Heterocaryosis and protoplasmic incompatibility in *Neurospora crassa. Proc. nat. Acad. Sci., Wash.* **42,** 613

[192] GAUGER W.L. (1961) The germination of zygospores of *Rhizopus stolonifer. Amer. J. Bot.* **48,** 427

[193] GIESY R.M. & DAY P.R. The septal pores of *Coprinus lagopus* (Fr) *sensu* Buller in relation to nuclear migration. *Amer. J. Bot.,* in the press

[194] GILES N.H. (1951) Studies on the mechanism of reversion in biochemical mutants of *Neurospora crassa. Cold Spring Harbor Symp. Quant. Biol.* **16,** 283

[195] GILES N.H. (1956) Forward and back mutation at specific loci in Neurospora. In *Mutation, Brookhaven Symp. in Biol.* **8,** 103

[196] GILES N.H. (1959) Mutations at specific loci in Neurospora. *Proc. X Internat. Congr. Genetics,* I, 261. (Montreal 1958), Univ. Toronto Press

[197] GILES N.H., PARTRIDGE C.W.H. & NELSON N.J. (1957) The genetic control of adenylosuccinase in *Neurospora crassa. Proc. nat. Acad. Sci., Wash.* **43,** 305

[198] GILES N.H. (1963) Genetic fine structure in relation to function in Neurospora. *Proc. XI Int. Congr. Genetics,* (The Hague), Vol. 2, Pergamon Press

[199] GIRBARDT M. (1961) Licht- und elektronmikroskopische Untersuchungen an *Polystictus versicolor.* II. Die Feinstruktur von Grundplasma und Mitochondrien. *Archiv für Mikrobiologie* **39,** 351

[200] GOODGAL S.H. (1950) The effect of photoreactivation on the frequency of ultraviolet-induced morphological mutations in the microconidial strain of *Neurospora crassa. Genetics* **35,** 667

[201] GOWDRIDGE B. (1956) Heterocaryons between strains of *Neurospora crassa* with different cytoplasms. *Genetics* **41,** 780

[202] GREGG M. (1957) Germination of the oospores of *Phytophthora erythroseptica. Nature, Lond.* **180,** 150

[203] GREGORY K.F. & SHYU W.-J. (1961) Apparent cytoplasmic inheritance of tyrosinase competence in *Streptomyces scabies. Nature, Lond.* **191,** 465

[204] GRIFFITHS D.J. & CARR A.J.H. (1961) Induced mutation for pathogenicity in *Puccinia coronata avenae. Trans. Brit. mycol. Soc.* **44,** 601

[205] GRIGG G.W. (1958) Competitive suppression and the detection of mutations in microbial populations. *Austral. J. biol. Sci.* **11,** 69

[206] GRINDLE M. (1963) Heterokaryon incompatibility in unrelated strains in the *Aspergillus nidulans* group. *Heredity* **18**, 191

[207] GRINDLE M. (1963) Heterokaryon compatibility of closely related wild isolates of *Aspergillus nidulans*. *Heredity* **18**, 397

[208] GROSS S.R. (1958) The enzymatic conversion of 5-dehydroshikimic acid to protocatechuic acid. *J. biol. Chem.* **233**, 1146

[209] GROSS S.R. (1962) On the mechanism of complementation at the *leu-2* locus of *Neurospora crassa*. *Proc. Nat. Acad. Sci., Wash.* **48**, 922

[210] GROSS S.R. & FEIN A. (1960) Linkage and function in Neurospora. *Genetics* **45**, 885

[211] GROSS S.R. & WEBSTER R.E. (1963) Some aspects of inter-allelic complementation involving leucine biosynthetic enzymes of Neurospora. *Cold Spring Harb. Symp. Quant. Biol.* **28**, 543

[212] GUTZ H. (1961) Distribution of X-ray and nitrous acid-induced mutations in the genetic fine structure of the *ad-7* locus of *Schizosaccharomyces pombe*. *Nature, Lond.* **191**, 1125

[213] HALISKY P.M. (1956) Inheritance of pathogenicity in *Ustilago avenae*. *Research Studies State Coll. Wash.* **24**, 348

[214] HARDER R. (1927) Zur Frage nach der Rolle von Kern und Protoplasma im Zellegschehen und bei der übertragung von Eigenschaften. *Z. Botan.* **19**, 337

[215] HARTMAN P.E., LOPER J.C. & SERMAN D. (1960) Fine structure mapping by complete transduction between histidine-requiring Salmonella mutants. *J. gen. Microbiol.* **22**, 323

[216] HASTIE A. C. (1962) Genetic recombination in the hop-wilt fungus *Verticillium albo-atrum*. *J. gen. Microbiol.* **27**, 373

[217] HAWKES J.G. (1958) Significance of wild species and primitive forms for potato breeding. *Euphytica* **7**, 257

[218] HAWTHORNE D.C. (1955) The use of linear asci for chromosome mapping in Saccharomyces. *Genetics* **40**, 511

[219] HAWTHORNE D.C. (1963) Directed mutation of the mating type alleles as an explanation of homothallism in yeast. *Proc. XI Int. Congr. Genetics (The Hague)* **1**, 34 (Pergamon Press)

[220] HAWTHORNE D.C. & MORTIMER R.K. (1960) Chromosome mapping in Saccharomyces: centromere linked genes. *Genetics* **45**, 1084

[221] HAWTHORNE D.C. & MORTIMER R.K. (1963) Super-suppressors in yeast. *Genetics* **48**, 617

[222] HEAGY F.C. & ROPER J.A. (1952) Deoxyribonucleic acid content of haploid and diploid Aspergillus conidia. *Nature* **170**, 713

[223] HELINSKI D.R. & YANOFSKY C. (1963) A genetic and biochemical analysis of second-site reversion. *J. Biol. Chem.* **238**, 1043

[224] HEMMONS L., PONTECORVO, G. & BUFTON A.W.J. (1952) Perithecium analysis in *Aspergillus nidulans*. *Heredity* **6**, 135

[225] HERRIOTT R.M. (1961) Formation of heterozygotes by annealing a mixture of transforming DNAs. *Proc. nat. Acad. Sci., Wash.* **47**, 146

[226] HERSHEY A.D. & CHASE M. (1952) Independent functions of viral protein and nucleic acid in growth of bacteriophage. *J. gen. Physiol.* **36**, 39

[227] HESLOT H. (1960) *Schizosaccharomyces pombe:* un nouvel organisme pour l'étude de la mutagenèse chimique. *Erwin-Baur-Gedachtnisvorlesungen* I, 98. Akademie-Verlag, Berlin

[228] HESLOT H. (1962) Étude quantitative de réversions biochimiques induites chez la levure *Schizosaccharomyces pombe* par des radiations et des substances radiomimétiques. *Erwin-Baur-Gedachtnisvorlesungen* II, 193. Akademie-Verlag, Berlin

294 FUNGAL GENETICS

[229] HÍRS C.H.W., MOORE S. & STEIN W.H. (1960) The sequence of amino acid residues in performic acid-oxidised ribonuclease. *J. biol. Chem.* **235**, 633

[230] HOAGLAND M.B., STEPHENSON M.L., SCOTT J.F., HECHT L.I. & ZAMECNIK P.C. (1958) A soluble ribonucleic acid intermediate in protein synthesis. *J. biol. Chem.* **231**, 241

[231] HOLLIDAY R. (1956) A new method for the identification of biochemical mutants of micro-organisms. *Nature, Lond.* **178**, 987

[232] HOLLIDAY R. (1961) The genetics of *Ustilago maydis. Genet. Res., Camb.* **2**, 204

[233] HOLLIDAY R. (1961) Induced mitotic crossing-over in *Ustilago maydis. Genet. Res., Camb.* **2**, 231

[234] HOLLIDAY R. (1962) Mutation and replication in *Ustilago maydis. Genet. Res., Camb.* **3**, 472

[235] HOLLIDAY R. (1964) The induction of mitotic recombination by mitomycin C in Ustilago and Saccharomyces. *Genetics*, in the press

[236] HOLLIDAY R. (1964) A mechanism for gene conversion in fungi. *Genet. Res., Camb.* **5**, 282

[237] HOLLOWAY B.W. (1955) Genetic control of heterocaryosis in *Neurospora crassa. Genetics* **40**, 117

[238] HOLTON C.S. (1959) Genetic controls of host-parasite interactions in smut diseases. Chapter 15 in *Plant Pathology: Problems and Progress, 1908–1958.* (Ed. by Holton C.S. *et al.*) Univ. Wisconsin Press, Madison

[239] HOLTON C.S. & HALISKY P.M. (1960) Dominance of avirulence and monogenic control of virulence in race hybrids of *Ustilago avenae. Phytopathology* **50**, 766

[240] HOROWITZ N.H., FLING M., MACLEOD H.L. & SUEOKA N. (1960) Genetic determination and enzymatic induction of tyrosinase in Neurospora. *J. mol. Biol.* **2**, 96

[241] HOROWITZ N.H., FLING M., MACLEOD H. & SUEOKA N. (1961) A genetic study of two new structural forms of tyrosinase in Neurospora. *Genetics* **46**, 1015

[242] HOROWITZ N.H., HOULAHAN M.B., HUNGATE M.G. & WRIGHT B. (1946) Mustard gas mutations in Neurospora. *Science* **104**, 233

[243] HOROWITZ N.H. & MACLEOD H. (1960) The DNA content of Neurospora nuclei. *Microbiol Genet. Bull.* **17.** 6

[244] HOWE H.B., Jr. (1956) Crossing over and nuclear passing in *Neurospora crassa. Genetics* **41**, 610

[245] HOWE H.B. Jr. & TERRY C.E. Jr. (1962) Genetic studies of resistance to chemical agents in *Neurospora crassa. Canad. J. Genet. Cytol.* **4**, 447

[246] HSU K.S. (1963) The genetic basis of actidione resistance in *Neurospora crassa. J. gen. Microbiol.* **32**, 341

[247] HUEBSCHMAN C. (1952) A method for varying the average number of nuclei in the conidia of *Neurospora crassa. Mycologia* **44**, 599

[248] ISHIKAWA T. (1962) Genetic studies of *ad-8* mutants in *Neurospora crassa* I. Genetic fine structure of the *ad-8* locus. *Genetics* **47**, 1147

[249] ISHIKAWA T. (1962) Genetic studies of *ad-8* mutants in *Neurospora crassa.* II. Inter-allelic complementation at the *ad-8* locus. *Genetics* **47**, 1755

[250] ISHITANI C. (1956) A high frequency of heterozygous diploids and somatic recombination produced by U.V. in imperfect fungi. *Nature, Lond.* **178**, 706

[251] ISHITANI C., IKEDA Y. & SAKAGUCHI K. (1956). Hereditary variation and genetic recombination in Koji-molds (*Aspergillus oryzae* and *A. sojae*). VI. Genetic recombination in heterozygous diploids. *J. gen. appl. Microbiol.* **2**, 401

[252] ITANO H.A. & ROBINSON E.A. (1960) Genetic control of the α- and β-chains of haemoglobin. *Proc. nat. Acad. Sci., Wash.* **46**, 1492

[253] JACOB F. & MONOD J. (1961) Genetic regulatory mechanisms in the synthesis of proteins. *J. mol. Biol.* **3,** 318

[254] JACOB F., SCHAEFFER P. & WOLLMAN E. L. (1960) Episomic elements in bacteria. *Symp. Soc. gen. Microbiol.* **10,** 67

[255] JAMES A. P. & SPENCER P. E. (1958) The process of spontaneous extranuclear mutation in yeast. *Genetics* **43,** 317

[256] JEFFREY S.I.B., JINKS J.L. & GRINDLE M. (1962) Intraracial variation in *Phytophthora infestans* and field resistance in potato blight. *Genetics* **32,** 323

[257] JENSEN K.A., KIRK I., KÖLMARK G. & WESTERGAARD M. (1951) Chemically induced mutations in Neurospora. *Cold Spr. Harb. Sym. quant. Biol.* **16,** 245

[258] JINKS J.L. (1954) Somatic selection in fungi. *Nature, Lond.* **174,** 409

[259] JINKS J.L. (1956) Naturally occurring cytoplasmic changes in fungi. *C.R. Lab. Carlsberg, Ser. physiol.* **26,** 183

[260] JINKS J. L. (1957) Selection for cytoplasmic differences. *Proc. roy. Soc.* B. **146,** 527

[261] JINKS J.L. (1958) Cytoplasmic differentiation in fungi. *Proc. roy. Soc.,* B. **148,** 314

[262] JINKS J.L. (1959) Selection for adaptability to new environments in *Aspergillus glaucus. J. gen. Microbiol.* **20,** 223

[263] JINKS J.L. (1959) The genetic basis of 'duality' in imperfect fungi. *Heredity* **13,** 525

[264] JINKS J.L. (1959) Lethal suppressive cytoplasms in aged clones of *Aspergillus glaucus. J. gen. Microbiol.* **21,** 397

[265] JINKS J.L. (1963) Cytoplasmic inheritance in fungi. In *Methodology in Basic Genetics* (Ed. W.J. Burdette). Holden-Day, Inc., San Francisco

[266] JINKS J. L. & GRINDLE M. (1963) Changes induced by training in *Phytophthora infestans. Heredity* **18,** 245

[267] JINKS J.L. & GRINDLE M. (1963) The genetical basis of heterokaryon incompatibility in *Aspergillus nidulans. Heredity* **18,** 407

[268] JOHNSON T. (1953) Variation in the rusts of cereals. *Biol. Rev.* **28,** 105

[269] JOHNSON T. (1954) Selfing studies with physiological races of wheat stem rust, *Puccinia graminis* var. *tritici. Canad. J. Bot.* **32,** 506

[270] JOHNSON T. (1960) Genetics of pathogenicity. In *Plant Pathology.* Ed. by Horsfall J.G. & Dimond, A.E. Acad. Press, New York. Vol. 2., ch. 11, p. 407

[271] JOSSE J. (1961) Studies in the mechanism of DNA synthesis. *5th Int. Biochem. Congr., Moscow.* (Pergamon Press), Symposium No. 1

[272] KAFER E. (1958) An 8-chromosome map of *Aspergillus nidulans. Advanc. Genet.* **9,** 105

[273] KÄFER E. (1960) High frequency of spontaneous and induced somatic segregation in *Aspergillus nidulans. Nature, Lond.* **186,** 619

[274] KÄFER E. (1961) The processes of spontaneous recombination in vegetative nuclei of *Aspergillus nidulans. Genetics* **46,** 1581

[275] KAKAR S.N. (1963) Allelic recombination and its relation to recombination of outside markers in yeast. *Genetics* **48,** 957

[276] KAKAR S.N. & WAGNER R.P. (1964) Genetic and biochemical analysis of isoleucine-valine mutants of yeast. *Genetics* **49,** 213

[277] KAPULER A.M. & BERNSTEIN H. (1963) A molecular model for an enzyme based on a correlation between the genetic and complementation maps of the locus specifying the enzyme. *J. Mol. Biol.* **6,** 443

[278] KEITT G.W. (1952) Inheritance of pathogenicity in *Venturia inaequalis* (Cke.) Wint. *Amer. Nat.* **86,** 373

20

296 FUNGAL GENETICS

[279] KEITT G.W. & BOONE D.M. (1954) Induction and inheritance of mutant characters in *Venturia inaequalis* in relation to its pathogenicity. *Phytopathology* 44, 362

[280] KEITT G.W. & BOONE D.M. (1956) Use of induced mutations in the study of host-pathogen relationships. *Genetics in Plant Breeding, Brookhaven Symp. in Biol.* 9, 209

[281] KEITT G.W., BOONE D.M. & SHAY J.R. (1959) Genetic and nutritional controls of host-parasite interactions in apple scab. Ch. 16, p. 157 in *Plant Pathology: Problems and Progress 1908–1958*. Ed. by Holton C.S. *et al.* Univ. Wisconsin Press, Madison

[282] KEITT G.W. & LANGFORD M.H. (1941) *Venturia inaequalis* (Cke.) Wint. I. A ground work for genetic studies. *Amer. J. Bot.* 28, 805

[283] KENDREW J.C., DICKERSON R.E., STRANDBERG B.E., HART R.G., DAVIES D.R., PHILLIPS D.C. & SHORE V.C. (1960) Structure of myoglobin. A three-dimension fourier synthesis at 2 Å resolution. *Nature, Lond.* 185, 422

[284] KITANI Y., OLIVE L.S. & EL ANI A.S. (1961) Transreplication and crossing-over in *Sordaria fimicola*. *Science* 134, 668

[285] KITANI Y., OLIVE L.S. & EL ANI A.S. (1962) Genetics of *Sordaria fimicola*. V. Aberrant segregation at the g locus. *Amer. J. Bot.* 49, 697

[286] KLEIN R.M. & KLEIN D.T. (1962) Interaction of ionizing and visible radiation in mutation induction in *Neurospora crassa*. *Amer. J. Bot.* 49, 870

[287] KLINE D.M., BOONE D.M.. & KEITT G.W. (1957) *Venturia inaequalis* (Cke.) Wint. XIV. Nutritional control of pathogenicity of certain induced biochemical mutants. *Amer. J. Bot.* 44, 797

[288] KNIEP H. (1923) Über erbliche Änderungen von Geschlechtsfaktoren bei Pilzen. *Z. indukt. Abstamm.-u. Vereblehre* 31, 170

[289] KÖHLER F. (1935) Genetische Studien an *Mucor mucedo* Brefeld. *Z. indukt. Abstamm.-u. Vereblehre* 70, 1

[290] KÖLMARK G. (1953) Differential response to mutagens as studied by the Neurospora reverse mutation test. *Hereditas* 39, 270

[291] KÖLMARK G. (1956) Mutagenic properties of certain esters of inorganic acids investigated by the Neurospora back-mutation test. *C.R. Lab. Carlsberg, Sér. physiol.* 26, 205

[292] KÖLMARK G. & WESTERGAARD M. (1953) Further studies on chemically induced reversions at the adenine locus of Neurospora. *Hereditas* 39, 209

[293] LACOUR L.F. & PELC S.R. (1959) Effect of colchicine on the utilisation of thymidine labelled with tritium during chromosomal reproduction. *Nature, Lond.* 183, 1455

[294] LACOUR L.F. & RUTISHAUSER A. (1953) Chromosome breakage experiments with endosperm sub-chromatid breakage. *Nature, Lond.* 172, 502

[295] LANGE M. (1952) Species concept in the genus Coprinus, a study on the significance of inter-sterility. *Dansk. Bot. Ark.* 14, 1

296] LEBEN C., BOONE D.M. & KEITT G.W. (1955) *Venturia inaequalis* IX. Search for mutants resistant to fungicides. *Phytopathology* 45, 467

[297] LEDERBERG J. (1955) Recombination mechanisms in bacteria. *J. cell. comp. Physiol.* 45, suppl. 2, 75

[298] LEDERBERG J. (1959) Genes and antibodies. *Science* 129, 1649

[299] LEDERBERG J. & LEDERBERG E.M. (1952) Replica plating and indirect selection of bacterial mutants. *J. Bact.* 63, 399

[300] LEHMAN I.R., ZIMMERMAN S.B., ADLER J., BESSMAN M.J., SIMMS E.S. & KORNBERG A. (1958). Enzymatic synthesis of deoxyribonucleic acid. V. Chemical composition of enzymatically synthesized deoxyribonucleic acid. *Proc. nat. Acad. Sci., Wash.* 44, 1191

[301] LEIN J., MITCHELL H.K. & MITCHELL M.B. (1948) A method for the selection of biochemical mutants of Neurospora. *Proc. nat. Acad. Sci., Wash.* **34**, 435

[302] LESTER H.E. & GROSS S.R. (1959) Efficient method for selection of auxotrophic mutants of Neurospora. *Science* **129**, 572

[303] LEUPOLD U. (1950) Die vererbung von Homothallie und Heterothallie bei *Schizosaccharomyces pombe. C.R. Lab. Carlsberg, Ser. Physiol.* **24**, 381

[304] LEUPOLD U. (1955) Methodisches zur Genetik von *Schizosaccharomyces pombe. Schweiz. Z. Allg. Path. Bakt.* **18**, 1141

[305] LEUPOLD U. (1956) Tetraploid inheritance in Saccharomyces. *J. Genet.* **54**, 411

[306] LEUPOLD U. (1956) Tetrad analysis of segregation in autopolyploids. *J. Genet.* **54**, 427

[307] LEUPOLD U. (1958) Studies on recombination in *Schizosaccharomyces pombe. Cold Spring Harb. Symp. Quant. Biol.* **23**, 161

[308] LEVI J.D. (1956) Mating reaction in yeast. *Nature, Lond.* **177**, 753

[309] LEVINTHAL C. (1959) Bacteriophage genetics. In *The Viruses*, Vol. 2 (F.M. Burnet & W.M. Stanley, eds.), Academic Press, N.Y.

[310] LEWIS D. (1961) Genetical analysis of methionine suppressors in Coprinus. *Genet. Res., Camb.* **2**, 141

[311] LEWIS D. (1963) Structural gene for the methionine-activating enzyme and its mutation as a cause of resistance to ethionine. *Nature, Lond.* **200**, 151

[312] LEWIS R.W. (1953) An outline of the balance hypothesis of parasitism. *Amer. Nat.* **87**, 273

[313] LHOAS P. (1961) Mitotic haploidization by treatment of *Aspergillus niger* diploids with *para*-fluorophenylalanine. *Nature* **190**, 744

[314] LINDEGREN C.C. (1932) The genetics of Neurospora II. The segregation of sex factors in asci of *Neurospora crassa, N. sitophila* and *N. tetrasperma. Bull. Torrey. Bot. Club* **59**, 119

[315] LINDEGREN C.C. (1949) *The yeast cell, its genetics and cytology.* Educational Publishers Inc., St. Louis

[316] LINDEGREN C.C. (1952) Gene conversion in Saccharomyces. *J. Genet.* **51**, 625

[317] LINDEGREN C.C. & HINO S. (1957) The effect of anaerobiosis on the origin of respiratory-deficient yeast. *Exp. Cell Res.* **12**, 163

[318] LINDEGREN C.C. & LINDEGREN G. (1941). X-ray and ultra-violet induced mutations in Neurospora, I. X-ray mutations; II. ultra-violet mutations. *J. Hered.* **32**, 405 & 435

[319] LINDEGREN C.C. & LINDEGREN G. (1942). Locally-specific patterns of chromatid and chromosome interference in Neurospora. *Genetics* **27**, 1

[320] LINDEGREN C.C. & LINDEGREN G. (1944) Sporulation in *Saccharomyces cerevisiae. Bot. Gaz.* **105**, 304

[321] LINDEGREN C.C. & LINDEGREN G. (1951) Tetraploid Saccharomyces. *J. gen. Microbiol.* **5**, 885

[322] LINDEGREN C.C., SHULT E. & DESBOROUGH S. (1960) The induction of respiratory deficiency by adaptational stress. *Canad. J. Genet. Cytol.* **2**, 1

[323] LINDEGREN C.C., WILLIAMS M.A. & MCCLARY D.O. (1956) The distribution of chromatin in budding yeast cells. *Antonie van Leeuwenhoek* **22**, 1

[324] LISSOUBA P. (1960) Mis en evidence d'une unité génétique polarisée et essai d'analyse d'uncas d'interférence négative. *Ann. des Sc. Nat., Bot.*, 12th ser., 644

[325] LISSOUBA P., MOUSSEAU J., RIZET G. & ROSSIGNOL J.L. (1962) Fine structure of genes in the Ascomycete *Ascobolus immersus. Advanc. Genet.* **11**, 343

[326] LISSOUBA P. & RIZET G. (1960) Sur l'existence d'une unité génétique polarisée ne subissant que des échanges non reciproques. *C.R. Acad. Sci., Paris*, **250**, 3408

[327] LOEGERING W.Q. & POWERS H.R. Jr. (1962) Inheritance of pathogenicity in a cross of physiological races III and 36 of *Puccinia graminis f. sp. tritici.* *Phytopathol.* **52,** 547

[328] MCDONALD K.D. & PONTECORVO G. (1953) "Starvation" technique. In Pontecorvo *et al.*, ref. [424]

[329] MACHLIS L. (1958) Evidence for a sexual hormone in Allomyces. *Physiol. Plant.* **11,** 181

[330] MACHLIS L. (1958) A study of sirenin, the chemotactic sexual hormone from the water mold Allomyces. *Physiol. Plant.* **11,** 845

[331] MACHLIS L. & CRASEMANN J.M. (1956) Physiological variation between the generations and among the strains of watermolds in the subgenus Euallomyces. *Amer. J. Bot.* **43,** 601

[332] MCCLARY D.O., NULTY W.L. & MILLER G.R. (1959) Effect of potassium versus sodium in the sporulation of Saccharomyces. *J. Bact.* **78,** 362

[333] MCCLINTOCK B. (1945) Preliminary observations of the chromosomes of *Neurospora crassa. Amer. J. Bot.* **32,** 671

[334] MACKINTOSH M.E. & PRITCHARD R.H. (1963) The production and replica plating of micro-colonies of *Aspergillus nidulans. Genet. Res., Camb.* **4,** 320

[335] MADELIN M.F. (1956) Studies on the nutrition of *Coprinus lagopus* Fr. especially as affecting fruiting. *Ann. Bot., N.S.* **20,** 307

[336] MADSEN N.B. & GURD F.R.N. (1957) The interaction of muscle phosphorylase with p-chloromercuribenzoate. III. The reversible dissociation of phosphorylase. *J. biol. Chem.* **223,** 1055

[337] MAHONEY M. & WILKIE D. (1962) Nucleo-cytoplasmic control of perithecial formation in *Aspergillus nidulans. Proc. Roy. Soc.,* B **156,** 524

[338] MALING B. (1959) Linkage data for group IV markers in Neurospora. *Genetics* **44,** 1215

[339] MALING B. (1960) Replica plating and rapid ascus collection of Neurospora. *J. gen. Microbiol.* **23,** 257

[340] MANNEY T.R. (1964) Action of a super-suppressor in yeast in relation to allelic mapping and complementation. *Genetics* **50,** 109

[341] MANNEY T.R. & MORTIMER R.K. (1964) Allelic mapping in yeast using X-ray induced mitotic reversion. *Science* **143,** 581

[342] MARKERT C.L. (1953) Lethal and mutagenic effects of ultraviolet radiation in Glomerella conidia. *Exp. Cell. Res.* **5,** 429

[343] MARKERT C.L. (1956) Response of Glomerella conidia to irradiation by X-rays and fast neutrons. *Papers Mich. Acad. Sci., Arts & Letters* **91,** 27

[344] MARMUR J. & GROSSMAN L. (1961) Ultraviolet light induced linking of deoxyribonucleic acid strands and its reversal by photoreactivating enzyme. *Proc. nat. Acad. Sci., Wash.* **47,** 778

[345] MARTIN P.G. (1959) Apparent self-fertility in *Neurospora crassa. J. gen. Microbiol.* **20,** 213

[346] MATHER K. (1938) *The Measurement of Linkage in Heredity,* pp. 132, Methuen, London

[347] MATHER K. & BEALE G.H. (1942) The calculation and precision of linkage values from tetrad analysis. *J. Genet.* **43,** 1

[348] MATHIESON M.J. (1952) Ascospore dimorphism and mating type in *Chromocrea spinulosa. Ann. Bot., N.S.* **16,** 449

[349] MATHIESON M.J. (1956) Polarized segregation in *Bombardia lunata. Ann. Bot., N.S.* **20,** 623

[350] MATHIESON M.J. & CATCHESIDE D.G. (1955) Inhibition of histidine uptake in *Neurospora crassa., J. gen. Microbiol.* **13,** 72

[351] MESELSON M. & STAHL F.W. (1958) The replication of DNA in *Escherichia coli. Proc. nat. Acad. Sci., Wash.* **44,** 671

[352] MESELSON M. & WEIGLE J.J. (1961) Chromosome breakage accompanying genetic recombination in bacteriophage. *Proc. nat. Acad. Sci., Wash.* **47,** 857

[353] MICHAELIS P. (1958) Cytoplasmic inheritance and the segregation of plasmagenes. *Proc.* 10th *Int. Congr. Genet.* (Montreal), **1,** 375. Univ. of Toronto Press

[354] MIDDLEKAUFF J.E., HINO S., YANG S.-P., LINDEGREN G. & LINDEGREN C.C. (1957) Gene control of resistance versus sensitivity to actidione in Saccharomyces. *Genetics* **42,** 66

[355] MINAMI Z. & IKEDA Y. (1962) Heterocaryosis observed in *Rhizopus javanicus J. gen. appl. Microbiol,* **8,** 92.

[356] MITCHELL H.K. (1957) Crossing over and gene conversion in Neurospora. In *The Chemical Basis of Heredity,* 94 (W.D. McElroy & B. Glass, eds.), Johns Hopkins Press, Baltimore

[357] MITCHELL M.B. (1955) Aberrant recombination of pyridoxine mutants of Neurospora. *Proc. nat. Acad. Sci., Wash.* **41,** 215

[358] MITCHELL M.B. (1955) Further evidence of aberrant recombination in Neurospora. *Proc. nat. Acad. Sci., Wash.* **41,** 935

[359] MITCHELL M.B. & MITCHELL H.K. (1950) The selective advantage of an adenineless double mutant over one of the single mutants involved. *Proc. nat. Acad. Sci., Wash.* **36,** 115

[360] MITCHELL M.B. & MITCHELL H.K. (1952) Observations on the behaviour of suppressors in Neurospora. *Proc. nat. Acad. Sci., Wash.* **38,** 205

[361] MITCHELL M.B. & MITCHELL H.K. (1952) A case of "maternal" inheritance in *Neurospora crassa. Proc. nat. Acad. Sci., Wash.* **38,** 442

[362] MITCHELL M.B. & MITCHELL H.K. (1956) A nuclear gene suppressor of a cytoplasmically inherited character in *Neurospora crassa. J. gen. Microbiol.* **14,** 84

[363] MITCHELL M.B., MITCHELL H.K. & TISSIERES A. (1953) Mendelian and nonmendelian factors affecting the cytochrome system in *Neurospora crassa. Proc. nat. Acad. Sci., Wash.,* **39,** 606

[364] MOAT A.G., PETERS N. Jr. & SRB A.M. (1959) Selection and isolation of auxotrophic yeast mutants with the aid of antibiotics. *J. Bacteriol.* **7,** 673

[365] MOORE R.T. & MCALEAR J.H. (1962) Fine structure of Mycota 7. Observations on septa of Ascomycetes and Basidiomycetes. *Amer. J. Bot.* **49,** 86

[366] MORPURGO G. (1962) A new method of estimating forward mutations in fungi: resistance to 8-azaguanine and p-fluorophenylalanine. *Sci. Repts. Ist. Super. Sanità* **2,** 9

[367] MOSEMAN J. G. (1959) Host-pathogen interaction of the genes for resistance in *Hordeum vulgare* and for pathogenicity in *Erysiphe graminis f. sp. hordei. Phytopathology* **49,** 469

[368] MOUTSCHEN-DAHMEN J., MOUTSCHEN-DAHMEN M. & LOPPES R. (1963) Differential mutagenic activity of 1(+) and d(−) diepoxybutane. *Nature* **199,** 406

[369] MURRAY N.E. (1960) Complementation and recombination between *methionine-2* alleles in *Neurospora crassa. Heredity* **15,** 207

[370] MURRAY N.E. (1963) Polarized recombination and fine structure within the *me-2* gene of *Neurospora crassa. Genetics* **48,** 1163

[371] MYERS J.W. & ADELBERG E.A. (1954) The biosynthesis of isoleucine and valine, I. Enzymatic transformation of the dihydroxy acid precursors to the keto acid precursors. *Proc. nat. Acad. Sci., Wash.* **40,** 493

300 FUNGAL GENETICS

[372] NAKAMURA K. & EGASHIRA T. (1961) Genetically mixed perithecia in Neurospora. *Nature, Lond.* **190**, 1129

[373] NELSON R.R. (1956) Transmission of factors for urediospore color in *Puccinia graminis var. tritici* by means of nuclear exchange getween vegetative hyphae. *Phytopathology* **46**, 538

[374] NELSON R.R. (1961) Evidence of gene pools for pathogenicity in species of Helminthosporium. *Phytopathology* **51**, 736

[375] NELSON R.R., WILCOXSON R.D. & CHRISTENSEN J.J. (1955) Heterocaryosis as a basis for variation in *Puccinia graminis var. tritici. Phytopathology* **45**, 639

[376] NEWMEYER D. (1957) Arginine synthesis in *Neurospora crassa:* genetic studies. *J. gen. Microbiol.* **16**, 449

[377] NEWMEYER D. (1962) Genes influencing the conversion of citrulline to argininosuccinate in *Neurospora crassa. J. gen. Microbiol.* **28**, 215

[378] NEWMEYER D. & TATUM E.L. (1953) Gene expression in Neurospora mutants requiring nicotinic acid or tryptophan. *Amer. J. Bot.* **40**, 393

[379] NEWTON M., JOHNSON T. & BROWN A.M. (1930) A preliminary study on the hybridisation of physiologic forms of *Puccinia graminis tritici. Sci. Agr.* **10**, 721

[380] NIEDERHAUSER J.S. (1956) The blight, the blighter and the blighted. *Trans. N.Y. Acad. Sci., Ser. II,* **19**, 55

[381] NIRENBERG M.W. & MATTHAEI J.H. (1961) The dependence of cell-free protein synthesis in *Escherichia coli* upon naturally occurring or synthetic polyribonucleotides. *Proc. nat. Acad. Sci., Wash.* **47**, 1588

[382] NOVICK A. & SZILARD L. (1950) Experiments with the chemostat on spontaneous mutations of bacteria. *Proc. nat. Acad. Sci., Wash.* **36**, 708

[383] NOVICK A. & WIENER M. (1957) Enzyme induction as an all or none phenomenon. *Proc. nat. Acad. Sci., Wash.* **43**, 553

[384] OFENGAND E.J., DIECKMANN M. & BERG P. (1961) The enzymic synthesis of amino acyl derivatives of ribonucleic acid III. Isolation of amino acid-acceptor ribonucleic acids from *Escherichia coli. J. biol. Chem.* **236**, 1741

[385] OGUR M., MINCKLER S., LINDEGREN G. & LINDEGREN C.C. (1952) The nucleic acids in a polyploid series of Saccharomyces. *Arch. Biochem. Biophys.* **40**, 175

[386] OHNISHI E., MACLEOD H. & HOROWITZ N.H. (1962) Mutants of Neurospora deficient in D-amino acid oxidase. *J. Biol. Chem.* **237**, 138

[387] OLIVE L.S. (1956) Genetics of *Sordaria fimicola* I. Ascospore color mutants. *Amer. J. Bot.* **43**, 97

[388] OLIVE L.S. (1958) On the evolution of heterothallism in fungi. *Amer. Nat.* **92**, 233

[389] PAPAZIAN H.P. (1951) The incompatibility factors and a related gene in *Schizophyllum commune. Genetics* **36**, 441

[390] PAPAZIAN H.P. (1958) The genetics of Basidiomycetes. *Advanc. Genet.* **9**, 41

[391] PARAG Y. (1962) Mutations in the B incompatibility factor of *Schizophyllum commune. Proc. nat. Acad. Sci., Wash.* **48**, 743

[392] PARAG Y. (1962) Studies in somatic recombination in dikaryons of *Schizophyllum commune. Heredity* **17**, 305

[393] PARTRIDGE C.W.H. (1961) Altered properties of the enzyme adenylosuccinase, produced by inter-allelic complementation at the *ad-4* locus in *Neurospora crassa. Biochem. Biophys. Res. Commun.* **3**, 613

[394] PATEMAN J.A. (1957) Back-mutation studies at the *am* locus in *Neurospora crassa. J. Genet.,* **55**, 444

[395] PATEMAN J.A. (1960) Inter-relationships at the *am* locus in *Neurospora crassa*. *J. gen. Microbiol.* **23**, 393

[396] PATEMAN J.A. (1960) High negative interference at the *am* locus in *Neurospora crassa*. *Genetics* **45**, 839

[396a] PATEMAN, J. A., COVE, D. J., REVER, B. M. & ROBERTS, D. B. (1964) A common co-factor for nitrate reductase and xanthine dehydrogenase which also regulates the synthesis of nitrate reductase. *Nature*, **201**, 58

[397] PATEMAN J.A. & FINCHAM J.R.S. (1958) Gene-enzyme relationships at the *am* locus in *Neurospora crassa*. *Heredity* **12**, 317

[398] PAXMAN G.J. (1963) Variation in *Phytophthora infestans*. *Eur. Potato J.* **6**, 14

[399] PERKINS D.D. (1955) Tetrads and crossing-over. *J. cell. comp. Physiol.* **45**, 119

[400] PERKINS D.D. (1959) New markers and multiple point linkage data in Neurospora. *Genetics* **44**, 1185

[401] PERKINS D.D. (1962) The frequency in Neurospora tetrads of multiple exchanges within short intervals. *Genet. Res., Camb.* **3**, 315

[402] PERKINS D.D. (1962) Crossing-over and interference in a multiply marked chromosome arm of Neurospora. *Genetics* **47**, 1253

[403] PERKINS D.D., EL-ANI A.S., OLIVE L.S. & KITANI Y. (1963) Interference between exchanges in tetrads of *Sordaria fimicola*. *Amer. Nat.* **47**, 249

[404] PERKINS D.D., GLASSEY M. & BLOOM B.A. (1962) New data on markers and rearrangements in Neurospora. *Canad. J. Genet. Cytol.* **4**, 187

[405] PERKINS D.D. & ISHITANI C. (1959) Linkage data for group III markers in Neurospora. *Genetics* **44**, 1209

[406] PERSON C. (1959) Gene-for-gene relationships in host-parasite systems. *Canad. J. Bot.* **37**, 1101

[407] PERUTZ M.F., ROSSMAN M.G., CULLIS A.F., MUIRHEAD H. & WILL G. (1960) Structure of haemoglobin. A three-dimensional Fourier synthesis at 5·5 Å resolution, obtained by X-ray analysis. *Nature, Lond.* **185**, 416

[408] PITTENGER T.H. (1954) The general incidence of pseudo-wild types in *Neurospora crassa*. *Genetics* **39**, 326

[409] PITTENGER T.H. (1956) Synergism of two cytoplasmically inherited mutants in *Neurospora crassa*. *Proc. nat. Acad. Sci., Wash.* **42**, 747

[410] PITTENGER T.H. & ATWOOD K.C. (1954) The relation of growth rate to nuclear ratio in Neurospora heterocaryons. *Genetics* **39**, 987. (Abstr.)

[411] PITTENGER T.H. & BRAWNER T.G. (1961) Genetic control of nuclear selection in Neurospora heterocaryons. *Genetics* **46**, 1645

[412] PITTENGER T.H. & COYLE M.B. (1963) Somatic recombination in pseudowild-type cultures of *Neurospora crassa*. *Proc. Nat. Acad. Sci., Wash.* **49**, 445

[413] PITTENGER T.H., KIMBALL A.W. & ATWOOD K.C. (1955) Control of nuclear ratios in Neurospora heterocaryons. *Amer. J. Bot.* **42**, 954

[414] PITTMAN D., SHULT E., ROSHANMANESH A. & LINDEGREN C.C. (1963) The procurement of biochemical mutants of *Saccharomyces* by the synergistic effect of ultraviolet radiation and 2,6-diaminopurine. *Canad. J. Microbiol.* **9**, 103

[415] PLUNKETT B.E. (1956) The influence of factors of the aeration complex and light upon fruit-body form in pure cultures of an agaric and a polypore. *Ann. Bot., N.S.* **20**, 563

[416] PLUNKETT B.E. (1958) Translocation and pileus formation in *Polyporus brumalis*. *Ann. Bot., N.S.* **22**, 237

[417] POMPER S. & BURKHOLDER P.R. (1949) Studies on the biochemical genetics of yeast. *Proc. nat. Acad. Sci., Wash.* **35**, 456

[418] PONTECORVO G. (1949) Auxanographic techniques in biochemical genetics. *J. gen. Microbiol.* **3**, 122

302   FUNGAL GENETICS

[419] PONTECORVO G. (1956) The parasexual cycle in fungi. *Annu. Rev. Microbiol.* 10, 393

[420] PONTECORVO G. (1963) Microbial genetics: retrospect and prospect. *Proc. Roy. Soc.*, B 158, 1

[421] PONTECORVO G. & KÄFER E. (1958) Genetic analysis by means of mitotic recombination. *Advanc. Genet.* 9, 71

[422] PONTECORVO G. & ROPER J.A. (1956) Resolving power of genetic analysis. *Nature, Lond.* 178, 83

[423] PONTECORVO G., ROPER J.A. & FORBES E. (1953) Genetic recombination without sexual reproduction in *Aspergillus niger*. *J. gen. Microbiol.* 8, 198

[424] PONTECORVO G., ROPER J. A., HEMMONS L. M., MACDONALD K. D. & BUFTON A. W.J. (1953) The genetics of *Aspergillus nidulans*. *Advanc. Genet.* 5, 141

[425] PONTECORVO G. & SERMONTI G. (1954) Parasexual recombination in *Penicillium chrysogenum*. *J. gen. Microbiol.* 11, 94

[426] PRITCHARD R.H. (1954) Ascospores with diploid nuclei. *Caryologia* 6 (suppl.), 1117

[427] PRITCHARD R.H. (1955) The linear arrangement of a series of alleles of *Aspergillus nidulans*. *Heredity* 9, 343

[428] PRITCHARD R.H. (1960) Localized negative interference and its bearing on models of gene recombination. *Genet. Res., Camb.* 1, 1

[429] PRITCHARD R. H. (1960) The bearing of recombination analysis at high resolution on genetic fine structure in *Aspergillus nidulans* and the mechanism of recombination in higher organisms. *Symposia Soc. gen. Microbiol.* 10, 155

[430] QUINTANILHA A. (1935) Cytologie et génétique de la sexualité chez les Hymenomycetes. *Bol. Soc. Broteriana* 10 (2nd series), 1

[431] QUINTANILHA A. (1944) La conduite sexuelle de quelques espèces d'Agaricacées. *Bol. Soc. Broteriana* 19 (2nd ser.), 39

[432] QUINTANILHA A. & PINTO-LOPES J. (1950) Aperçu sur l'état actuel de nos connaisances concernant la "conduite sexuelle" des espèces d'Hymenomycetes. I. *Bol. Soc. Broteriana* 24 (2nd ser.), 115

[433] RADHAKRISHNAN A.N., WAGNER R.P. & SNELL E.E. (1960) Biosynthesis of valine and isoleucine III. α-keto-β-hydroxy acid reductase and α-hydroxy-β-keto acid reductoisomerase. *J. biol. Chem.* 235, 2322

[434] RAISTRICK H. (1949) A region of biosynthesis. *Proc. Roy. Soc.*, 'B', 136, 481

[435] RAPER C.A. & RAPER J.R. (1964) Mutations affecting heterokaryosis in *Schizophyllum commune*. *Amer. J. Bot.*, 51, 503

[436] RAPER J.R. (1952) Chemical regulation of sexual processes in the Thallophytes. *Bot. Rev.* 18, 447

[437] RAPER J.R. (1960) The control of sex in fungi. *Amer. J. Bot.* 47, 794

[438] RAPER J.R., BAXTER M.G. & ELLINGBOE A.H. (1960) The genetic structure of the incompatibility factors of *Schizophyllum commune*: the A factor. *Proc. nat. Acad. Sci., Wash.* 46, 833

[439] RAPER J.R., BAXTER M.G & MIDDLETON R.B. (1958) The genetic structure of the incompatibility factors in *Schizophyllum commune*. *Proc. nat. Acad. Sci, Wash.* 44, 889

[440] RAPER J.R & ESSER K. (1961) Antigenic differences due to the incompatibility factors in *Schizophyllum commune*. *Z. Vererbunglehre* 92, 439

[441] RAPER J.R., KRONGELB G.S. & BAXTER M.G (1958) The number and distribution of incompatibility factors in Schizophyllum. *Amer. Nat.*, 92, 221

442] RAPER J.R. & MILES P.G. (1958) The genetics of *Schizophyllum commune*. *Genetics* 43, 530

[443] RAPER J.R. & RAPER C.A. (1962) Mutant modifiers of the A incompatibility factor in *Schizophyllum commune. Amer. J. Bot.* **49**, 667 (Abstr.)

[444] REAUME S.E. & TATUM E.L. (1949) Spontaneous and nitrogen mustard induced nutritional deficiencies in *Saccharomyces cerevisiae. Arch. Biochem.* **22**, 331

[445] REISSIG J.L. (1956) Replica plating with *Neurospora crassa. Microbial Genetics Bull.* **14**, 31

[446] REISSIG J.L. (1960) Forward and back mutation in the *pyr-3* region of Neurospora, I. Mutations from arginine dependence to prototrophy. *Genet. Res., Camb.* **1**, 356

[447] REISSIG J.L. (1963) Induction of forward mutants in the *pyr-3* region of Neurospora. *J. Gen. Microbiol.* **30**, 317

[448] REISSIG J.L. (1963) Spectrum of forward mutants in the *pyr-3* region of Neurospora. *J. Gen. Microbiol.* **30**, 327

[449] RIS, H. (1961) Ultrastructure and molecular organization of genetic systems. *Canad. J. Genet. Cytol.* **3**, 95

[450] RIZET G. (1952) Les phénomènes de barrage chez *Podospora anserina*. I. Analyse génétique des barrages entre souches *S* et *s. Rev. Cytol. Biol. Vég.* **13**, 51

[451] RIZET G. & ENGELMANN C. (1949) Contribution à l'étude génétique d'un Ascomycète tetrasporé: *Podospora anserina. Rev. Cytol. Biol. Vég.* **11**, 201

[452] RIZET G., LISSOUBA P. & MOUSSEAU J. (1960) Les mutations d'ascospore chez l'ascomycète *Ascobolus immersus* et l'analyse de la structure fine des gènes. *Bull. Soc. Franc. Physiol. Vég.* **6**, 175

[453] RIZET G., MARCOU D. & SCHECROUN J. (1958) Deux phénomènes d'heredité cytoplasmique chez l'ascomycete *Podospora anserina. Bull. Soc. Franc. Physiol. Vég.* **4**, 136

[454] RIZET G. & SCHECROUN J. (1959) Sur les facteurs cytoplasmiques associés au couple de gènes *S-s* chez le *Podospora anserina. C.R. Acad. Sci., Paris* **249**, 2392

[455] ROBERTS C.F. (1959) A replica plating technique for the isolation of nutritionally exacting mutants of a filamentous fungus (*Aspergillus nidulans*). *J. gen. Microbiol.* **20**, 540

[456] ROBERTS C.F. (1963) The genetic analysis of carbohydrate utilization in *Aspergillus nidulans. J. gen. Microbiol.* **31**, 45

[457] ROMAN H. (1955) A system selective for mutations affecting the synthesis of adenine in yeast. *C.R. Lab. Carlsberg, Sér. Physiol.* **26**, 299

[458] ROMAN H. (1956) Studies of gene mutation in Saccharomyces. *Cold Spring Harb. Symp. Quant. Biol.* **21**, 175

[459] ROMAN H., PHILLIPS M.M. & SANDS S.M. (1955) Studies of polyploid Saccharomyces I. Tetraploid segregation. *Genetics* **40**, 546

[460] ROPER J.A. (1952) Production of heterozygous diploids in filamentous fungi. *Experientia* **8**, 14

[461] ROPER J.A. (1958) Nucleo-cytoplasmic interactions in *Aspergillus nidulans. Cold Spring Harb. Symp. Quant. Biol.* **23**, 141

[462] ROPER J.A. & KÄFER, E. (1957) Acriflavine-resistant mutants of *Aspergillus nidulans. J. gen. Microbiol.* **16**, 660

[463] ROPER J.A. & PRITCHARD R.H (1955) Recovery of complementary products of mitotic crossing-over. *Nature, Lond.* **175**, 639

[464] ROSSIGNOL J.-L. (1964) Phenomènes de recombinaison intragénique et unité functionelle d'un locus chez l'*Ascobolus immersus*. Doctoral Thesis, Paris

[465] ROWELL J.B. (1955) Segregation of sex factors in a diploid line of *Ustilago zeae* induced by alpha radiation. *Science* **121**, 304

[466] ROWELL J.B. (1955) Functional role of compatibility factors and an *in vitro* test for sexual compatibility with haploid lines of *Ustilago zeae*. *Phytopathology* **45**, 370

[467] ROWELL J.B. & DE VAY J.E. (1954) Genetics of *Ustilago zeae* in relation to basic problems of its pathogenicity. *Phytopathology* **44**, 356

[468] ROWELL J.B., LOEGERING W.Q. & POWERS H.R. Jr. (1963) Genetic model for physiologic studies of mechanisms governing development of infection type in wheat stem rust. *Phytopathol.* **53**, 932

[469] RUPERT C.S. (1960) Photoreactivation of transforming DNA by an enzyme from baker's yeast. *J. gen. Physiol.* **43**, 573

[470] RYAN F.J., BEADLE G.W. & TATUM E.L. (1943) The tube method of measuring the growth rate of Neurospora. *Amer. J. Bot.* **30**, 784

[471] ST. LAWRENCE P. (1956) The *q* locus of *Neurospora crassa*. *Proc. nat. Acad. Sci., Wash.* **42**, 189

[472] SAMBORSKI D.J. (1963) A mutation in *Puccinia recondita* Rob. ex Desm. f. sp. *tritici* to virulence on Transfer, Chinese Spring × *Aegilops umbellate* Zhuk. *Canad. J. Bot.* **41**, 475

[473] SANSOME E.R. (1946) Maintenance of heterozygosity in a homothallic species of the *Neurospora tetrasperma* type. *Nature, Lond.* **157**, 484

[474] SANSOME E.R. (1961) Meiosis in the oogonium and antheridium of *Pythium debaryanum*. *Nature, Lond.* **191**, 827

[475] SANSOME E.R. (1963) Meiosis in *Phythium debaryanum* Hesse and its significance in the life-history of the Biflagellatae. *Trans. Brit. Mycol. Soc.* **46**, 63

[476] SANSOME E.R., DEMEREC M. & HOLLAENDER A. (1945) Quantitative irradiation experiments with *Neurospora crassa*. I. Experiments with X-rays. *Amer. J. Bot.* **32**, 218

[477] SANSOME E.R. & HARRIS B.J. (1962) Use of camphor-induced polyploidy to determine the place of meiosis in fungi. *Nature, Lond.* **196**, 291

[478] SANWAL B.D & LATA M. (1961) The occurrence of two different glutamic dehydrogenases in Neurospora. *Canad. J. Microbiol.* **7**, 319

[479] SARACHEK A. & BISH J.T. (1963) Post-irradiation protein synthesis and the induction of cytoplasmic and genic mutations in Saccharomyces by ultra-violet irradiation. *Cytologia* **28**, 450

[480] SARACHEK A. & FOWLER G.L. (1959) The induction by allyl glycine of heritable respiratory deficiency in Saccharomyces and its reversal by sulphur amino acids. *Canad. J. Microbiol.* **5**, 584

[481] SASS J.E. (1929) The cytological basis for homothallism and heterothallism in the Agaicaceae. *Amer. J. Bot.* **16**, 663

[482] SCHLESINGER M.T. & LEVINTHAL C. (1963) Hybrid protein formation of *E. coli* alkaline phosphatase leading to *in vitro* complementation. *J. Mol. Biol.* **7**, 1

[483] SCHWINGHAMER E.A. (1959) The relation between radiation dose and the frequency of mutations for pathogenicity in *Melampsora lini*. *Phytopathology* **49**, 260

[484] SEQUEIRA L. (1954) Nuclear phenomena in the basidia and basidiospores of *Omphalia flavida*. *Mycologia* **46**, 470

[485] SETLOW R.B. & CARRIER W.L. (1964) The disappearance of thymine dimers from DNA: an error correcting mechanism. *Proc. Nat. Acad. Sci., Wash.* **51**, 226

[486] SHARPE H.S. (1958) A closed system of cytoplasmic variation in *Aspergillus glaucus*. *Proc. roy. Soc., 'B'* **148**, 355

[487] SHATKIN A.J. & TATUM E.L. (1959) Electron microscopy of *Neurospora crassa* mycelia. *J. biophys. biochem. Cytol.* **6**, 423

[488] SHATKIN A.J. & TATUM E.L. (1961) The relationship of *m*-inositol to morphology in *Neurospora crassa*. *Amer. J. Bot.* **48**, 760

[489] SHEAR C.L. & DODGE B.O. (1927) Life histories and heterothallism of the red bread mold fungi of the *Monilia sitophila* group. *J. Agr. Res.* **34**, 1019

[490] SHULT E.E., DESBROUGH S. & LINDEGREN C.C. (1962) Preferential segregation in Saccharomyces. *Genet. Res., Camb.* **3**, 196

[491] SHULT E.E. & LINDEGREN C.C. (1956) A general theory of crossing-over. *J. Genet.* **54**, 343

[492] SIDDIQI O.H. (1962) The fine genetic structure of the *paba*-1 region of *Aspergillus nidulans*. *Genet. Res., Camb.* **3**, 68

[493] SIDDIQI O.H. (1962) Mutagenic action of nitrous acid on *Aspergillus nidulans*. *Genet. Res., Camb.* **3**, 303

[494] SIMONS M.D. & MURPHY H.C. (1955) A comparison of certain combinations of oat varieties as crown rust differentials. *U.S.D.A. Tech. Bull. No.* 112, pp. 22

[495] SINGLETON J.R. (1953) Chromosome morphology and the chromosome cycle in the ascus of *Neurospora crassa*. *Amer. J. Bot.* **40**, 124

[496] SISLER H.D. & COX C.E. (1960) Physiology of fungitoxicity. In *Plant Pathology* Vol. II, 507. (J.G. Horsfall & A.E. Dimond, eds.), Academic Press, N.Y.

[497] SJÖWALL M. (1945). Studien uber sexualität, vererbung und zytologie bei einigen diözischen Mucoraceen. Pub. Lund: Carl Bloms: pp. 97

[498] SLONIMSKI P. & EPHRUSSI B. (1949) Action de l'acriflavine sur les levures V. Le système des cytochromes des mutants 'petite colonie.' *Ann. Inst. Pasteur* **77**, 47

[499] SMITH B.R. (1962) The location of 11 *am* alleles in linkage group 5. *Neurospora Newsletter* **1**, 18

[500] SMITH B.R. (1965) Interallelic recombination at the *histidine-5* locus in *Neurospora crassa. Heredity*, in the press

[501] SMITH H.H. & SRB A.M. (1951) Induction of mutations with *β*-propiolactone. *Science* **114**, 490

[502] SMOOT J.J., GOUGH F.J., LAMEY H.A., EICHENMULLER J.J. & GALLEGLY M.E. (1958) Production and germination of oospores of *Phytophothora infestans*. *Phytopathology* **48**, 165

[503] SOMERS C.E., WAGNER R.P. & HSU T.C. (1960) Mitosis in vegetative nuclei of *Neurospora crassa. Genetics* **45**, 801

[504] SORGER G.J. (1963) TPNH-cytochrome c reductase and nitrate reductase in mutant and wild type Neurospora and Aspergillus. *Biochem. Biophys. Res. Comm.* **12**, 395

[505] SPEYER J.F., LENGYEL P., BASILIO C. & OCHOA S. (1962) Synthetic polynucleotides and the amino acid code II. *Proc. nat. Acad. Sci., Wash.* **48**, 63

[506] SRB A.M. (1955) Spontaneous and chemically-induced mutations giving rise to canavanine resistance in yeast. *Compt.-rend. Lab. Carlsberg, Sér. physiol.* **26**, 363

[507] SRB A.M. (1958). Some consequences of nuclear cytoplasmic recombinations among various Neurosporas. *Cold Spring Harb. Symp. Quant. Biol.* **23**, 269

[508] SRB A.M., FINCHAM J.R.S. & BONNER D.M. (1950) Evidence from gene mutations in Neurospora for close metabolic relationships among ornithine, proline and α-amino-δ-hydroxyvaleric acid. *Amer. J. Bot.* **37**, 533

[509] SRB A.M. & HOROWITZ N.H. (1944) The ornithine cycle in Neurospora and its genetic control. *J. biol. Chem.* **154**, 129

[510] STADLER D.R. (1956) A map of linkage group VI of *Neurospora crassa. Genetics* **41**, 528

[511] STADLER D.R. (1956) Double crossing-over in Neurospora. *Genetics* **41**, 623

[512] STADLER D.R. (1959) The relationship of gene conversion to crossing-over in Neurospora. *Proc. nat. Acad. Sci., Wash.* **45**, 1625

[513] STADLER D.R. (1963) Genetic control of tryptophan accumulation in Neurospora. *Proc. XI Int. Congr. Genetics* (The Hague), **1**, 52, Pergamon Press

[514] STADLER D.R. & TOWE A.M. (1963) Recombination of allelic cysteine mutants in Neurospora. *Genetics* **48**, 1323

[515] STAHL F. (1961) A chain model for chromosomes. *J. Chimie Physique* **58**, 1072

[516] STAPLETON G.E., HOLLAENDER A. & MARTIN F.L. (1952) Comparative lethal and mutagenic effects of ionizing radiation on *Aspergillus terreus*. *J. cell. comp. Physiol.* **39**, *suppl.* 1, 87

[517] STEFFENSEN D. (1959) A comparative view of the chromosome. *Brookhaven Symposia in Biol.* **12**, 103

[518] STERN C. (1936) Somatic crossing-over and segregation in *Drosophila melanogaster*. *Genetics* **21**, 625

[519] STEVENS C.M. & MYLROIE A. (1953) Production and reversion of biochemical mutants of *Neurospora crassa* with mustard compounds. *Amer. J. Bot.* **40**, 424

[520] STRICKLAND W.N. (1948) Abnormal tetrads in *Aspergillus nidulans*. *Proc. roy. Soc. 'B'* **148**, 533

[521] STRICKLAND W.N. (1948) An analysis of interference in *Aspergillus nidulans*. *Proc. roy. Soc., 'B'* **149**, 82

[522] STRICKLAND W.N. (1960) A rapid method for obtaining unordered Neurospora tetrads. *J. gen. Microbiol.* **22**, 583

[523] STRICKLAND W.N. (1961) Tetrad analysis of short chromosome regions of *Neurospora crassa*. *Genetics* **46**, 1125

[524] STRICKLAND W.N., PERKINS D.D. & VEATCH C.C. (1959) Linkage group data for group V markers in Neurospora. *Genetics* **44**, 1221

[525] SURZYCKI S. & PASZEWSKI A. (1964) Non-random segregation of chromosomes in *Ascobolus immersus*. *Genet. Res., Camb.* **5**, 20

[526] SUSKIND S.R. & JORDAN E. (1959) Enzymatic activity of a genetically altered tryptophan synthetase in *Neurospora crassa*. *Science* **129**, 1614

[527] SUSKIND S.R. & KUREK L.I. (1959) On a mechanism of suppressor gene regulation of tryptophan synthetase activity in *Neurospora crassa*. *Proc. nat. Acad. Sci., Wash.* **45**, 193

[528] SUSKIND S.R., LIGON D.S. & CARSIOTIS M. (1962) Mutationally altered tryptophan synthetase in *Neurospora crassa*. In *Molecular Basis of Neoplasia*, 15th Ann. Symp. on Fundamental Cancer Research, Univ. of Texas M.D. Anderson Hospital & Tumor Institute

[529] SUSKIND S.R., YANOFSKY C. & BONNER D.M. (1955) Allelic strains of Neurospora lacking tryptophan synthetase : a preliminary immunochemical characterisation, *Proc. nat. Acad. Sci., Wash.* **41**, 577

[530] SUYAMA Y. (1960) Effects of pyridoxal phosphate and serine in conversion of indoleglycerol phosphate to indole by extracts from tryptophan mutants of *Neurospora crassa*. *Biochem. Biophys. Res. Comm.* **3**, 493

[531] SUYAMA Y. (1963) *In vitro* complementation in the tryptophan synthetase system of Neurospora. *Biochem. Biophys. Res. Comm.* **10**, 144

[532] SUYAMA Y. & BONNER D.M. (1964) Complementation between tryptophan synthetase mutants of *Neurospora crassa*. *Biochem. Biophys. Acta.* **81**, 565

[533] SUYAMA Y., MUNKRES K.D. & WOODWARD V.W. (1959) Genetic analysis of the *pyr-3* locus of *Neurospora crassa*: the bearing of recombination and gene conversion upon intra-allelic linearity. *Genetica* **30**, 293

[534] SWANSON C.P. (1957) *Genetics and Cytogenetics* pp. 596. MacMillan, London

[535] SWIEZYNSKI K.M. & DAY P.R. (1960) Heterokaryon formation in *Coprinus lagopus*. *Genet. Res., Camb.* **1**, 114

[536] Swiezynski K. M. & Day P. R. (1960) Migration of nuclei in *Coprinus lagopus*. *Genet. Res., Camb.* 1, 129

[537] Takahashi T. (1958) Complementary genes controlling homothallism in Saccharomyces. *Genetics* 43, 705

[538] Takahashi T. (1959) Filtration methods for selecting auxotrophic mutants of flocculent type yeast. *Report Kihara Inst. Biol. Res.* 10, 57

[539] Takahashi T., Saito H. & Ikeda Y. (1958) Heterothallic behaviour of a homthallic strain in Saccharomyces yeast. *Genetics* 43, 249

[540] Takemaru T. (1957) Genetics of *Collybia velutipes* V. Mating patterns between F1 mycelia of legitimate and illegitimate origins in the strain NL-55. *Bot. Mag. (Tokyo)* 70, 244

[541] Takemaru T. (1961) Genetical studies on fungi X. The mating system in Hymenomycetes and its genetical mechanism. *Biol. J. Okayama Univ.* 7, 133

[542] Tatum E. L., Barratt R. W. & Cutter, V. M. (1949) Chemical induction of colonial paramorphs in Neurospora and Syncephelastrum. *Science* 109, 509

[543] Tatum E. L., Barratt R. W., Fries N. & Bonner D. M. (1950) Biochemical mutant strains of Neurospora produced by physical and chemical treatment. *Amer. J. Bot.* 37, 38

[544] Taylor J. H. (1957) The time and mode of duplication of the chromosomes. *Amer. Nat.* 91, 209

[545] Taylor J. H. (1958) The organization and duplication of genetic material. *Proc. 10th Int. Congr. Genet.* I, 63. Univ. Toronto Press

[546] Taylor J. H., Woods P. S. & Hughes W. L. (1957) The organization and duplication of chromosomes as revealed by autoradiographic studies using tritium-labeled thymidine. *Proc. nat. Acad. Sci., Wash.* 43, 122

[547] Teas H. J., Horowitz N. H. & Fling M. (1948) Homoserine as a precursor of threonine and methionine in Neurospora. *J. biol. Chem.* 172, 651

[548] Terakawa H. (1960) The incompatibility factors in *Pleurotus ostreatus*. *Sci. Papers Coll. Gen. Educ. Univ. Tokyo* 10, 65

[549] Threlkeld S.F.H. (1962) Some asci with non-identical sister spores from a cross in *Neurospora crassa*. *Genetics* 47, 1187.

[550] Tinline R.D. (1962) *Cochliobolus sativus* V. Heterokaryosis and parasexuality. *Canad. J. Bot.* 40, 425

[551] Tinline R.D. (1963) *Cochliobolus sativus* VII. Nutritional control of the pathogenicity of some auxotrophs to wheat seedlings. *Canad. J. Bot.* 41, 489

[552] Tsugita A., Gish D. T., Young J., Fraenkel-Conrat, H. Knight, C. A. & Stanley W. M. (1960) The complete amino acid sequence of the protein of tobacco mosaic virus. *Proc. nat. Acad. Sci., Wash.* 46, 1463

[553] Vogel H. J. & Bonner D. M. (1954) On the glutamate-proline-ornithine interrelation in *Neurospora crassa*. *Proc. nat. Acad. Sci., Wash.* 40, 688

[554] Vogel R. H. & Kopac M. J. (1959) Glutamic γ-semialdehyde in arginine and proline synthesis of Neurospora: a mutant-tracer analysis. *Biochem. Biophys. Acta.* 36, 505

[555] Vogel R. H. & Vogel H. J. (1963) Evidence for acetylated intermediates of arginine synthesis in *Neurospora crassa*. *Genetics* 48, 914

[556] Wagner R.P., Bergquist A., Barbee T. & Kiritani K. (1964) Genetic blocks in the isoleucine-valine pathway of *Neurospora crassa*. *Genetics* 49, 865

[557] Wagner R. P., Somers C. E. & Bergquist A. (1960) Gene structure and function in Neurospora. *Proc. nat. Acad. Sci., Wash.* 46, 708

[558] Wallace M.E. & Michie D. (1953) Affinity, a new genetic phenomenon in the house mouse. *Nature, Lond.* 171, 27

[559] Watson I.A. (1957) Mutation for pathogenicity in *Puccinia graminis var. tritici*. *Phytopathology* 47, 507

[560] WATSON I.A. (1957) Further studies on the production of new races from mixtures of races of *Puccinia graminis var. tritici* on wheat seedlings. *Phytopathology* **47**, 510

[561] WATSON I.A. & LUIG N.H. (1958) Somatic hybridization in *Puccinia graminis var. tritici*. *Proc. Linn. Soc. N.S. Wales*, **83**, 190

[562] WATSON I.A. & LUIG N.H. (1962) Asexual intercrosses between somatic recombinants of *Puccinia graminis*. *Proc. Linn. Soc. N.S.W.* **87**, 99

[563] WATSON J.D. & CRICK F.H.C. (1953) A structure for deoxyribose nucleic acid. *Nature, Lond.* **171**, 737

[564] WATSON J.D. & CRICK F.H.C. (1953) Genetical implications of the structure of deoxyribonucleic acid. *Nature, Lond.* **171**, 964

[565] WEBBER B.B. (1960) Genetical and biochemical studies of histidine-requiring mutants of *Neurospora crassa*, II. Evidence concerning heterogeneity among *hist*-3 mutants. *Genetics* **45**, 1617

[566] WEBBER B.B. & CASE M.E. (1960) Genetical and biochemical studies of histidine-requiring mutants of *Neurospora crassa*. *Genetics* **45**, 1605

[567] WEISS S.B. & NAKAMOTO T. (1961) On the participation of DNA in RNA biosynthesis. *Proc. nat. Acad. Aci., Wash.* **47**, 694

[568] WEISS S.B. & NAKAMOTO T. (1961) The enzymatic synthesis of RNA: nearest-neighbour base frequencies. *Proc. nat. Acad. Sci., Wash.* **47**, 1400

[569] WESTERGAARD M. (1957) Chemical mutagenesis in relation to the concept of the gene. *Experientia* **13**, 224

[570] WESTERGAARD M. & MITCHELL H.K. (1947) Neurospora V. A synthetic medium favoring sexual reproduction. *Amer. J. Bot.* **34**, 573

[571] WHEELER H.E. (1954) Genetics and evolution of heterothallism in Glomerella. *Phytopathology* **44**, 342

[572] WHEELER H.E. (1956) Linkage groups in Glomerella. *Amer. J. Bot.* **43**, 1

[573] WHEELER H.E. & McGAHEN J.W. (1952) Genetics of Glomerella X. Genes affecting sexual reproduction. *Amer. J. Bot.* **39**, 110

[574] WHITEHOUSE H.L.K. (1949) Multiple allelomorph heterothallism in the fungi. *New Phytol.* **48**, 212

[575] WHITEHOUSE H.L.K. (1951) A survey of heterothallism in the Ustilaginales. *Trans. Brit. mycol. Soc.* **34**, 340

[576] WHITEHOUSE H.L.K. (1957) Mapping chromosome centromeres from tetratype frequencies. *J. Genet.* **55**, 348

[577] WHITEHOUSE H.L.K. (1963) A theory of crossing-over by means of hybrid deoxyribonucleic acid. *Nature* **199**, 1034

[578] WHITEHOUSE H.L.K. & HASTINGS P.J. (1964) The analysis of genetic recombination on the polaron hybrid DNA model. *Genet. Res., Camb.*, in the press

[579] WILCOXSON R.D., TUITE J.F. & TUCKER S. (1958) Urediospore germ tube fusions in *Puccinia graminis*. *Phytopathology* **48**, 358

[580] WILDE P. (1961) Ein Beitrag zur Kenntnis der Variabilität von *Phytophthora infestans*. *Archiv. f. Mikrobiol.* **40**, 163

[581] WILKIE D. & LEWIS D. (1963) The effect of ultraviolet light on recombination in yeast. *Genetics* **48**, 1701

[582] WILLIAMSON D.H. & SCOPES A.W. (1961) Synchronization of division in cultures of *Saccharomyces cerevisiae* by control of the environment. *Symp. Soc. gen. Microbiol., London* **11**, 217

[583] WILSON C.M. (1952) Meiosis in Allomyces. *Bull. Torrey Bot. Club*, **79**, 139

[584] WILSON J.F. (1961) Micrurgical techniques for Neurospora. *Amer. J. Bot.* **48**, 46

[585] WILSON J.F. (1963) Transplantation of nuclei in *Neurospora crassa*. *Amer. J. Bot.* **50**, 780

[586] Wilson J.F., Garnjobst L. & Tatum E.L. (1961) Heterocaryon incompatibility in *Neurospora crassa*—micro-injection studies. *Amer. J. Bot.* **48,** 299

[587] Winge Ø. & Roberts C. (1949) A gene for diploidization in yeast. *Compt.-rend. lab. Carlsberg, Sér. Physiol.* **24,** 314

[588] Witkin E.M. (1958) Post-irradiation metabolism and the timing of ultraviolet-induced mutations in bacteria. *Proc. 10th Int. Congr. Genet. (Montreal)* I, 280, *Univ. of Toronto Press*

[589] Woodward D.O. (1959) Enzyme complementation *in vitro* between adenylosuccinaseless mutants of *Neurospora crassa*. *Proc. nat. Acad. Sci., Wash.* **45,** 846

[590] Woodward D.O., Partridge C.W.H. & Giles N.H. (1958) Complementation at the *ad-4* locus in *Neurospora crassa*. *Proc. nat. Acad. Sci., Wash.* **44,** 1237

[591] Woodward D.O., Partridge C.W.H. & Giles N.H. (1960) Studies of adenylosuccinase in mutants and revertants of *Neurospora crassa*. *Genetics* **45,** 555

[592] Woodward V.W. (1956) Mutation rates of several gene loci in Neurospora. *Proc. nat. Acad. Sci., Wash.* **42,** 752

[593] Woodward V.W., De Zeeuw J.R. & Srb A.M. (1954) The separation and isolation of particular biochemical mutants of Neurospora by differential germination of conidia, followed by filtration and selective plating. *Proc. nat. Acad. Sci., Wash.* **40,** 192

[594] Woodward V.W. & Schwartz P. (1964) Neurospora mutants lacking ornithine transcarbamylase. *Genetics* **49,** 845

[595] Wright R.E. & Lederberg J. (1956) Extranuclear transmission in yeast heterocaryons. *Proc. nat. Acad. Sci., Wash.* **43,** 919

[596] Yanofsky C. (1952) The effects of gene change on tryptophan desmolase formation. *Proc. nat. Acad. Sci., Wash.* **38,** 215

[597] Yanofsky C. (1960) The tryptophan synthetase system. *Bact. Rev.* **24,** 221

[598] Yanofsky C., Helinski D.R. & Maling B.D. (1962) The effects of mutation on the composition and properties of the *A* protein of *Escherichia coli* tryptophan synthetase. *Cold Spring Harb. Symp. Quant. Biol.* **26,** 11

[599] Yanofsky C. & Bonner D.M. (1955) Gene interaction in tryptophan synthetase formation. *Genetics* **40,** 761

[600] Yielding K.L. & Tomkins G.M. (1960) Structural alterations in crystalline glutamic dehydrogenase induced by steroid hormones. *Proc. nat. Acad. Sci., Wash.* **46,** 1483

[601] Yotsuyanagi Y. (1955) Mitochondria and refractive granules in the yeast cell. *Nature, Lond.* **176,** 1207

[602] Yuasa A. & Lindegren C.C. (1959) The integrity of the centriole in Saccharomyces. *Antonie van Leeuwenhoek* **25,** 73

[603] Yura T. (1959) Genetic alteration of pyrroline-5-carboxylate reductase in *Neurospora crassa*. *Proc. nat. Acad. Sci., Wash.* **45,** 197

[604] Yu-Sun C.C.C. (1964) Nutritional studies of *Ascobolus immersus*. *Amer. J. Bot.* **51,** 231

[605] Zalokar M. (1959) Growth and differentiation in Neurospora hyphae. *Amer. J. Bot.* **46,** 602

[606] Zetterberg G. (1961) A specific and strong mutagenic effect of N-nitroso-N-methyl urethan in Ophiostoma. *Hereditas* **47,** 295

[607] Zetterberg G. (1962) On the specific mutagenic effect of N-nitroso-N-methylamine in *Ophiostoma*. *Hereditas* **48,** 371

AUTHOR INDEX

GENERAL INDEX

permeases 62 255
   for amino acids in Neurospora 62
Peronosporales 25–6
petite-colony yeast
   biochemical characteristics 239 253
   neutral cytoplasmic type 239–41 253
   segregational type 239–40
   suppressive cytoplasmic type 240–1
phenylalanine requiring mutants 180–1
Pholiota mutabilis 248
photoreactivation 69
Phycomyces blakesleeanus 24
P. nitens 208
Phycomycetes 20–6 208
physiologic races of pathogens 258 *et seq.*
   basis 272–3
   possible origins 264–72
Phytophthora cactorum 26
P. erythroseptica 26
P. infestans
   germination of oospores 26 265
   heterothallism 25 265
   physiologic races 259 264–7 272
pigment mutations 34
plasmagenes 252
Plectomycetes 29–32
pleiotropism 175 181–2 210
Pleurotus ostreatus 216
Podospora anserina 32 33 95
   barrage phenomenon 246–7 256
   heterogenic incompatibility in 221
   secondary homothallism 32 85 219–21
   senescence 251
Poisson distribution 91–2 94 96
poky Neurospora and similar mutants 241–3 249
   modifier of 243
polarity in intragenic recombination 150 *et seq.*
polarization in the ascus 10 100
polaron 150–4 156
polynucleotides 274 276 278–9
polyploidy 124 *et seq.* 171
   in Allomyces 22 124
Polyporus betulinus 216
P. brumalis 43
post-meiotic segregation—see *half-chromatid*
potatoes, blight resistant varieties 266

proline requiring mutants 54 177–9 189–90 192
proline biosynthesis 177–9
promycelium 38–9
prophase 6
β-propiolactone as a mutagen 70
protein structure 183–4
   genetic control of 190 *et seq.*
protocatechuic acid oxidase 181
protoperithecium 3 33–4 224
prototrophs 46
   formed by mutation 44 60–1
   from heteroallelic crosses 140–4 200
pseudowild types in Neurospora 120–3
   as test for complementation 134
   origin 121–2
   recombination in 122
   self-fertility in 121
Puccinia coronata avenae
   mutation to virulence 271
   physiologic races 264–5
P. graminis tritici
   life cycle 37
   mitotic recombination 264 269
   mutation to virulence 269
   physiologic races 264
P. recondita tritici
pycniospore 37
pycnium 37–8 268
pyr-3 locus of Neurospora 173
   system for selecting mutants at 61
Pyrenomycetes 29 32–4 223
pyrimidine requiring mutants 53
pyridoxal phosphate 185
pyridoxin requiring fungi 19 53
pyrroline-5-carboxylate reductase 177–9 192
Pythium debaryanum 26

quadrivalents 124 128

random spore analysis 31 80–1 84
recessive 283
recessive lethals 62–3
reciprocal crosses, test for extranuclear inheritance 235 241 246–8
recombination 283
recombination fraction 79
   limiting value 92–3
   relation to exchange frequency 94
red cytoplasmic variant in Aspergillus 252–3
regulatory genes 203–5

two strand double exchanges 14 15 85
possible excess 89–91 164
tyrosinase 190 192 205–6
tyrosine requiring mutant 181

ultraviolet light
as a mutagen 55–6 67–9 72 74–7
effect in increasing mitotic crossing-
over 113 270
effect on DNA 69
inducing chromosome aberrations
77
mutant yield and dosage 68
unbalanced growth 58–9
univalent 124 126 128
uracil 274–278
Uredinales 36–9 268–72
uredospore 37–9
mutations affecting pigment 39 269
Ustilaginales 36 39–40 267–8
Ustilago avenae 264 268
U. maydis
auxotrophic mutants 53–4
DNA content 106 171
diploids 171 106 268
heterothallism 267–8
life cycle 39–40
linkage groups 103
mitotic segregation 107 113–4
nutrition 19
pathogenicity 39 106 267–8
replica plating 51 107

vegetative death in *Aspergillus glaucus*
250–1

vegetative segregation—see mitotic segregation
vegetative selection of cytoplasmic differences 236–7
Venturia inaequalis 33–4 75
chromosomes 103
inheritance of virulence 261
isolation of mutants 52–4 61 71
linkage groups 103
making of crosses 34
nutrition 19 272–3
pathogenicity 272
physiologic races 261 264 272–3
Verticillium albo-atrum 44 264
virulence 258 *et seq.*
mutations affecting 264 266 269–72
viruses, analogy with extra-nuclear inheritance 253
visible mutations 46 132

X-rays
and chromosome breakage 76–7 270
272
increasing segregation frequency in
heteroallelic diploids 115 140
mutagenic effect 67–9 75–6
use in fine structure mapping in yeast
115 140

yeasts—see *Saccharomyces, Schizosaccharomyces, Hansenula*

zoospores 25
zygospore 23–5
germination 23–4
zygote 22 283
zygotene 4 5